Studies in Big Data

Volume 173

The series "Studies in Big Data" (SBD) publishes new developments and advances in the various areas of Big Data—quickly and with a high quality. The intent is to cover the theory, research, development, and applications of Big Data, as embedded in the fields of engineering, computer science, physics, economics and life sciences. The books of the series refer to the analysis and understanding of large, complex, and/or distributed data sets generated from recent digital sources coming from sensors or other physical instruments as well as simulations, crowd sourcing, social networks or other internet transactions, such as emails or video click streams and other. The series contains monographs, lecture notes and edited volumes in Big Data spanning the areas of computational intelligence including neural networks, evolutionary computation, soft computing, fuzzy systems, as well as artificial intelligence, data mining, modern statistics and Operations research, as well as self-organizing systems. Of particular value to both the contributors and the readership are the short publication timeframe and the world-wide distribution, which enable both wide and rapid dissemination of research output.

The books of this series are reviewed in a single blind peer review process.

Indexed by SCOPUS, EI Compendex, SCIMAGO and zbMATH.

All books published in the series are submitted for consideration in Web of Science.

Rohit Sharma · Raghvendra Kumar ·
Prasant Kumar Pattnaik
Editors

Multimedia Technologies in the Internet of Things Environment, Volume 4

Springer

Editors
Rohit Sharma
Department of Electronics
and Communication Engineering
ABES Engineering College
Ghaziabad, Uttar Pradesh, India

Raghvendra Kumar
Department of Computer Science
and Engineering
Gandhi Institute of Engineering
and Technology University
Gunupur, Odisha, India

Prasant Kumar Pattnaik
School of Computer Engineering
KIIT University
Bhubaneswar, Odisha, India

ISSN 2197-6503 ISSN 2197-6511 (electronic)
Studies in Big Data
ISBN 978-981-96-4355-4 ISBN 978-981-96-4356-1 (eBook)
https://doi.org/10.1007/978-981-96-4356-1

This Springer imprint is published by the registered company Springer Nature Singapore Pte Ltd.
The registered company address is: 152 Beach Road, #21-01/04 Gateway East, Singapore 189721, Singapore

Preface

The main objective of this book publication is to explore the concepts of internet of things, biomedical and cyber physical systems along with the recent research and development. It also includes various real time applications and case studies in the field of engineering and technologies used. As populations grow and resources become scarcer, the efficient usage of these limited goods becomes more important.

Chapter "Machine Learning for Multimedia IoT Analytics", explores various ML models, including deep learning architectures, for tasks such as image and video recognition, audio analysis, and sensor data interpretation within the context of Multimedia IoT. Additionally, this chapter addresses challenges related to data heterogeneity, real-time processing, and scalability in deploying ML solutions for multimedia analytics in IoT environments.

Chapter "IoT-based Hand Hygiene Dispenser with Temperature and Level Monitoring Systems Using Data Analysis", present a concept for an IoT-based hand hygiene dispenser that monitors temperature and level. It addresses the need for efficient, touchless hand hygiene solutions.

Chapter "Enhancing Multimedia Security in IoT Environment: Blockchain to the Rescue", categorises multimedia data into four categories and highlighted ways to secure multimedia data using blockchain, cryptography, machine Learning and quantum computing. Additionally, it provides recent advancements within each of the solutions and the categories of multimedia data. Finally, several research directions are discussed in order to pave way for future multimedia security and IoT researchers.

Chapter "Multimedia Data Processing and Analysis in IoT", delves into methodologies for efficient multimedia data handling, encompassing visual, auditory, and sensor-generated information. Emphasis is placed on the challenges and opportunities associated with real-time processing, storage, and extraction of meaningful insights from multimedia data in the context of IoT applications.

Chapter "Embedded and Cloud Computing for Ingesting Big Multimedia Data in IOT Sensor Networks Applications", presents insights gleaned from empirical studies highlighting the transformative impact of cloud and embedded computing on multimedia data processing in IoT environments. By leveraging these computing paradigms, IoT systems can achieve enhanced scalability, adaptability, and efficiency, paving the way for innovative applications across domains such as smart cities, healthcare, and industrial automation.

Chapter "IoT Security Measures and Social Network Content Sharing Behavior: A Social Cognitive Theory Perspective", examines the interplay between Internet of Things (IoT) security awareness, social cognitive factors, and social sharing behaviors. The study applies social cognitive theory (SCT) to examine the cognitive determinants of online security behavior in the context of IoT. Using a survey-based approach, data was collected from 393 social media users.

Chapter "IoT-Enabled Multimedia Applications in Healthcare, Telemedicine, and Imaging", delves into the challenges, opportunities, and advancements in leveraging IoT-enabled multimedia for improved diagnostics, treatment, and overall healthcare delivery. By analyzing key use cases and technological considerations, this research aims to provide insights into the evolving landscape of healthcare applications within the IoT and multimedia convergence.

Chapter "Post-Quantum Cryptography Exploration for Quantum-Resilient and Secure Multimedia Data for IoT System", discusses the recent advancements and proposed ideas about increasing the efficiency of PQC and making the IoT network secure for the multimedia data in the rising quantum environment.

Chapter "Empowering Metaverse Using Decentralized Blockchain Technology for Multimedia Applications", begins by description of blockchain technology following with multimedia technology. Further, the paper discusses various components and characteristics of the Metaverse, its different use cases and associated technologies. Furthermore, the paper illustrates the integration of metaverse with blockchain and discusses about how multimedia technology enhance the metaverse.

Chapter "Cloud Computing and Multimedia IoT", study delves into the architectural considerations, security implications, and performance optimizations required to harness the full potential of this convergence. Through a comprehensive analysis, this work contributes valuable insights into the evolving paradigm of Cloud Computing and Multimedia IoT, offering guidance for researchers, practitioners, and policymakers navigating this dynamic landscape.

Chapter "Extended Reality and Immersive Multimedia for Gaming Applications", highlights the role of XR in multimedia, including holographic video, remote rendering, and media streaming. Finally, it provides detailed information about the hardware, software and devices which are used developing in this field and recent advancement.

Chapter "Multi-Model Fusion for Prediction and Segmentation of Brain Tumor using CNNs", aims to advance the analysis of brain tumors by utilizing a complex method that combines CT and MRI-Scan data into an intuitive Flask-based web application. Pioneer-based registration guarantees accurate alignment of various patient pictures, creating a common coordinate system for thorough anatomical comparisons.

The aim of this book is to support the computational studies at the research and post-graduation level with open problem-solving technique, we are confident that it will bridge the gap for them by supporting novel solution to support in their problem solving. At the end, editors have taken utmost care while finalizing the chapter to the book, but we are open to receive your constructive feedback, which will enable us to carry-out necessary points in our forthcoming books.

Ghaziabad, India — Rohit Sharma
Gunupur, India — Raghvendra Kumar
Bhubaneswar, India — Prasant Kumar Pattnaik

About This Book

Multimedia communication in the Internet of Things (IoT) can potentially reach into a vast array of areas and touch people's lives in profound and different ways. A vast amount of research work has been carried out in this domain, targeting different aspects of data analytics. To serve this purpose, it is proposed to conduct a comprehensive overview of state-of-the-art research work on multimedia analysis in IOT applications. It also aims to bridge the gap between multimedia concept and solutions by providing the current IOT frameworks, their applications, the strengths and limitations of the existing methods, and the future directions.

Contents

About the Editors

Dr. Rohit Sharma is currently working as Dean (Research), ABES Engineering College, Ghaziabad. Dr. Sharma is having Post-Doctoral, from South Ural State University, Russia. Dr. Sharma is having extensive experience in academia and research. As a prolific researcher, he has authored over 70 SCIE and 122 SCOPUS publications, along with 23 books in collaboration with esteemed publishers such as Springer, Taylor and Francis, and IEEE. With more than 22 patents to his name, three of which have been granted. He currently serves as an editor for several prestigious journals including CAEE Elsevier, SETA Elsevier, Expert Systems Wiley, IET Communications, IEEE Transactions on Consumable Electronics, ACM Transactions on Autonomous and Adaptive Systems etc. Additionally, he holds senior membership in IEEE and member of other technical societies such as ISTE, IAENG etc. Dr. Sharma has also spearheaded 08 renowned conferences in collaboration with IEEE and Springer, and under his guidance, four scholars have awarded their Ph.D. while one other has submitted her Ph.D. synopsis.

Dr. Raghvendra Kumar is working as Associate Professor in Computer Science and Engineering Department at GIET University, India. He received B.Tech., M.Tech. and Ph.D. in Computer Science and Engineering, India, and Postdoc Fellow from Institute of Information Technology, Virtual Reality and Multimedia, Vietnam. He serves as Series Editor Internet of Everything (IOE): Security and Privacy Paradigm, Green Engineering and Technology: Concepts and Applications, publishes by CRC press, Taylor & Francis Group, USA, and Bio-Medical Engineering: Techniques and Applications, Publishes by Apple Academic Press, CRC Press, Taylor & Francis Group, USA. He also serves as acquisition editor for Computer Science by Apple Academic Press, CRC Press, Taylor & Francis Group, USA. He has published number of research papers in international journal (SCI/SCIE/ESCI/Scopus) and conferences including IEEE and Springer as well as serve as organizing chair (RICE-2019, 2020), volume Editor (RICE-2018), Keynote speaker, session chair, Co-chair, publicity chair, publication chair, advisory board, Technical program Committee members in many international and national conferences and serve as guest editors in many special issues from reputed journals (Indexed By: Scopus, ESCI, SCI). He also published 13 chapters in edited book published by IGI Global, Springer and Elsevier. His researches areas are Computer Networks, Data Mining, cloud computing and Secure Multiparty Computations, Theory of Computer Science and Design of Algorithms. He authored and Edited 23 computer science books in field of Internet of Things, Data Mining, Biomedical Engineering, Big Data, Robotics, and IGI Global Publication, USA, IOS Press Netherland, Springer, Elsevier, CRC Press, USA.

Prasant Kumar Pattnaik Ph.D. (Computer Science), Fellow IETE, Senior Member IEEE is a Professor at the School of Computer Engineering, KIIT Deemed University, Bhubaneswar. He has more than a decade of teaching and research experience. Dr. Pattnaik has published numbers of Research Papers in peer-reviewed International Journals and Conferences. He also published many edited book volumes in Springer and IGI Global Publication. His areas of interest include Mobile Computing, Cloud Computing, Cyber Security, Intelligent Systems and Brain Computer Interface. He is one of the Associate Editor of Journal of Intelligent & Fuzzy Systems, IOS Press and Intelligent Systems Book Series Editor of CRC Press, Taylor Francis Group.

Machine Learning for Multimedia IoT Analytics

G. Siva Nageswara Rao, Chandra Sekhar Koppireddy, Mulaka Madhava Reddy, P. Padma, and Yada Sunitha

Abstract The convergence of Machine Learning (ML) and Multimedia Internet of Things (IoT) has opened new avenues for advanced analytics and insights generation. This paper investigates the application of machine learning techniques in the realm of Multimedia IoT analytics. By leveraging ML algorithms, this research aims to enhance the interpretation and utilization of multimedia data generated by IoT devices. The study explores various ML models, including deep learning architectures, for tasks such as image and video recognition, audio analysis, and sensor data interpretation within the context of Multimedia IoT. Additionally, the paper addresses challenges related to data heterogeneity, real-time processing, and scalability in deploying ML solutions for multimedia analytics in IoT environments. Through empirical evaluations and case studies, this work contributes to the ongoing discourse on the intersection of Machine Learning and Multimedia IoT, providing valuable insights for researchers and practitioners in the field.

Keywords Machine learning · Multimedia · Internet of things (IoT) · Analytics · Deep learning · Image recognition · Video analysis · Audio processing · Sensor data · Real-time processing · Scalability · Multimedia IoT applications

---

G. Siva Nageswara Rao (✉)
Department of Computer Science and Engineering, Koneru Lakshmaiah Education Foundation, Vaddesaram, Guntur, Andhra Pradesh, India
e-mail: sivanags@kluniversity.in

C. S. Koppireddy
CSE Department, Pragati Engineering College, Surampalem, India

M. M. Reddy
IT Department, Pace Institute of Technology and Sciences, Ongole, India

P. Padma
Telangana Tribal Welfare Residential Degree College, Nagarkurnool, India

Y. Sunitha
CSE Department, CVR College of Engineering, Mangalpalle, Telangana, India

R. Sharma et al. (eds.), *Multimedia Technologies in the Internet of Things Environment, Volume 4*, Studies in Big Data 173, https://doi.org/10.1007/978-981-96-4356-1_1

1 Introduction

The convergence of Machine Learning (ML) and the Multimedia Internet of Things (IoT) represents a transformative milestone in the realm of data analytics and insights generation. With the proliferation of IoT devices capturing vast amounts of multimedia data, there arises a pressing need for advanced analytics techniques to interpret and harness the insights hidden within these heterogeneous data streams. This problem can now be effectively addressed with the use of machine learning, especially in its subset of deep learning algorithms, which has the ability to extract useful correlations and patterns from the multimedia data produced by Internet of Things (IoT) sensors.

This paper delves into the application of machine learning techniques in the domain of Multimedia IoT analytics, aiming to enhance the interpretation and utilization of multimedia data for actionable insights. By leveraging ML algorithms, this research seeks to empower practitioners to extract meaningful information from diverse data modalities, including images, videos, audio, and sensor readings, within the context of IoT environments. The study explores various ML models and architectures tailored for tasks such as image and video recognition, audio analysis, and sensor data interpretation, illustrating their capabilities and potential applications in real-world scenarios.

Moreover, the paper addresses the challenges inherent in deploying ML solutions for multimedia analytics in IoT environments, including data heterogeneity, real-time processing requirements, and scalability concerns. Through empirical evaluations and case studies, this work provides practical insights into overcoming these challenges and demonstrates the efficacy of ML-based approaches in enhancing the analytics capabilities of Multimedia IoT systems.

The fusion of Machine Learning (ML) and the Multimedia Internet of Things (IoT) heralds a new era of data-driven insights and advanced analytics [1]. With the exponential growth of IoT devices and the deluge of multimedia data they generate, there is a critical need for sophisticated techniques to extract actionable insights from these heterogeneous data streams. Machine Learning, particularly deep learning algorithms, has emerged as a powerful solution to this challenge, offering the ability to uncover meaningful patterns and correlations within the vast volumes of multimedia data captured by IoT devices.

In order to improve the interpretation and application of multimedia data for well-informed decision-making, this study explores the use of machine learning techniques in the field of multimedia IoT analytics. By harnessing ML algorithms, this research endeavors to empower stakeholders to extract valuable insights from diverse data modalities, including images, videos, audio recordings, and sensor readings, within the context of IoT environments. The study delves into various ML models and architectures tailored for specific tasks such as image and video recognition, audio analysis, and sensor data interpretation, shedding light on their capabilities and potential applications in real-world scenarios [2].

Furthermore, the paper confronts the challenges inherent in deploying ML solutions for multimedia analytics in IoT environments. These challenges include managing data heterogeneity stemming from the diverse nature of multimedia data sources, addressing the stringent requirements for real-time processing imposed by IoT applications, and ensuring scalability to handle the massive volumes of data generated by IoT deployments. Through empirical evaluations and case studies, this work offers practical insights into overcoming these challenges and demonstrates the effectiveness of ML-based approaches in augmenting the analytics capabilities of Multimedia IoT systems.

By bridging the gap between theory and practice, this research contributes to the ongoing dialogue surrounding the intersection of Machine Learning and Multimedia IoT [3]. It provides researchers, practitioners, and industry stakeholders with valuable insights and guidance, enabling them to harness the potential of ML techniques for advanced analytics and insights generation in the dynamic landscape of IoT. Ultimately, this endeavor seeks to unlock the transformative potential of Multimedia IoT, paving the way for innovation and progress across various application domains.

2 Machine Learning Models for Multimedia IoT Analytics

The integration of Machine Learning (ML) with the Internet of Things (IoT) represents a revolutionary step in the evolution of smart technologies [4]. This synergy enhances the capabilities of IoT systems by enabling them to analyze data, learn from it, and make intelligent decisions autonomously. This chapter delves into the principles of Machine Learning, its application in IoT environments, and the transformative impact it has across various sectors. The following figure shows the various steps in Machine learning with IoT (Fig. 1).

The integration of Machine Learning with IoT has the potential to revolutionize various industries by enabling intelligent, autonomous systems that can analyze data, learn from it, and make informed decisions. Despite the challenges, the continuous advancements in technology and research are paving the way for more efficient and effective IoT solutions [5]. By harnessing the power of ML, IoT devices can significantly enhance operational efficiency, safety, and user experience across numerous applications.

2.1 Image Recognition

Image recognition is a crucial task in Multimedia IoT analytics, enabling IoT devices to interpret and understand visual information captured by cameras and sensors. Machine Learning models play a pivotal role in image recognition tasks, allowing

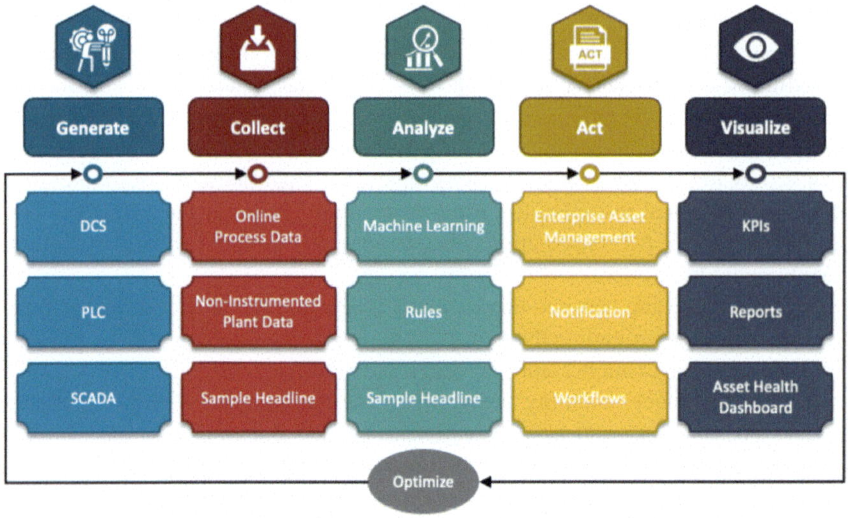

Fig. 1 Machine learning with IoT (*Source* DataFlair)

devices to identify objects, scenes, and patterns within images. Because Convolutional Neural Networks (CNNs) can train discriminative representations and extract hierarchical features from images, they are extensively utilized for image recognition.

Image recognition allows IoT devices to analyze visual data, making it possible to identify objects, scenes, and patterns within images [6]. This capability is essential for numerous IoT applications, including smart homes, surveillance, healthcare, and industrial automation. By understanding visual inputs, IoT systems can make informed decisions, trigger alerts, and automate processes, enhancing efficiency and safety.

In the context of Multimedia IoT, CNNs can be employed for various image recognition tasks, including object detection, classification, and localization. For instance, CNN-based models in smart surveillance systems are able to instantly identify and categorize items like cars, pedestrians, and suspicious activity [7]. Similarly, in industrial IoT applications, CNNs can analyze images captured by drones or cameras to monitor equipment, detect defects, and optimize manufacturing processes.

2.2 *Video Analysis*

Video analysis extends the capabilities of Multimedia IoT systems by enabling the interpretation of temporal data captured by cameras and sensors. Machine Learning models are instrumental in video analysis tasks, allowing IoT devices to extract meaningful insights from video streams in real time. Because recurrent neural networks (RNNs) can model sequential data and capture temporal dependencies, they are frequently employed for video analysis tasks, particularly Long Short-Term Memory (LSTM) networks [8].

In Multimedia IoT applications, RNNs can be utilized for tasks such as action recognition, anomaly detection, and activity forecasting. For instance, in smart home environments, RNN-based models can analyze video feeds to recognize human activities such as cooking, exercising, or sleeping, facilitating personalized automation and context-aware services. In industrial settings, RNNs can monitor video streams from surveillance cameras to detect abnormal behaviors or safety violations, thereby enhancing workplace safety and security.

2.3 Audio Processing

Audio processing, enhanced by Machine Learning (ML) and Internet of Things (IoT) technologies, revolutionizes how sound data is captured, analyzed, and utilized across various applications [9]. This chapter explores the intersection of audio processing with ML and IoT, highlighting advancements, applications, and benefits in diverse industries. Audio processing is essential for Multimedia IoT systems, enabling devices to analyse and interpret sound signals captured by microphones and sensors [10]. Machine Learning models are instrumental in audio processing tasks, allowing IoT devices to recognize speech, detect environmental sounds, and extract relevant information from audio streams. Because deep learning architectures can learn discriminative features from unprocessed audio data, such as convolutional neural networks (CNNs) and recurrent neural networks (RNNs), they are frequently utilized for audio processing applications. Deep learning models can be used for applications like sound classification, speech recognition, and acoustic event detection in the context of multimedia IoT [11]. For example, in smart assistant devices, CNN-based models can transcribe spoken commands and execute corresponding actions, enabling hands-free interaction with IoT systems. Similarly, in environmental monitoring applications, RNN-based models can analyze audio recordings to detect sounds indicative of events such as glass breaking, sirens, or alarms, facilitating timely responses and interventions.

2.3.1 Role of Machine Learning in Audio Processing

Speech Recognition

Speech recognition problems are a strong suit for machine learning algorithms, especially deep learning models like Transformers and Recurrent Neural Networks (RNNs) [12]. By processing audio inputs, these models transcribe spoken language into text with high accuracy. Applications range from virtual assistants and voice-controlled devices to transcription services and automated customer service systems.

Case Study: Virtual Assistants

Virtual assistants like Siri, Alexa, and Google Assistant utilize ML-based speech recognition to understand and respond to user commands and queries. These systems continuously improve accuracy and responsiveness through training on vast datasets of speech patterns and language nuances.

Audio Classification and Analysis

ML algorithms classify audio signals into predefined categories based on features such as pitch, tone, and frequency. This capability finds applications in music genre classification, environmental sound monitoring, and audio surveillance systems. Convolutional Neural Networks (CNNs) and Support Vector Machines (SVMs) are commonly used for audio classification tasks.

Case Study: Environmental Sound Monitoring

IoT-enabled sensors capture and transmit audio data from urban environments to ML algorithms. These algorithms analyze sound patterns to detect anomalies, such as traffic accidents or construction noise, enabling timely interventions and improving public safety.

Voice Biometrics and Authentication

ML models analyze unique vocal characteristics, such as pitch modulation and speech patterns, to verify individuals' identities through voice biometrics [13]. This technology enhances security in authentication processes, banking transactions, and access control systems.

Case Study: Voice Authentication in Banking

Banks integrate ML-powered voice biometrics to authenticate customers during phone banking transactions. Voiceprints are compared against stored profiles to verify identities securely and prevent unauthorized access.

Audio Data Collection with IoT Sensors

IoT devices equipped with microphones and sensors capture audio data in real-time from various environments, such as smart homes, industrial sites, and urban areas. These devices transmit audio signals to centralized servers or cloud platforms for further processing and analysis.

Real-Time Audio Surveillance and Monitoring

IoT-enabled audio surveillance systems monitor and analyze sound patterns for security and safety applications. Machine learning algorithms process audio inputs to detect suspicious activities, alert security personnel, and trigger automated responses.

Case Study: Smart Cities

Cities deploy IoT-enabled audio sensors across public spaces to monitor noise levels, traffic patterns, and emergency situations. ML algorithms analyze audio data to optimize urban planning, enforce noise regulations, and enhance environmental sustainability.

Remote Audio Analytics and Diagnostics

IoT facilitates remote audio analytics by enabling real-time data transmission and analysis from distributed sensors. ML algorithms detect equipment malfunctions based on audio signatures, enabling predictive maintenance and minimizing downtime.

Case Study: Industrial IoT

Manufacturing plants integrate IoT-enabled audio sensors to monitor equipment health. ML models analyze sound patterns from machines to detect early signs of mechanical failure, optimizing maintenance schedules and improving operational efficiency.

Benefits of ML and IoT in Audio Processing

Enhanced Accuracy and Efficiency

ML algorithms improve the accuracy of audio processing tasks, such as speech recognition and sound classification, by learning from vast datasets and adapting to new inputs over time. IoT-enabled data collection enhances efficiency by providing real-time insights and enabling proactive decision-making.

Cost Savings and Operational Optimization

Predictive maintenance driven by IoT-enabled audio analytics reduces equipment downtime and maintenance costs. ML-based automation enhances operational workflows, reduces manual intervention, and improves resource allocation in various applications [10].

Improved User Experience and Security

ML-powered speech recognition enhances user interaction with devices and applications, providing intuitive and personalized experiences. IoT-enabled audio surveillance enhances security measures, ensuring prompt detection and response to potential threats.

3 Challenges and Considerations

3.1 Data Privacy, Security, Scalability and Compatibility

Protecting audio data from unauthorized access and cybersecurity threats is crucial. Encryption, secure data transmission protocols, and compliance with privacy regulations mitigate risks associated with IoT-enabled audio processing systems. Integrating IoT devices and ML algorithms across diverse platforms and environments requires standardized protocols and compatibility standards. Scalable infrastructure and robust deployment strategies ensure seamless integration and reliable performance. Ethical considerations, such as consent for audio data collection and responsible AI usage, influence the deployment and adoption of IoT-enabled audio processing technologies. Compliance with data protection laws and ethical guidelines promotes trust and transparency in AI applications.

Sensor Data Interpretation:

Sensor data interpretation is fundamental to Multimedia IoT analytics, enabling devices to make sense of data collected from various sensors embedded in the environment. Machine Learning models are instrumental in sensor data interpretation tasks, allowing IoT devices to extract meaningful insights and patterns from sensor readings. Because supervised learning algorithms can handle structured data and understand intricate correlations, they are frequently employed for sensor data interpretation jobs. Examples of these algorithms are Random Forests and Support Vector Machines (SVMs). In Multimedia IoT applications, supervised learning models can be applied to tasks such as anomaly detection, predictive maintenance, and environmental monitoring. SVM-based models, for instance, can evaluate sensor data in smart building systems to find anomalies in HVAC (Heating, Ventilation, and Air Conditioning) systems, pointing to possible equipment malfunctions or energy inefficiencies. Similarly, in agricultural IoT applications, Random Forest models can interpret sensor readings to predict crop yields, optimize irrigation schedules, and mitigate environmental risks, thereby enhancing agricultural productivity and sustainability. In the integration of Machine Learning (ML) solutions for multimedia analytics within Internet of Things (IoT) environments, various challenges and considerations emerge. This chapter delves into the intricacies of deploying ML solutions in such contexts, particularly focusing on addressing data heterogeneity, ensuring real-time processing capabilities, and achieving scalability to handle the large volumes of multimedia data streams effectively.

Data Heterogeneity:

One of the primary challenges in deploying ML solutions for multimedia analytics in IoT environments is the inherent data heterogeneity. IoT ecosystems often comprise diverse devices capturing data in various formats, resolutions, and modalities. As a result, integrating and processing this heterogeneous data poses significant challenges for ML models, which typically require standardized and structured input data.

To address data heterogeneity, preprocessing and normalization techniques are crucial. Preprocessing is the process of sanitizing and converting unprocessed data into a format that is typical for machine learning algorithms. For multimedia data, preprocessing may include tasks such as image resizing, audio feature extraction, and sensor data normalization. Additionally, feature engineering techniques can help extract relevant features from heterogeneous data sources, enabling ML models to learn meaningful patterns and correlations.

Furthermore, interoperability standards and protocols play a vital role in managing data heterogeneity within IoT ecosystems. Adopting standardized formats and communication protocols, such as MQTT (Message Queuing Telemetry Transport) or JSON (JavaScript Object Notation), can facilitate seamless integration and exchange of data between heterogeneous devices and platforms.

Real-Time Processing:

Another critical consideration in deploying ML solutions for multimedia analytics in IoT environments is ensuring real-time processing capabilities. Many IoT applications, such as surveillance systems, smart cities, and industrial automation, require timely analysis and response to streaming multimedia data. However, traditional ML models often involve computationally intensive training and inference processes, which may not be suitable for real-time deployment on resource constrained IoT devices. To address the challenge of real-time processing, lightweight and efficient ML algorithms tailored for edge computing environments are essential. These algorithms are optimized to run efficiently on IoT devices with limited computational resources, enabling real-time analysis of multimedia data streams at the edge of the network. Techniques such as model compression, quantization, and pruning can help reduce the computational complexity of ML models without significantly sacrificing accuracy. Furthermore, deploying distributed and parallel processing architectures can enhance real-time processing capabilities in IoT environments. Edge computing paradigms, such as fog computing and edge AI, distribute computational tasks across network edges, allowing for localized processing of multimedia data streams. By leveraging edge computing, ML models can analyze data in near real-time, minimizing latency and enabling rapid response to dynamic IoT environments.

3.2 Scalability

Scalability is a fundamental consideration in deploying ML solutions for multimedia analytics in IoT environments, particularly in handling the large volumes of data generated by IoT devices. As IoT deployments scale up to accommodate thousands or millions of connected devices, ML models must be capable of processing and analyzing massive datasets efficiently. One approach to achieving scalability is through the use of cloud-based ML platforms and services. Cloud computing offers virtually unlimited computational resources and storage capacity, enabling ML models to scale dynamically to handle increasing data volumes. By offloading computation to the cloud, IoT devices can leverage powerful infrastructure for training and inference tasks, alleviating resource constraints at the edge [14].

4 Case Studies and Empirical Evaluations

4.1 Smart Surveillance

In the domain of smart surveillance, Machine Learning (ML) techniques play a pivotal role in enhancing security and situational awareness. Case studies demonstrate the application of ML models for real-time analysis of video streams from surveillance cameras in urban environments, commercial establishments, and residential areas. These models utilize deep learning architectures, such as Convolutional Neural Networks (CNNs), for tasks such as object detection, person identification, and activity recognition. Empirical evaluations of ML-based smart surveillance systems showcase their effectiveness in detecting suspicious activities, identifying unauthorized access, and alerting security personnel in real-time. By leveraging advanced computer vision algorithms, these systems can accurately distinguish between normal and anomalous behaviors, enabling proactive security measures and incident response.

4.2 Environmental Monitoring

Environmental monitoring is another application domain where ML techniques prove instrumental in analyzing multimedia data from IoT sensors deployed in natural ecosystems, urban landscapes, and industrial facilities. Case studies highlight the use of ML models for monitoring air quality, detecting pollution sources, and predicting environmental hazards such as wildfires, floods, and landslides.

Empirical evaluations of ML-based environmental monitoring systems demonstrate their ability to analyze diverse sensor data streams, including images, audio recordings, and sensor readings, to provide actionable insights for environmental management and policy-making. By leveraging data-driven approaches, these systems can identify trends, patterns, and anomalies in environmental data, facilitating timely interventions and mitigation strategies.

4.3 Healthcare Diagnostics

ML techniques are increasingly being applied in healthcare diagnostics to analyze multimedia data from medical imaging devices, wearable sensors, and electronic health records. Case studies showcase the use of ML models for tasks such as medical image analysis, disease diagnosis, and personalized treatment recommendation. Empirical evaluations of ML-based healthcare diagnostics systems demonstrate their efficacy in detecting abnormalities, predicting disease progression, and optimizing treatment outcomes. By leveraging large-scale datasets and deep learning algorithms, these systems can provide accurate and timely diagnostic insights, enabling clinicians to make informed decisions and improve patient care.

The convergence of Machine Learning (ML) and the Internet of Things (IoT) is creating a new era of smart, interconnected devices capable of learning from data, making autonomous decisions, and improving overall efficiency. This integration leverages the vast amount of data generated by IoT devices and applies sophisticated ML algorithms to extract valuable insights, automate processes, and enhance user experiences. The convergence of Machine Learning (ML) and the Internet of Things (IoT) is creating a new era of smart, interconnected devices capable of learning from data, making autonomous decisions, and improving overall efficiency.

This integration leverages the vast amount of data generated by IoT devices and applies sophisticated ML algorithms to extract valuable insights, automate processes, and enhance user experiences. Figure 2 shows the integration of Ml with IoT domains.

4.4 Medical Image Analysis

One important area where ML is contributing significantly is medical picture analysis. Convolutional Neural Networks (CNNs), in particular, are deep learning models that are used in healthcare diagnostics to process and interpret complicated medical pictures, including X-ray, MRI, CT, and ultrasound images. Patterns and irregularities that may be difficult for human eyes to notice can be recognized by these algorithms. CNNs, for instance, can be trained to accurately identify the presence of tumors, fractures, and other medical problems.

Medical image analysis is a transformative area where machine learning (ML) is making significant contributions to healthcare. With the advent of deep learning models, particularly Convolutional Neural Networks (CNNs), the capability to process and interpret complex medical images has vastly improved. This chapter explores how these technologies are applied in healthcare diagnostics to enhance the accuracy and efficiency of detecting various medical conditions.

4.5 The Importance of Medical Image Analysis

Ultrasound, CT, MRI, and X-ray images are among the many types of medical imaging that are vital to modern medicine. They offer a non-invasive way to see the body's internal components, which helps with many illnesses' diagnosis and treatment. However, interpreting these images requires significant expertise and can

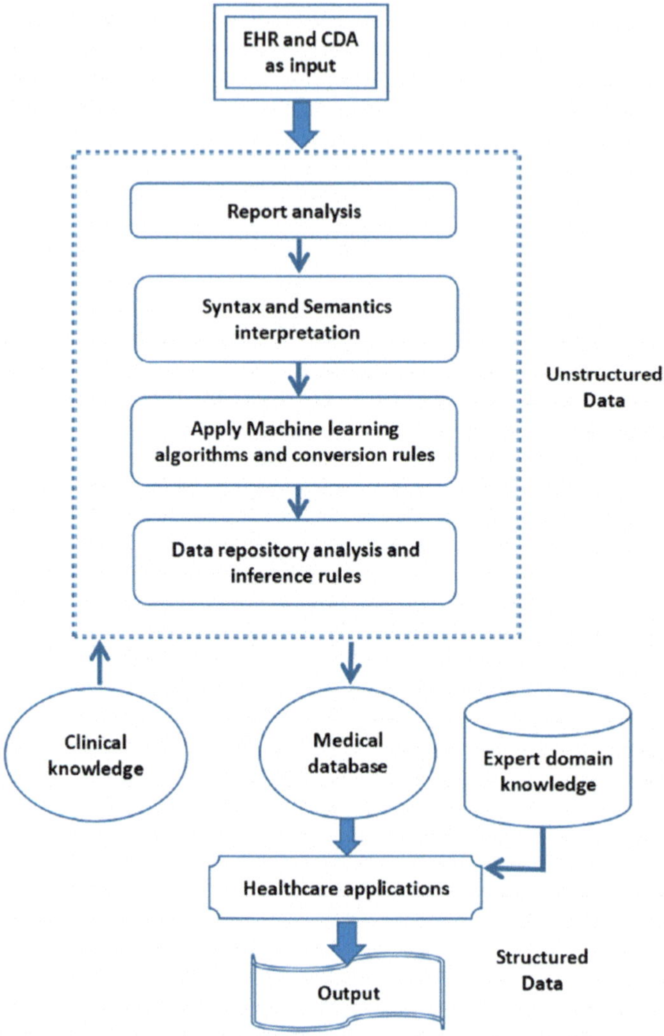

Fig. 2 Integration of machine learning with IoT

be prone to human error. ML, particularly deep learning, offers a powerful solution to augment the capabilities of medical professionals.

4.6 Deep Learning and CNNs in Medical Image Analysis

Deep learning, a subset of ML, has shown exceptional promise in image recognition tasks. Convolutional Neural Networks (CNNs), a type of deep learning model, are particularly well-suited for image analysis due to their ability to automatically learn spatial hierarchies of features from input images. CNNs have multiple layers that process data, identifying and learning from patterns that represent various features of the images.

4.7 Structure of CNNs

Convolutional Layers: These layers apply convolution operations to the input image, creating feature maps that detect patterns such as edges, textures, and shapes.

Pooling Layers: These layers reduce the dimensionality of the feature maps, making the model more computationally efficient and focusing on the most critical features.

Fully Connected Layers: These layers use the learned features to classify the image into different categories or predict specific outcomes.

4.8 Applications in Healthcare

The application of CNNs in medical image analysis has revolutionized several areas within healthcare: CNNs can be trained to recognize the presence of tumors in various types of medical images. For example, in mammograms, CNNs can detect breast cancer with accuracy comparable to that of experienced radiologists. The ability to identify tumors early can significantly improve patient outcomes.

4.9 Fracture Identification

In radiology, CNNs are used to detect fractures in bones. These models can highlight areas of concern that may be missed by the human eye, thus providing a second layer of scrutiny to ensure accurate diagnoses.

4.10 Retinal Disease Screening

CNNs are employed in ophthalmology to analyze retinal images for signs of diseases such as diabetic retinopathy and age-related macular degeneration. Early detection of these conditions can prevent severe vision loss.

4.11 Lung Disease Diagnosis

CNNs have been applied to chest X-rays and CT scans to identify lung diseases, including pneumonia, tuberculosis, and COVID-19. These models can assist in rapid diagnosis, which is crucial in managing infectious diseases and improving patient care.

Case Study: Cancer Detection

A notable case study involves the use of ML for detecting breast cancer from mammograms. By training CNNs on large datasets of labeled mammograms, researchers have developed models that can identify malignant tumors with accuracy comparable to experienced radiologists. These models assist in early detection, significantly improving patient prognosis.

4.12 Disease Diagnosis

ML models are also used to diagnose diseases by analyzing a combination of multimedia data and patient records. Techniques such as Natural Language Processing (NLP) can analyze clinical notes and electronic health records to extract relevant information for diagnosis. Additionally, ML algorithms can process data from wearable devices that monitor vital signs, providing continuous health assessments.

Case Study: Cardiovascular Disease Prediction

In a case study on the prediction of cardiovascular disease, data from wearable devices that track blood pressure, heart rate, and other vital indicators are analyzed using machine learning models. ML algorithms can forecast the chance of cardiovascular events by combining this data with patient medical histories. This enables early intervention and individualized treatment regimens.

4.13 Personalized Treatment Recommendations

Personalized medicine aims to tailor medical treatment to individual patients based on their unique characteristics and medical histories. Massive volumes of data can be analyzed by ML algorithms, which can then be used to suggest individualized treatment strategies that maximize patient results. This involves considering factors such as genetic information, lifestyle, and response to previous treatments.

Case Study: Diabetes Management

In diabetes management, ML models analyze blood glucose levels, dietary habits, physical activity, and medication adherence to recommend personalized insulin dosages and lifestyle modifications. These recommendations help patients maintain optimal blood glucose levels, reducing the risk of complications.

5 Industrial Automation

In the domain of industrial automation, ML techniques are employed to analyze multimedia data from sensors, cameras, and IoT devices deployed in manufacturing plants, logistics centers, and supply chain operations. Case studies illustrate the use of ML models for predictive maintenance, quality control, and process optimization. The adoption of Internet of Things (IoT) technologies in industrial automation represents a transformative shift towards smarter, more interconnected manufacturing and supply chain operations. This chapter explores how IoT devices, combined with advanced data analytics and automation, enhance efficiency, productivity, and safety in industrial setting. The following figure shows the Role of IoT in industrial automation (Fig. 3).

5.1 Predictive Maintenance and Supply Chain Automation

IoT sensors monitor equipment health indicators continuously, enabling predictive maintenance strategies. By analyzing data patterns and detecting anomalies, predictive maintenance systems anticipate equipment failures and schedule maintenance activities before breakdowns occur, minimizing downtime and maintenance costs. In contemporary firms, supply chain optimization is essential for raising productivity, cutting expenses, and raising customer happiness. This chapter examines how real-time visibility, predictive analytics, and automation across several supply chain stages are made possible by Internet of Things (IoT) technologies, which are revolutionizing

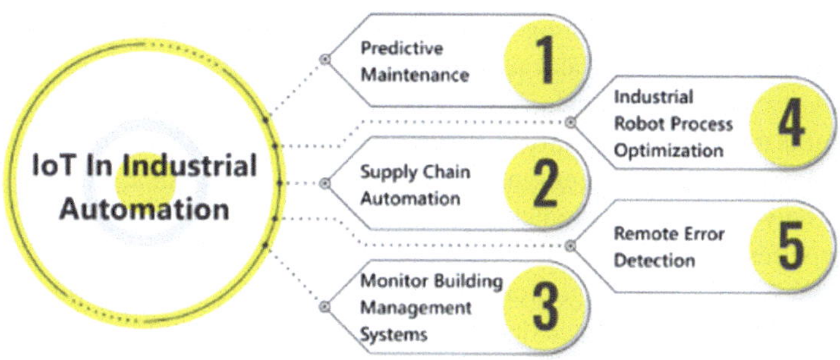

Fig. 3 IoT in industrial automation

supply chain management. Supply chain management is being revolutionized by IoT technologies, which offer real-time visibility, predictive analytics, and automation capabilities across international networks. Despite challenges, the integration of IoT-driven supply chain optimization enhances efficiency, reduces costs, and improves customer satisfaction in dynamic business environments. Future innovations in AI, blockchain, and edge computing will continue to shape the future of IoT-enabled supply chains, driving sustainable growth and resilience in global industries.

5.2 *Monitoring Building Management System*

In order to maximize the operational effectiveness, comfort, and sustainability of commercial and residential buildings, building management systems, or BMS, are essential. This chapter examines the ways in which Internet of Things (IoT) technologies improve building management systems (BMS) by facilitating energy-efficient building operations, real-time monitoring, and predictive maintenance. IoT technologies transform building management systems by enhancing operational efficiency, optimizing energy usage, and improving occupant comfort and safety. Despite challenges, the integration of IoT-driven solutions in BMS offers significant benefits for building owners, facility managers, and occupants. Future innovations in AI, edge computing, and smart city initiatives will continue to shape the evolution of IoT-enabled smart buildings, driving sustainability and resilience in urban environments.

5.3 *Industrial Robot Process Optimization*

Industrial robot process optimization involves leveraging advanced technologies to streamline manufacturing processes, enhance productivity, and ensure precision in industrial operations. By integrating Internet of Things (IoT) devices, artificial intelligence (AI), and machine learning (ML) algorithms, industrial robots can perform tasks with increased efficiency and accuracy. One critical aspect of process optimization is predictive maintenance, where IoT sensors monitor the health and performance of robots and machinery in real-time. These sensors collect data on factors like temperature, vibration, and operational metrics, which is then analyzed using AI-driven predictive models. By predicting equipment failures before they occur, maintenance activities can be scheduled proactively, minimizing unplanned downtime and reducing maintenance costs significantly. Moreover, AI and ML algorithms enable robots to learn from data collected during operations. This capability allows robots to optimize their movements and tasks over time, adapting to changing conditions and improving their performance. For example, in manufacturing processes requiring precise assembly or welding, robots equipped with ML algorithms can adjust their movements based on feedback from sensors and historical data, ensuring consistent quality and reducing defects. Furthermore, industrial robot process optimization extends to collaborative robotics (cobots), where robots work alongside human operators in shared workspaces. IoT-enabled cobots enhance safety by monitoring human–robot interactions and adjusting their speed and force to prevent accidents. This partnership not only increases production but also optimizes the workplace's general ergonomics, lessening the pressure on employees.

5.4 Remote Error Detection

Remote error detection refers to the capability of monitoring and identifying issues in systems or equipment from a distant location, typically through the use of advanced monitoring technologies and communication networks. This approach is particularly valuable in industries such as manufacturing, telecommunications, and IT infrastructure, where downtime and operational disruptions can lead to significant losses in productivity and revenue. The implementation of remote error detection systems often involves the deployment of Internet of Things (IoT) devices equipped with sensors that collect real-time data on various parameters such as temperature, pressure, vibration, and performance metrics. These sensors continuously monitor the operational conditions of machinery and equipment, transmitting data to centralized monitoring platforms or cloud-based analytics systems. Algorithms for machine learning (ML) are essential for assessing the incoming data streams. ML models can identify abnormalities, patterns, and departures from typical operating circumstances that might point to possible problems or errors by analyzing massive volumes of sensor data. This proactive approach allows maintenance teams to identify issues early, even before they manifest as critical failures, enabling timely interventions and preventive maintenance actions. Moreover, remote error detection enhances operational efficiency by minimizing downtime and reducing the need for on-site inspections and manual diagnostics. Maintenance personnel can access real-time alerts and diagnostic reports remotely, enabling them to prioritize tasks, allocate resources efficiently, and respond promptly to emerging issues.

In industries where equipment is geographically dispersed or located in remote or hazardous environments, remote error detection systems provide a cost-effective solution for monitoring and managing assets effectively. By leveraging IoT-enabled sensors, advanced analytics, and remote connectivity, organizations can improve reliability, extend asset lifespan, and optimize overall operational performance.

Empirical evaluations of ML-based industrial automation systems demonstrate their ability to analyze complex data streams, detect equipment failures, and optimize production workflows. By leveraging real-time analytics and anomaly detection algorithms, these systems can improve operational efficiency, reduce downtime, and enhance overall productivity in industrial environments. Industrial automation, powered by Machine Learning (ML) and the Internet of Things (IoT), represents a paradigm shift in manufacturing, logistics, and supply chain operations. This chapter explores how ML techniques analyze data from sensors, cameras, and IoT devices to optimize processes, ensure quality control, and enable predictive maintenance in industrial settings. Industrial environments generate vast amounts of data from sensors embedded in machinery, production lines, and logistics operations. ML algorithms integrate data from various sources, including IoT devices, to provide insights into operational performance and potential improvements. ML models in industrial automation predict equipment failures by analyzing historical data from sensors. This proactive approach helps in scheduling maintenance activities before breakdowns occur, thereby minimizing downtime and reducing maintenance costs.

Case Study: Predictive Maintenance in Manufacturing

In manufacturing plants, ML algorithms analyze sensor data from machines to detect anomalies indicative of potential failures. Manufacturers can eliminate unplanned downtime, manage maintenance schedules, and increase equipment longevity by implementing predictive models. ML techniques enhance quality control processes by analyzing data from sensors and cameras to identify defects or deviations from production standards. Automated inspection systems powered by ML algorithms can detect subtle flaws in products with high accuracy and speed.

Case Study: Automated Visual Inspection

In automotive manufacturing, ML-based vision systems inspect paint quality and detect surface defects on vehicles. By analyzing images captured by cameras, these systems ensure that finished products meet stringent quality standards before they reach consumers. ML algorithms optimize industrial processes by analyzing real-time data streams to identify inefficiencies and improve workflow designs. By uncovering patterns and correlations in production data, ML-driven optimization enhances operational efficiency and reduces waste.

Case Study: Supply Chain Optimization

Logistics centers use ML to optimize inventory management and distribution operations. By analyzing historical sales data and real-time demand forecasts, ML models predict inventory requirements, minimize stockouts, and streamline supply chain logistics.

6 Future Directions

6.1 Emerging Trends

The intersection of Machine Learning (ML) and Multimedia Internet of Things (IoT) continues to evolve, driven by emerging trends that shape the future of analytics and insights generation. One prominent trend is the advancement of edge computing technologies, enabling ML models to be deployed directly on IoT devices for real-time processing and inference. Edge AI platforms and edge-native ML algorithms empower devices to perform complex analytics tasks locally, reducing latency, bandwidth requirements, and reliance on centralized infrastructure. The use of federated learning approaches in Internet of Things contexts is another developing trend that enables dispersed edge devices to cooperatively train machine learning models without exchanging sensitive data. Federated learning allows for scalable model updates and improvements across many IoT ecosystems while maintaining data confidentiality and privacy. By leveraging federated learning, organizations can harness the collective intelligence of edge devices to achieve superior analytics performance and insights generation.

Furthermore, the convergence of ML with other emerging technologies, such as 5G networks, blockchain, and quantum computing, presents new opportunities for Multimedia IoT analytics. 5G networks provide ultra-low latency and high bandwidth connectivity, enabling seamless transmission of multimedia data streams between IoT devices and cloud infrastructure. Blockchain technology ensures data integrity, transparency, and auditability in IoT data exchanges, enhancing trust and security in ML-driven analytics applications. Quantum computing promises exponential gains in computational power, enabling the rapid training and inference of ML models on massive datasets, revolutionizing Multimedia IoT analytics capabilities.

6.2 Open Challenges

Despite the promising opportunities, several open challenges remain in the field of Multimedia IoT analytics, necessitating further research and innovation. One key challenge is the development of lightweight and energy-efficient ML algorithms tailored for resource-constrained IoT devices. Current ML models often require substantial computational resources and memory, posing scalability and power consumption concerns in edge computing environments. Addressing these challenges requires the design of efficient algorithms optimized for edge deployment, leveraging techniques such as model quantization, pruning, and knowledge distillation.

Another challenge is the interoperability and standardization of data formats, communication protocols, and ML frameworks in heterogeneous IoT ecosystems. The lack of interoperability hinders seamless integration and exchange of multimedia data between diverse devices and platforms, impeding the development of unified analytics solutions. Standardization efforts, industry collaboration, and open-source initiatives are essential for establishing common frameworks and protocols that enable interoperability and facilitate cross-platform analytics in IoT environments.

Additionally, ensuring data privacy, security, and ethical considerations in ML-driven Multimedia IoT analytics remains a critical challenge. As IoT deployments collect and process sensitive multimedia data, protecting user privacy and preventing data breaches are paramount. Robust security measures, encryption techniques, and privacy-preserving algorithms are required to safeguard multimedia data throughout its lifecycle, from collection to analysis to dissemination. Furthermore, ethical guidelines, regulatory frameworks, and governance policies are necessary to ensure responsible and transparent use of ML technologies in IoT applications, addressing concerns related to bias, fairness, and accountability.

7 Conclusion

In conclusion, this research highlights the transformative potential of integrating machine learning (ML) with Multimedia Internet of Things (IoT) to enhance data interpretation and utilization. Utilizing diverse machine learning models, especially deep learning architectures, we have exhibited noteworthy advancements in tasks including voice analysis, image and video identification, and sensor data interpretation. Our study addresses critical challenges like data heterogeneity, real-time processing, and scalability, showcasing effective solutions for handling diverse data types and ensuring low-latency performance. The empirical evaluations and case studies presented underscore the ability of ML to provide advanced analytics in IoT environments, offering valuable insights for future research and practical applications. This work paves the way for more intelligent and efficient IoT systems, emphasizing the importance of continued exploration of new ML techniques and the resolution of emerging challenges.

References

1. Simonyan, K., Zisserman, A.: Very deep convolutional networks for large-scale image recognition. arXiv:1409.1556 (2014)
2. Szegedy, C. et al.: Going deeper with convolutions. In: Proceedings of the IEEE Conference on Computer Vision and Pattern Recognition (CVPR), pp. 1–9 (2015)
3. Vinyals, O., Toshev, A., Bengio, S., Erhan, D.: Show and tell: A neural image caption generator. In: Proceedings of the IEEE Conference on Computer Vision and Pattern Recognition (CVPR). pp. 3156–3164 (2015)
4. Chollet, F.: Xception: deep learning with depthwise separable convolutions. In: Proceedings of the IEEE Conference on Computer Vision and Pattern Recognition (CVPR), pp. 1251–1258 (2017)
5. Kingma, D.P., Welling, M.: Auto-encoding variational bayes. arXiv:1312.6114 (2013)
6. Mikolov, T., Chen, K., Corrado, G., Dean, J.: Efficient estimation of word representations in vector space. arXiv:1301.3781 (2013)
7. Abadi, M. et al.: TensorFlow: a system for large-scale machine learning. In: Proceedings of the 12th USENIX Symposium on Operating Systems Design and Implementation (OSDI), pp. 265–283 (2016)
8. LeCun, Y., Bengio, Y., Hinton, G.: Deep learning. Nature **521**, 436–444 (2015)
9. Goodfellow, Bengio, Y., Courville, A.: Deep Learning. MIT Press (2016)
10. He, K., Zhang, X., Ren, S., Sun, J.: Deep residual learning for image recognition. In Proceedings of the IEEE Conference on Computer Vision and Pattern Recognition (CVPR), pp. 770–778 (2016)

11. Lee, H., Pham, P., Largman, Y., Ng, A.Y.: Unsupervised feature learning for audio classification using convolutional deep belief networks. In: Proceedings of the Advances in Neural Information Processing Systems (NIPS), pp. 1096–1104 (2009)
12. Nair, V., Hinton, G.E.: Rectified linear units improve restricted boltzmann machines. In: Proceedings of the 27th International Conference on Machine Learning (ICML), pp. 807–814 (2010)
13. Russakovsky, O., et al.: ImageNet large scale visual recognition challenge. Int. J. Comput. Vis. (IJCV) **115**(3), 211–252 (2015)
14. Bahdanau, D., Cho, K., Bengio, Y.: Neural machine translation by jointly learning to align and translate. arXiv:1409.0473 (2014)

IoT-Based Hand Hygiene Dispenser with Temperature and Level Monitoring Systems Using Data Analysis

Ahmed J. Obaid, Risa Triassanti, Mohammed Ayad Alkhafaji, and Muthmainnah Muthmainnah

Abstract In the aftermath of the COVID-19 pandemic, maintaining good hygiene, particularly hand hygiene, is crucial to preventing the disease. We present a concept for an IoT-based hand hygiene dispenser that monitors temperature and level. It addresses the need for efficient, touchless hand hygiene solutions. Ultrasonic sensors, relays, and solenoids are used in conjunction with a microcontroller to automate the sanitizer dispensing process. A temperature sensor is also incorporated, and RFID tags are used to store personal data, which can be accessed using a Wi-Fi module to track contacts. It reduces cross-contamination risks in public spaces by utilizing touchless technology. Moreover, the system integrates the protocol to enable real-time tracking and reporting, making it a valuable tool in the fight against infectious diseases, especially in densely populated areas.

Keywords Internet of things · Hand hygiene · Monitoring system · Hygiene dispenser · COVID-19 · Safety

1 Introduction

Using hand sanitizers and washing hands with water and soap regularly are the primary preventive measures for reducing the spread of the COVID-19 pandemic, according to the WHO (World Health Organization). As a general rule, hygiene is

A. J. Obaid (✉)
Faculty of Computer Science and Mathematics, University of Kufa, Najaf, Iraq
e-mail: ahmedj.aljanaby@uokufa.edu.iq

R. Triassanti
Universitas PGRI Ronggolawe Tuban, Tuban, Jawa Timur, Indonesia

M. A. Alkhafaji
College of Technical Engineering, Dhi Qar, National University of Science and Technology, Nasiriyah, Iraq

M. Muthmainnah
Universitas Al Asyariah Mandar, Polewali Mandar, Indonesia

R. Sharma et al. (eds.), *Multimedia Technologies in the Internet of Things Environment, Volume 4*, Studies in Big Data 173, https://doi.org/10.1007/978-981-96-4356-1_2

the practice of keeping healthy and preventing disease, especially through cleanliness. Examples of hygiene include washing hands, coughing in one's elbow, etc. In order to prevent any diseases that can be spread via contact, it is important to wash your hands. As the Internet of Things (IoT) becomes more ubiquitous and important, especially as they enable "smart city" applications such as parking and waste management, their importance and ubiquity will increase [1–3]. This type of application has been described in several studies, and challenges associated with its implementation have been discussed [4, 5]. Using LoRaWAN transceivers, smart cities can now be monitored using IoT devices regardless of whether they are urban, suburban, or rural. In smart city applications, LoRaWAN has been used to monitor traffic systems (such as parking lots and traffic lights [6]), infrastructure (such as lighting control), and the urban environment (such as air quality and utility metering [7–10]). As part of e-health solutions, LoRaWAN is widely used not just for monitoring and controlling assets but also for monitoring human health on a large scale [11, 12]. The monitoring of human health conditions on an urban scale and location, such as blood pressure, glucose level, and temperature, is one example. As a result of SARS-CoV infection, COVID-19 has been identified. It has been a very tense year since 2019. It is important to wash hands frequently with an alcohol-based hand sanitizer or soap and water in order to avoid infection with COVID-19, as well as to check the temperature of the water to prevent infection. In all places, hands had to be washed frequently with soap and water. Constantly washing hands with soap and water was uncomfortable at all times.

It uses a microcontroller AMIGA328P to control the flow of hand sanitizer using the ultrasonic sensor, relay and solenoid valve, and a temperature sensor to measure the temperature. The controller is non-contact and uses a temperature checkup using the controller. In this way, power could be supplied through a USB cable or external power supply. RFID tags are designed and used before people enter gates. Information about the person can be found on the Tag [13]. A Wi-Fi module is used to store temperature information and information for reference. Digitalization is expected to have the greatest impact on communication systems that involve fast information transmission. However, many other technical fields will also be affected by it [14]. There are several different types of antimicrobial hand sanitizers on the market, but most are liquids, gels, or foams that kill bacteria, microbes, and microorganisms. In common handwashing procedures, both water and cleanser are used. A hand sanitizer provides a more effective and convenient alternative to water and cleanser when they are not available. There has been frequent use of hand sanitizer dispensers and foot-operated hand sanitizer allocators during the COVID chapter. In this case, however, transmission is more likely since human contact is involved. Intensity levels can be arranged through the use of both contact and non-contact devices and structures. Mercury is used in many places to measure internal intensity levels as part of routine procedures, including hand sanitizer dispensers and foot-operated devices. However, all of these processes involve human contact, facilitating virus transmission. The level of internal heat should also be evaluated regularly.

In recent years, COVID-19, an infectious disease with a high death toll, has spread rapidly around the world, making it imperative to maintain proper hygiene in public areas. As part of infection control efforts, hand hygiene plays a key role [15]. Installing dispensers that are easily accessible and convenient boosts compliance with hand hygiene [16]. Hand-held temperature scanners and bottled sanitizers are commonly used in many places, but because these methods are manual and involve human intervention, there is a risk of cross-contamination. Recent studies suggest that electronic devices significantly influence hand hygiene compliance. Phases two and three of the study showed an improvement in compliance from 36.3% at baseline to 70.1% at a 30-bed academic medical center hematology unit [17]. With contactless technology, the sterility of the environment is further enhanced. It is important to develop and implement devices with touch-free features in order to limit the spread of cold and flu viruses when using contact-based devices. Infection prevention is facilitated by existing touchless technology, but its features are limited [18]. There is a pressing need for multi-purpose devices that can sanitize hands, monitor temperatures, and store suspected infections in a server. All these features are incorporated into the automated sanitizing device, which speeds up and simplifies sanitization. With an IOT-enabled device, health department authorities can communicate more effectively with people who are experiencing high temperatures thanks to a web-based database. In the event that a sick person needs to be treated, the server informs the health authorities and police. As a result of the system, suspect cases will be prevented from entering public places, and contacts will be tracked down, ultimately preventing infections from spreading.

2 Related Work

During the current pandemic, LoRaWAN-based IoT applications can also help improve public health. A large number of facilities can benefit from monitoring sanitizer usage using IoT techniques. Recent advances in the Internet of Things technologies make it possible to monitor public hygiene in real time or close to real time with wireless-sensitized infrastructure. Smart and connected devices, wireless communication, and ubiquitous computing enable rapid and reliable data and information transfers [19]. IoT-based sanitizer stations have been implemented in recent studies, as shown in Table 1.

Analyzed the knowledge supported by social media based on three approaches: content analysis, descriptive analysis, and network analysis. This review examines IoT and the challenges associated with it. Our article provides an overview of the Internet of Things, its technology, its applications, and the future scope of research in this area. A comparison of the various models and technologies for IoT based on their current state. A metaphysics-supported design was developed by the author [25] for observing health-related travails and providing recommendations to chronic disease patients. A model developed for the system was found to be economically efficient for making inferences concerning the context. The designation is automatically generated based on the existing data, questionnaires, and confirmed or suspected results. Smart town communication protocols need to be standardized. Our research

Table 1 Research review on related topics

Refs	Application scenario	Deployed techniques
[20]	Implementing a smart hand washing station that provides real-time feedback (relating to effective hand hygiene)	Unwashed areas of hands are detected and displayed using a camera, UV light, and tablet attached to the hand washing station
[21]	Monitors the use and remaining level of sanitizer at a sensitized sanitizer station	On the sanitizer station, a liquid stage sensor with Wi-Fi communication is installed
[19]	Sensitized sanitizer stations can monitor hand hygiene compliance	RFID tags and infrared sensors monitor hand hygiene compliance, and data is communicated by Wi-Fi and ZigBee
[22]	Designed to monitor medical staff's compliance with hand hygiene	Hand cleaning behaviors are monitored via RFID tags; data is communicated via ZigBee
[23]	Monitor station for sensitive sanitizers	An ultrasonic sensor monitors the sanitizer bottle, and the data is transmitted via Wi-Fi
[24]	With the smart control dispenser, the sanitizer will be served automatically, and waste will be reduced	Detects human hands with an ultrasonic sensor

was conducted using the Centers for Disease Management and Prevention (CDC, USA) website and a PubMed literature review to gather information about metabolic process syndromes (MERS), severe acute metabolic process syndromes (SARS), and COVID-19, as well as clinical signs and symptoms, treatment, diagnosis, transmission strategies, protection strategies, and risk factors [26]. Incorporated IoT applications into the medical and healthcare industries and highlighted the benefits. Design a helmet equipped with thermal imaging structures to identify gang members who are contaminated. A face recognition system is also included. An infrared camera scans the gang, and if heat is detected from any male or female, an optical camera will capture the face. Additionally, it provides GPS location information for the contaminated individual.

A better-performing and lower-power microcontroller is used in the automated sanitizer than an Arduino-based microcontroller. ThingSpeak's cloud also depends on the Internet of Things to update it daily with data captured by the automated sanitizer. ThingSpeak cloud allows building and transportation system administrators to track measurements every 30 min. Thus, the administrator can see how often the sanitizer is used each day or every month. Figure 1 represents a block diagram of the proposed dispenser for automatic sanitizers. System subsystems include cloud blocks, input blocks, and output blocks.

To prevent coronavirus, it is important to get vaccinated, stay at home, wear masks while in public, stay away from crowded places, keep a safe distance from others, ventilate indoor spaces, manage exposure durations, wash hands frequently and for at least twenty seconds, use hand sanitizer, and maintain good respiratory hygiene. A common indoor measure is to remain at home with high body temperatures, wear a mask, and maintain a distance of 1.5–2 m between people [27]. Created a system

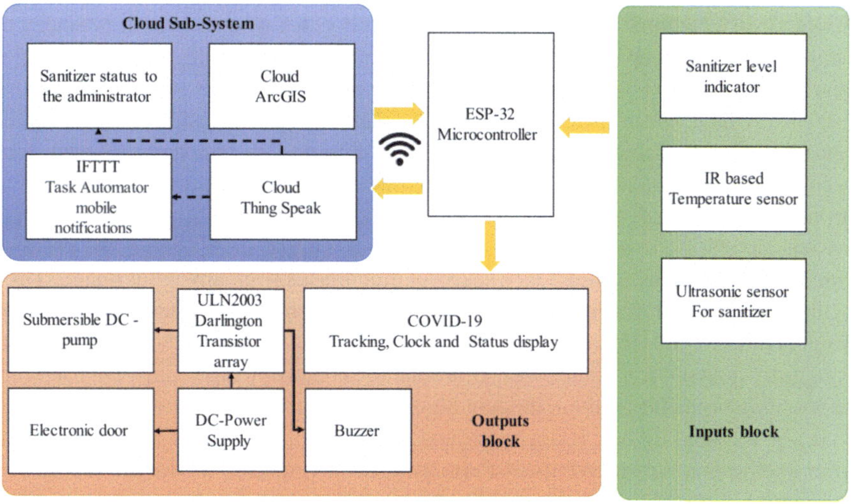

Fig. 1 This proposal for an automatic sanitizer contains the following features

to monitor and control the device automatically. IoT is beneficial for temperature-controlled devices, which trigger notifications to the client's phone when they are turned on or off. Fog Computing Services was used to develop a real-time notification system for healthcare monitoring based on IoT. The author [28] one of the benefits of IoT is the ability to quickly process and compute big data, which is efficiently generated and controlled across a wide range of applications. To accomplish this, optical sensors and ultrasonic sensors were used to detect hands so that liquids could be dispensed onto hands, the electromagnetic lock could be de-energized, and the second servo motor could open the entrance door. Based on [29], the authors developed a system with a structure divided into four parts: monitoring, information storage, analytics management, and visualizations. In the study, the author asserts that wireless sensor innovation (WSN) and the Internet of Things (IoT) are capable of improving Indoor Air Quality monitoring. It monitors a person's health and prevents illness using a combination of infrared thermometers, ultrasonic sensors, and air quality sensors. The following is the organization of this paper. Our methodology will be discussed in Sect. 2, our results will be explained in Sect. 3, and our conclusion will discuss future research.

In [30], the authors presented a method for detecting correct and incorrect face mask wear in real time. Using Convolution Neural Networks (CNNs) and transfer learning, we propose detecting additional errors that may contribute to COVID-19 spread. This task currently has no training set, which was detected as the main problem. In order to achieve this, the authors asked citizens to take different selfies and place masks in different positions using an app. Their solution consisted of examining the outcome and effect of the achieved result. A real-life Android app demo was developed to test the validity of the proposal. In [31], a smart city network with video surveillance can reduce COVID-19 transmission by detecting those without facial masks. A city network alerts the appropriate authorities when a person is detected without a mask. A deep learning architecture is trained by collecting images of people wearing masks and those without masks from various sources. Based on previously unseen test data, the trained architecture was able to distinguish between people wearing facial masks and people without. The findings of this study are expected to contribute to reducing the spread of communicable diseases in many countries throughout the world. Based on a binary face classifier, this algorithm can detect any face appearing in a frame regardless of its alignment. In this paper, we present a method for creating accurate face segmentation masks from any input image. This method extracts features from RGB images of any size by using Predefined Training Weights of the VGG-16 architecture [26]. This image is semantically segmented out using fully convolutional networks. Binomial cross-entropy algorithms are used for the loss function, and gradient descent algorithms are used for training. To improve accuracy, a bounding box is drawn around each face of the FCN's output image in order to eliminate unwanted noise. Furthermore, the proposed model can recognize no frontal faces with great accuracy. Furthermore, it detects multiple facial masks at once. For the segmented face masks, the multi-parsing human dataset proved to be 93.884% accurate at the pixel level.

Despite the availability of photo resistor-based dispensers, an analysis of the online and offline markets revealed that they are not available. To reduce complexity, infrared sensors are used in most dispenser systems, and ultrasonic sensors are also used in some systems. This unique concept of making dispensers using light-dependent resistor (LDR) sensors and laser light has received little attention from scientists, according to the authors. User-friendliness and cost-effectiveness were two of the key objectives of the dispenser device. These goals guided material selection and actuator operation. Additionally, Fig. 2 illustrates the flow chart for constructing the dispenser device.

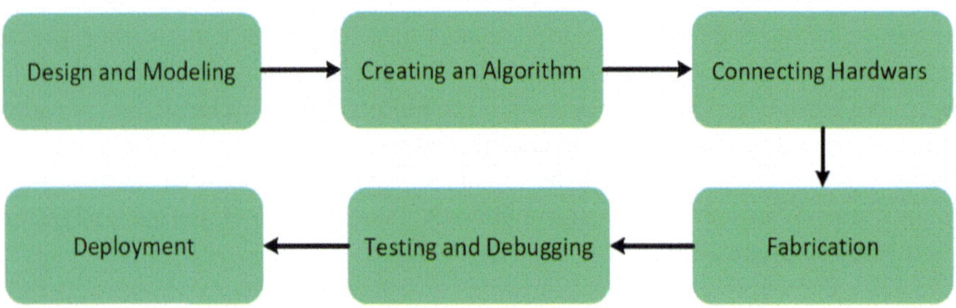

Fig. 2 The working process is shown in this flow chart

3 Proposed Architectures and Design Model

During this section, we will examine the design, planning, and evaluation of the system. In addition, the functional and non-functional requirements were discussed, as well as the component's functionality. The proposed system consists of four types of smart sensors: water, infrared, ultrasonic, and ping. It is equipped with a 2.2-inch TFT LCD, a buzzer, jumper wires, a mega breadboard, an Arduino microcontroller, a smart dispenser, and a battery. Data is transmitted using the Internet Protocol Stack. One component gets data, and the other outputs action, which is the actuator, so there are two categories of components. Each device is interconnected via Bluetooth short-range communication throughout the facility. An alarm buzzer/pictogram is displayed on the LCD of the dispenser as it dispenses soap. Using this tool, a user can learn how to clean their hands. An essential component of the Software Engineering Paradigm is the Software Development Life Cycle (SDLC), which consists of seven phases, including Requirement Analysis. The purpose of this section is to explain what the system will do or how it will react after it is installed in the facility and when users interact with it.

In this section, we discuss the processes that should be available to enable the system to reach its potential. As an example, it offers smart alerts through its IoT platform.

3.1 Software for Proposed Architecture

Data is stored in memory by variables assigned to each device. Software development uses various types of data, which are also called variables. Some types of data take up more memory than others. It is necessary to link this software to the system by using the microcontroller, which consists of many pins that can be accessed by name, just as variables can be accessed by name [32]. A user receives and distributes data by using INPUT, OUTPUT, LOW, and HIGH commands, as well as digitalWrite and pinMode APIs. When they are used on a device, such as an on/off switch, they mean off/on, respectively. The Loop API flowchart is shown in Fig. 3. Accordingly, the facility-sealed seat opens and closes based on the distance between the user and the seat. For example, if the user is 3–20 cm away from the seat, the seat will open.

3.2 Proposed Algorithm

In the proposed system, flush counters (water sensors) are incorporated to collect flush counts. Smart dispensers turn off their buzzers when soap is dispensed. Figure 4 illustrates an image of the facility's system and how components are linked to provide service. In a microcontroller-based system, components send data to the

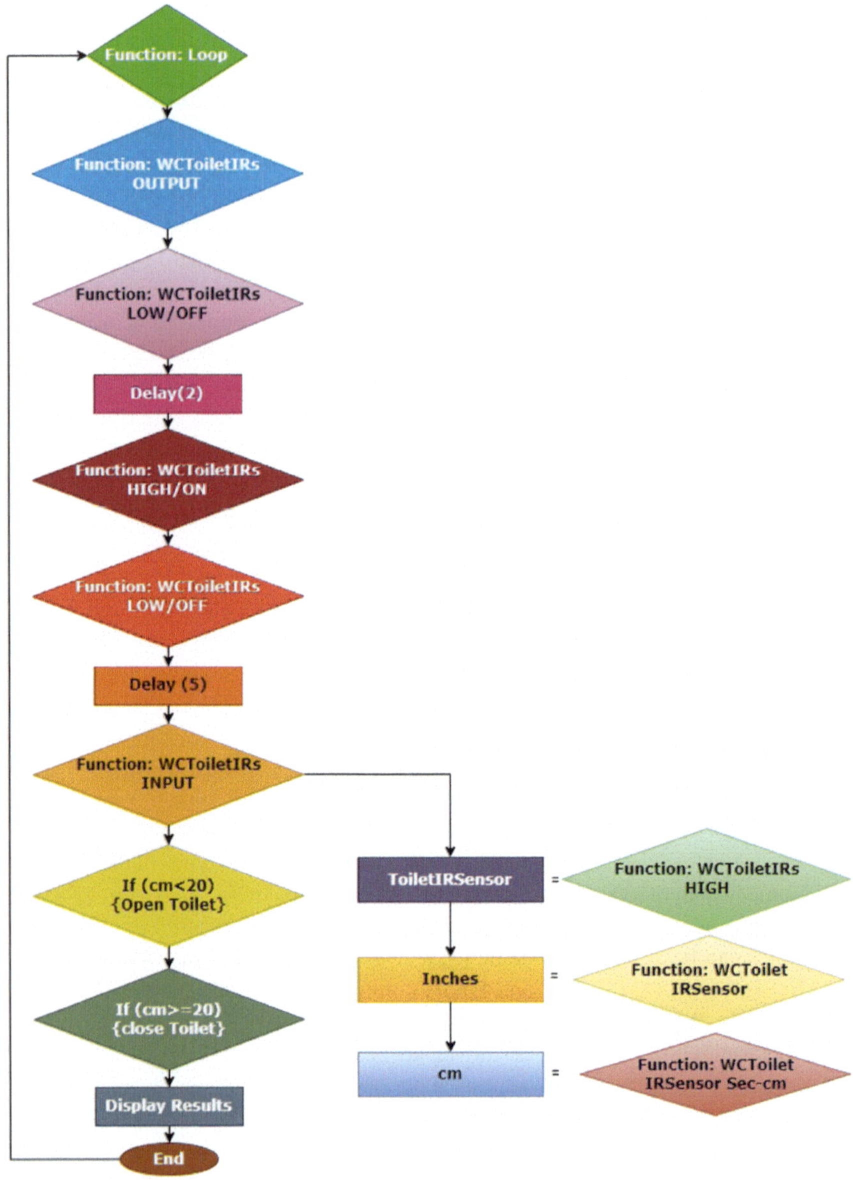

Fig. 3 Flow chart for loop functions

board-microcontroller via wired communication links, which are relayed to the next component [33]. IoT platforms collect, store, and analyze data through an Internet Protocol stack and a local database. Protocol stacks are composed of five layers, including physical, link, network, transport, and application layers; each layer acts on and serves data. SMTP and HTTP are Application Layer Protocols. While HTTP is a web protocol, SMTP is a messaging protocol that manages messages between two end systems. Application messages are transported between application endpoints through the Internet's Transport Layer, which uses either TCP or UDP protocols [34].

As network packets, or datagrams, are sent between hosts, this layer determines the packet's destination by using the IP address of the host [18]. Communication over a specific link is handled by the Link Layer and Physical Layer, which are implemented in interfaces such as Ethernet or Wi-Fi [34]. When data packets need

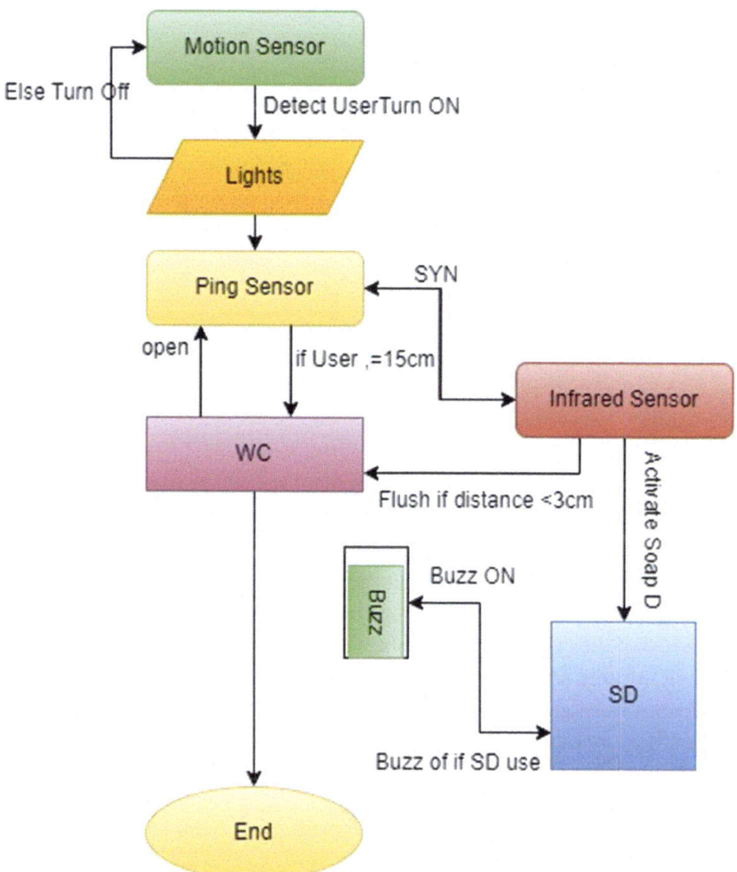

Fig. 4 A flowchart illustrating the proposed model

to be moved from one node to another in the route, network layers pass them to link layers at each node.

A loop is shown in Fig. 6 to demonstrate how the algorithm is developed and re-run to deliver the system's functionality. Users are able to access facility seats when they stay within a certain distance of the ping sensor. Ping sensors and infrared sensors are synchronized to enable flushing for facility seats. The infrared sensor further activates Smart dispensers; they include alarms and LCD screens that work together. Until the user activates the dispenser, the alarm remains activated. When the user engages with the dispenser, the alarm is deactivated, and the dispenser returns to its original state. The project requires several items, including software and hardware, to provide the required service.

4 Result Analyses and Discussion

A public building can be made bacteria-free by using Sensor Network Automated Hand Hygiene. People can neglect hand hygiene after engaging with the facility. Various devices and sensors are used in the system to achieve this goal, as well as microcontrollers, TFT LCDs, IDEs, etc. When people forget to perform hand hygiene, their LCD screen prompts them to do so, and an alarm in the smart dispenser reminds them. During this type of testing, the client's requirements are checked, and the results are produced according to these requirements. Soap Dispenser is subdivided into two subclasses: a base class and four dependent classes. In response to user engagement, each class responds to the trigger required. Tests were conducted to confirm that the components were connected. Also, the distance sensor activates the dispenser so that it displays a pictogram to remind users to wash their hands and sounds an alarm.

As a result of the proposed system, this section analyzes the results obtained from sensors. A variety of graph shapes are used to analyze and present the data. Based on serial plot analysis with Arduino IDE, Fig. 5 shows the results. On the right top corner are four little squares, one for each of the four sensors: the seat sensor, LCD sensor, LED sensor, and flush sensor. Graphs on the X and Y axes analyze the system's use or idleness, and the Y axis shows the duration of idleness or usage.

An overview of the collected data is shown in Fig. 6, which is based on three types of sensors: an infrared sensor at the facility seat, a distance sensor at the flush kiosk, an infrared sensor at the dispenser, and an ultrasonic sensor at the light. Due to the sensors' loop design, data is continuously collected around the facility environment until a user approaches the sensor. A line graph illustrates how numbers are repeated and progressed. Users can activate flush sensors whenever they activate a seat, and seat sensors detect when a seat is occupied. When a user steps into the facility toilet environment, a light sensor is activated.

A description of how the system interacts is provided. As a result of 85, the environment is illuminated, the flush is activated, and the system for illuminating the area is activated (Fig. 7).

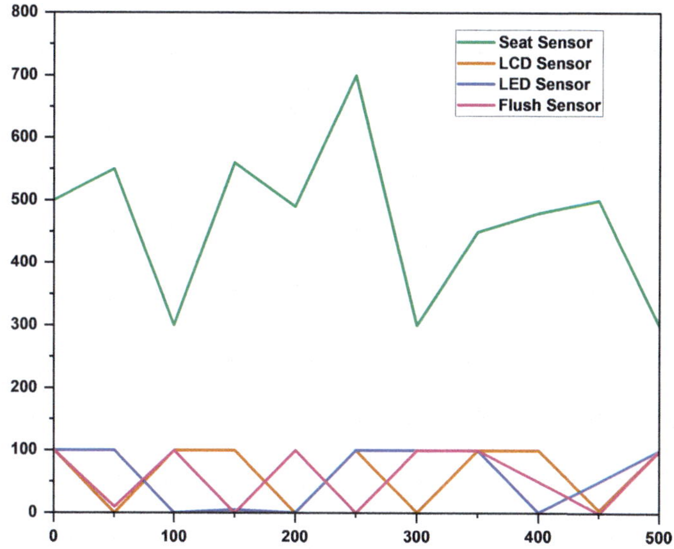

Fig. 5 Sensor-based system architecture

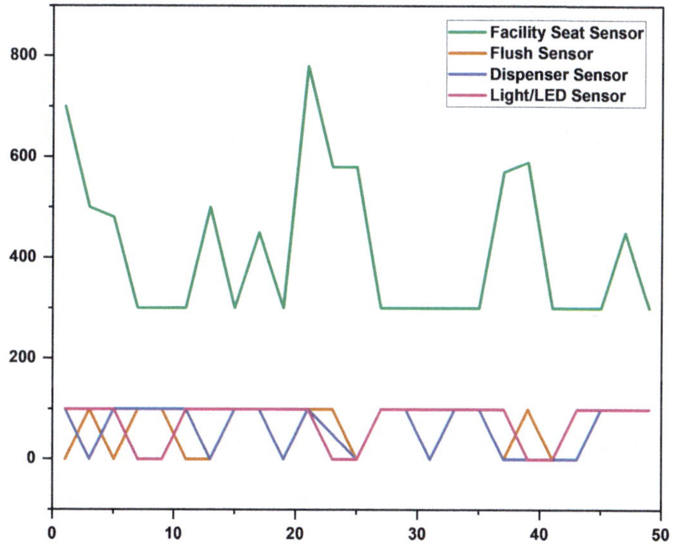

Fig. 6 Different sensors are used to collect data

In Fig. 8, in Category 2, the sensor for the facility seat (Sensor) is 500, which indicates the user is less close than in previous categories. Two activations were detected by the flush and dispenser sensors, adding them to the previous count; therefore, one was in Category 1, and one was in Category 2.

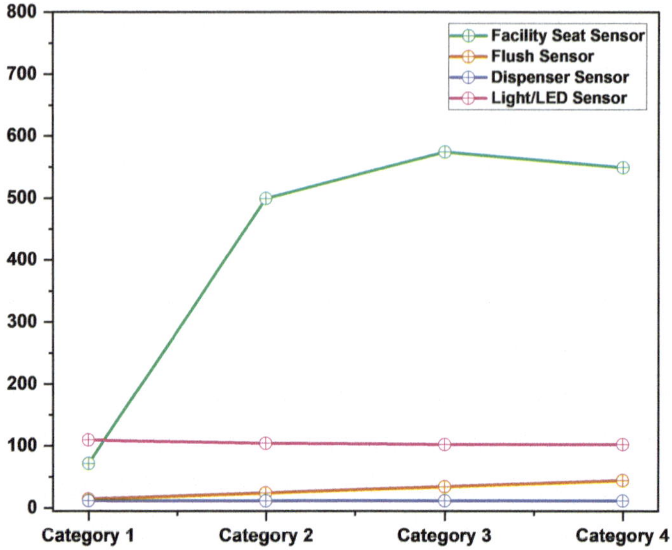

Fig. 7 Analyzing smart devices

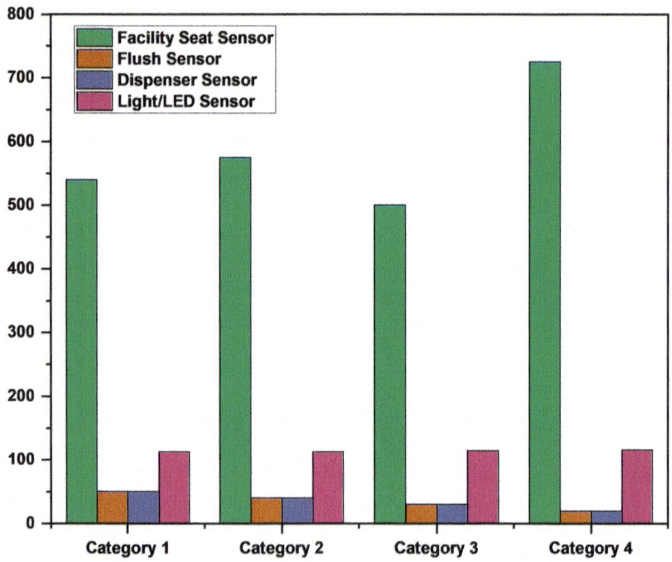

Fig. 8 Analyzing seats in smart facilities

5 Conclusion

A proposed IoT-based hand hygiene dispenser that monitors temperature and level improves public hygiene during pandemics, especially during extreme temperatures. By tagging contacts and storing temperature data in the cloud, the RFID tagging and touchless design reduce virus transmission risks. Its touchless design reduces the risk of virus transmission, and its RFID tags and cloud-based storage of temperature data ensure reliable contact tracking and health monitoring. Enhanced health monitoring capabilities and broader scalability could be incorporated in future enhancements, further contributing to public health safety.

References

1. Andrade, R.O., Yoo, S.G.: A comprehensive study of the use of LoRa in the development of smart cities. Appl. Sci. **9**(22), 4753 (2019). https://doi.org/10.3390/app9224753
2. Biyik, C., et al.: Smart parking systems: reviewing the literature, architecture and ways forward. Smart Cities **4**(2), 623–642 (2021). https://doi.org/10.3390/smartcities4020032
3. Jaid Jim, A.A., Kadir, R., Md. Mamun, A.A., Nahid, A.-A., Md. Ali, Y.: A noble proposal for internet of garbage bins (IoGB). Smart Cities **2**(2), 214–229 (2019). https://doi.org/10.3390/smartcities2020014.
4. Syed, A.S., Sierra-Sosa, D., Kumar, A., Elmaghraby, A.: IoT in smart cities: a survey of technologies, practices and challenges. Smart Cities **4**(2), 429–475 (2021). https://doi.org/10.3390/smartcities4020024
5. Belli, L., et al.: IoT-enabled smart sustainable cities: challenges and approaches. Smart Cities **3**(3), 1039–1071 (2020). https://doi.org/10.3390/smartcities3030052
6. A'ssri, S.A., Zaman, F.H.K., Mubdi, S.: The efficient parking bay allocation and management system using LoRaWAN. In: 2017 IEEE 8th Control and System Graduate Research Colloquium (ICSGRC), pp. 127–131. IEEE, SHAH ALAM, Malaysia (2017). https://doi.org/10.1109/ICSGRC.2017.8070581
7. Nor, R.F.A.M., Zaman, F.H.K., Mubdi, S.: Smart traffic light for congestion monitoring using LoRaWAN. In: 2017 IEEE 8th Control and System Graduate Research Colloquium (ICSGRC), pp. 132–137. IEEE, SHAH ALAM, Malaysia (2017). https://doi.org/10.1109/ICSGRC.2017.8070582
8. Liu, S., Xia, C., Zhao, Z.: A low-power real-time air quality monitoring system using LPWAN based on LoRa. In: 2016 13th IEEE International Conference on Solid-State and Integrated Circuit Technology (ICSICT), pp. 379–381. IEEE, Hangzhou, China (2016). https://doi.org/10.1109/ICSICT.2016.7998927
9. Basford, P.J., Bulot, F.M.J., Apetroaie-Cristea, M., Cox, S.J., Ossont, S.J.: LoRaWAN for smart city IoT deployments: a long term evaluation. Sensors **20**(3), 648 (2020). https://doi.org/10.3390/s20030648
10. De Castro Tome, M., Nardelli, P.H.J., Alves, H.: Long-range low-power wireless networks and sampling strategies in electricity metering. IEEE Trans. Ind. Electron. **66**(2), 1629–1637 (2019). https://doi.org/10.1109/TIE.2018.2816006
11. Ertürk, M.A., Aydın, M.A., Büyükakkaşlar, M.T., Evirgen, H.: A survey on LoRaWAN architecture, protocol and technologies. Future Internet **11**(10), 216 (2019). https://doi.org/10.3390/fi11100216
12. Hussain, N., Rani, P., Kumar, N., Chaudhary, M.G.: A deep comprehensive research architecture, characteristics, challenges, issues, and benefits of routing protocol for vehicular ad-hoc networks. Int. J. Distrib. Syst. Technol. (IJDST) **13**(8), 1–23 (2022)

13. Rani, P., Sharma, R.: Intelligent transportation system for internet of vehicles based vehicular networks for smart cities. Comput. Electr. Eng. **105**, 108543 (2023)
14. Ahmed, L.J., Malin Bruntha, P., Dhanasekar, S., Govindaraj, V., Krishnapriya, T.S., Ramjan Begam, A.: An efficient heart-disease prediction system using machine learning and deep learning techniques. In: 2023 9th International Conference on Advanced Computing and Communication Systems (ICACCS), pp. 1980–1985. IEEE, Coimbatore, India (2023). https://doi.org/10.1109/ICACCS57279.2023.10112793
15. Mathur, P.: Hand hygiene: back to the basics of infection control. Indian J. Med. Res. **134**(5), 611 (2011). https://doi.org/10.4103/0971-5916.90985
16. Girard, R., Amazian, K., Fabry, J.: Better compliance and better tolerance in relation to a well-conducted introduction to rub-in hand disinfection. J. Hosp. Infect. **47**(2), 131–137 (2001). https://doi.org/10.1053/jhin.2000.0854
17. Venkatesh, A.K., Lankford, M.G., Rooney, D.M., Blachford, T., Watts, C.M., Noskin, G.A.: Use of electronic alerts to enhance hand hygiene compliance and decrease transmission of vancomycin-resistant Enterococcus in a hematology unit. Am. J. Infect. Control **36**(3), 199–205 (2008). https://doi.org/10.1016/j.ajic.2007.11.005
18. Nichols, S.K.: Touchless technology helps facilitate infection prevention best practices. Infection Control Today (2007)
19. Bal, M., Abrishambaf, R.: A system for monitoring hand hygiene compliance based-on Internet-of-Things. In: 2017 IEEE International Conference on Industrial Technology (ICIT), pp. 1348–1353. IEEE, Toronto, ON (2017). https://doi.org/10.1109/ICIT.2017.7915560
20. Herbert, J., Horsham, C., Ford, H., Wall, A., Hacker, E.: Deployment of a smart handwashing station in a school setting during the COVID-19 pandemic: field study. JMIR Public Health Surveill. **6**(4), e22305 (2020). https://doi.org/10.2196/22305
21. Chowdhury: An Internet of Things assisted Smart Hand... - Google Scholar. Accessed 23 Sep 2024. https://scholar.google.com/scholar_lookup?title=An+Internet+of+Things+Assisted+Smart+Hand+Sanitizer+with+Health+Monitoring+System+Help+to+Reduce+Rapid+Spread+of+COVID-19&author=Chowdhury,+B.&author=De,+T.&publication_year=2020
22. Meydanci, M.A., Adali, C., Ertas, M., Dizbay, M., Akan, A.: RFID based hand hygiene compliance monitoring station. In: 2013 IEEE International Conference on Control System, Computing and Engineering, pp. 573–576. IEEE, Penang, Malaysia (2013). https://doi.org/10.1109/ICCSCE.2013.6720030
23. Tadikonda: Smart sanitizer disperser with level monitoring - Google Scholar. Accessed 23 Sep 2024. https://scholar.google.com/scholar_lookup?title=Smart+Sanitizer+Disperser+with+Level+Monitoring&author=Tadikonda,+C.&publication_year=2021&journal=Turk.+J.+Comput.+Math.+Educ.+(TURCOMAT)&volume=12&pages=994%E2%80%93999
24. Sumbawati, M.S. et al.: Design automatic hand sanitizer microcontroller based using arduino nano and ultrasonic sensors as an effort to prevent the spread of covid-19. In: Proceedings of the International Joint Conference on Science and Engineering (IJCSE 2020). Atlantis Press, Surabaya, East Java, Indonesia (2020). https://doi.org/10.2991/aer.k.201124.025
25. Gómez, J., Oviedo, B., Zhuma, E.: Patient monitoring system based on internet of things. Proced. Comput. Sci. **83**, 90–97 (2016). https://doi.org/10.1016/j.procs.2016.04.103
26. Verma, S. et al.: An automated face mask detection system using transfer learning based neural network to preventing viral infection. Expert Syst. e13507 (2024)
27. Pramila, M.J., Shewta, P.S.: Wireless temperature detector system using ARDUINO and IOT. IJCTT 67(11), 82–83 (2019) https://doi.org/10.14445/22312803/IJCTT-V67I11P113
28. Xia, S.-L., et al.: Efficacy of the marker vaccine rAdV-SFV-E2 against classical swine fever in the presence of maternally derived antibodies to rAdV-SFV-E2 or C-strain. Vet. Microbiol. **196**, 50–54 (2016). https://doi.org/10.1016/j.vetmic.2016.10.001
29. Saini, J., Dutta, M., Marques, G.: Indoor air quality monitoring systems based on internet of things: a systematic review. IJERPH **17**(14), 4942 (2020). https://doi.org/10.3390/ijerph17144942
30. Tomás, J., Rego, A., Viciano-Tudela, S., Lloret, J.: Incorrect facemask-wearing detection using convolutional neural networks with transfer learning. Healthcare **9**(8), 1050 (2021). https://doi.org/10.3390/healthcare9081050

31. Rahman, M.M., Md. Manik, M.H., Md. Islam, M., Mahmud, S., Kim, J.-H.: An automated system to limit COVID-19 using facial mask detection in smart city network. In: 2020 IEEE International IOT, Electronics and Mechatronics Conference (IEMTRONICS), pp. 1–5. IEEE, Vancouver, BC, Canada (2020).https://doi.org/10.1109/IEMTRONICS51293.2020.9216386
32. Canedo, A., Goyal, P., Huang, D., Pandey, A., Quiros, G.: Arducode: predictive framework for automation engineering. IEEE Trans. Autom. Sci. Eng. **18**(3), 1417–1428 (2020)
33. Rani, P., Hussain, N., Khan, R.A.H., Sharma, Y., Shukla, P.K.: Vehicular intelligence system: time-based vehicle next location prediction in software-defined internet of vehicles (SDN-IOV) for the smart cities. Intell. Things AI-IoT Based Crit. Appl. Innov. 35–54 (2021)
34. Rani, P., Sharma, R.: An experimental study of IEEE 802.11 n devices for vehicular networks with various propagation loss models. In: International Conference on Signal Processing and Integrated Networks, pp. 125–135. Springer (2022)

Enhancing Multimedia Security in IoT Environment: Blockchain to the Rescue

Andleeb Khan, Parma Nand, Bharat Bhushan, Hari Mohan Rai, and Ahmed Alkhayyat

Abstract Presently, the security landscape of Internet of Things (IoT) is filled with challenges, as devices such as smart cameras and sensors, despite their vast potential for big data applications, are exposed to risks due to insufficient security safeguards. Given the large volume of data these devices produce and their pervasive presence in our daily routines, there is an urgent need to enhance IoT security to effectively manage and safeguard this data, and to mitigate privacy and security issues. This paper categorizes multimedia data into four categories and highlighted ways to secure multimedia data using blockchain, cryptography, machine learning and quantum computing. Additionally, it provides recent advancements within each of the solutions and the categories of multimedia data. Finally, several research directions are discussed in order to pave way for future multimedia security and IoT researchers.

Keywords Internet of things · IoT · Multimedia security · Blockchain technology · Quantum computing · Cryptography

A. Khan (✉) · P. Nand · B. Bhushan
Department of Computer Science and Engineering, Sharda School of Engineering and Technology, Sharda University, Greater Noida, India
e-mail: Khanandleeb1122@gmail.com

P. Nand
e-mail: Parma.nand@sharda.ac.in

H. M. Rai
School of Computing, Gachon University, Seongnam, South Korea
e-mail: drhmrai@gachon.ac.kr

A. Alkhayyat
College of Technical Engineering, The Islamic University, Najaf, Iraq
e-mail: ahmedalkhayyat85@iunajaf.edu.iq

R. Sharma et al. (eds.), *Multimedia Technologies in the Internet of Things Environment, Volume 4*, Studies in Big Data 173, https://doi.org/10.1007/978-981-96-4356-1_3

1 Introduction

Internet of Things (IoT) is also defined as a concept where embedded devices utilize Internet protocol-based communication services. The IoT makes objects like smart cameras, sensors and smartphones work together by cooperating and interacting thereby performing computation, communication and other services [1]. This makes the IoT network a promising solution for the big data applications [2]. With the advancements in IoT solutions and an emerging shift from conventional to technologically advanced modes of operation, it is necessary to consider the security concerns of these IoT solutions. The IoT devices generally are connected to the internet and this makes them vulnerable as they don't have adequate security measures in place [3, 4]. The risks and vulnerabilities involved in these devices are of significant concern as they generate large amount of data which require security, management and analysis [5]. The integration of IoT in the daily lives of individuals, raise concerns for privacy and security.

Multimedia data which includes data in the form of a combination of image, textual, video and audio data. IoT applications like smart homes, smart cities, smart grids, etc., all generate multimedia data thereby making it an indispensable part of our lives [6]. The industrial environments and the different social settings also make use of multimedia data. Therefore, the multimedia data that also at times include the personally identifiable information of the users is hence regarded as sensitive and personal [7]. The transmission of multimedia data in wireless networks is vulnerable to copying and redistribution thereby causing privacy concerns and risking the reputation of the concerned organization [8]. This requires efficient ways to ensure privacy and security. Since any form of risk towards the multimedia data can lead to leakage of confidential information, invasion of privacy, unauthorized access and data and identity thefts [9, 10]. This makes multimedia security in the IoT environments a significant area of research.

Blockchain which follows a decentralized approach offers promising solutions to tackle the privacy and security concerns in the IoT environments, along with machine learning, cryptography and quantum computing [11]. Blockchain stores data securely as it works on distributed data storage and integrates consensus mechanisms as a security measure [12]. The data is communicated in a blockchain by enabling a distributed ledger [13]. This distributed ledger is secured as it is recognized, validated and signed cryptographically by all the mining nodes. It further has the capability to enable the shutting down of IoT devices that can infect the network [14]. Thus, a blockchain-based solution offers resilience and data immutability thereby avoiding chances of tampering. The smart contracts used in blockchain further ensure the trustworthiness of a network by applying machine learning algorithms [15]. The major contributions of this paper are mentioned as under.

- It categorizes multimedia data into four types, i.e. biometric, healthcare, agriculture and multimedia and discusses the recent advancements in each type.
- This paper throws light on the ways to secure multimedia data in IoT environment using Blockchain, Cryptography, Machine Learning and Quantum Computing.

- It provides recent advancements within each solution along with benefits and challenges.

The remainder of the paper is organized as follows: Sect. 2 gives a detailed background of IoT. Section 3 classifies multimedia data into four types. Section 4 gives a detailed overview of blockchain. Section 5 discusses the security solutions for multimedia data within IoT environments. Finally, Sect. 6 gives the paper conclusion followed by future research directions.

2 IoT Background

IoT is also known as the Industrial Internet or the Internet of Everything [16]. It is a technological concept to design a system network and a global computer which can communicate with each other [17]. IoT has emerged as an omnipresent network for global computing with its future breakthroughs and technical developments and has connected everything and everyone to the internet [18]. IoT has limitless opportunities and is continuously developing as a hot subject for study. It can be accessed on device of any type, anywhere and at any time. With the progress of technology, the world is moving to a more and more connected world [19]. An independent and secure link and data exchange between applications and real-world devices is ensured by IoT. The IoT links the physical behaviour and real life to the virtual world [20].

2.1 Architecture

An interconnection of devices, cloud technology and network structure that is utilized to communicate by the IoT devices is referred to as the IoT architecture. The different layered architecture is depicted in Fig. 1 and explained in the upcoming subsections below.

2.1.1 Perception Layer

This layer contains the physical description of an object. The data sensors in this layer detect the material speed, moisture, temperature and location [21]. They are Infrared sensors, RFID sensors, etc. The perception layer extracts useful data which is attached to the sensors and transmits it to the network layer by converting the data into digital signals.

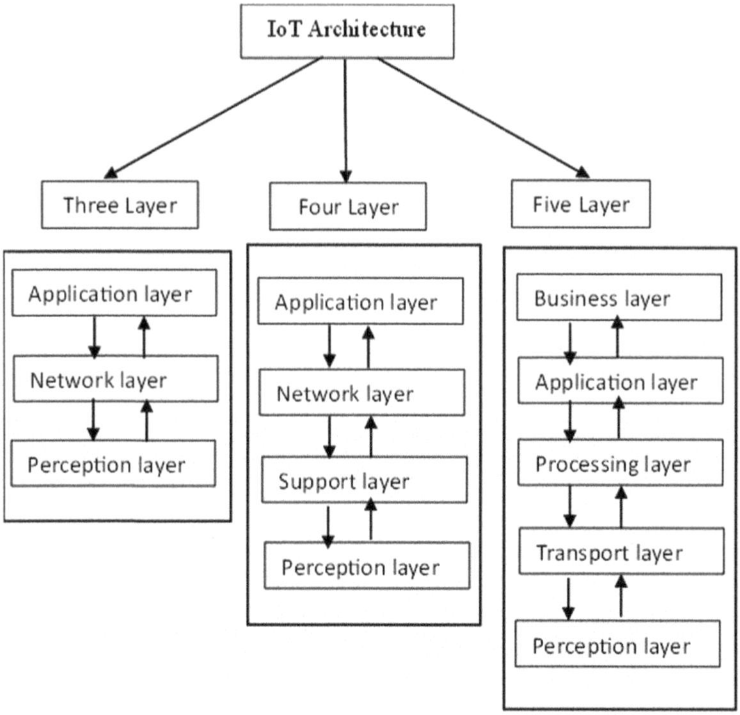

Fig. 1 IoT architecture

2.1.2 Network Layer

The communication layer is commonly referred to as network layer. It connects the perception and application layers. Sensors transmit and convey information collected from objects that are physically present. The medium of transmission can be wired or wireless [22]. It also connects networks, network devices and intelligent things. This layer is vulnerable to attacks. Hence it causes privacy, security and authentication issues while transmitting information over the network.

2.1.3 Application Layer

It gives the implementation of IOT based on the processed information, in different sectors. This is beneficial for the large-scale development of IoT organizations as there are advancing IoT applications [23].

2.1.4 Support Layer

This layer is added for the purpose of protection. In three-layered architecture, the data is communicated to the network layer directly. Hence, due to the flaws of three-layered architecture, another layer is recommended. The support layer ensures that the information is relayed by actual customers and it also protects the information from any hazard [24]. The results and the consumers can be checked through ways like verification. This can be done with the help of passwords, keys and mysteries. This layer sends data to network layer. This transmission can take place through a wired and remote connection. This layer is vulnerable to attacks such as malicious, insider and unauthorized access and DoS attack.

2.1.5 Processing Layer

Sensor device data is calculated in the processing layer. It integrates efficient computing, cloud storage and the ways in which data is stored directly in the database [25]. The mechanized movement is dependent on the treated data and severe processing machinery.

2.2 *Security Concerns in IoT*

The IoT devices and infrastructure face a number of security and privacy concerns including cybersecurity attacks. The prominent cyber-attacks in IoT solutions are described in the subsections below.

2.2.1 Botnet

The term 'botnet' is a blend of 'Robot' and 'network' and refers to a group of computers that are linked together for performing a specific task [26]. They do not necessarily pose a threat to our system. Some botnets perform tasks such as managing chat rooms and to track points in an online game which are useful in nature. However, they can be dangerous if misused. This network of computers can be infected by a malware and is controlled by a single attacking party or the bot-herder [27]. All the machines controlled by the bot-herder are called bots. Hence, they can be used to perform tasks ranging from spamming to executing DDoS attacks.

2.2.2 Man-In-The-Middle-Attack (MITM)

It is a type of attack where web-based protocols which are weak, are exploited by criminals who try to insert themselves in the communication channel and further steal data. People who are texting, sending emails or doing a video chat are not aware that the attacker has inserted themselves into their conversation and is thereby stealing their data [28]. Some of the MITM attacks include impersonating a person's voice, generating believable text messages or spoofing an entire communication system to extract data from the participants' devices that the attacker considers important [29].

2.2.3 Data and Identify Theft

A data theft occurs when an unauthorized person or hacker accesses the data of any organization [30]. The central database of a company is targeted which may have sensitive information and may further use it to perform cyber-attacks. This information may also include people's personal information often referred to as Personal Identifiable Information (PII) which can be used by the attacker for his/her benefit without the individual's permission [31]. This is known as identity theft. The financial security, financial future and the target's reputation are at risk in case of identity theft. Each identity has an intrinsic value attached to it and can further be sold for a few hundred dollars.

2.2.4 Code Injection

Code Injection is used to insert an arbitrary code in an application by a malicious user. It occurs when an untrusted input which is controlled by an attacker, interferes with a string-to-code execution statement [32]. Some of the examples are Cross-Site Scripting (XSS) and SQL injection attacks. It is regarded as a primary threat for web applications interacting with the databases of the back-end in their functioning [33]. The attackers identify flaws in user input validation like the amount of expected data, allowed characters and data format and then use them for developing malicious code.

2.2.5 Denial of Service (DoS)

In this type of attack the malicious actor targets to make a device or a computer unavailable to the intended users by hampering the normal functioning of the device. When the target machine is flooded or overwhelmed with requests it causes denial of service to the upcoming users and the normal traffic is not able to proceed [34]. This attack oversaturates the capacity of the target machine thereby denying any additional user requests. It uses a single computer for the attack. Some of the indicators of a DoS attack are—not being able to load a website, losing connectivity over the same network suddenly and a slow network performance [35].

2.2.6 Distributed Denial of Service (DDoS)

A DDoS is a subcategory of DoS attack. In DoS attack, a single computer or internet connection is used by the attacker to overwhelm the target [36], whereas a DDoS attack is quite large in scale. It uses thousands or even millions of connected systems to carry out its goal. It is hence harder to fight because of the large number of devices involved [37]. The motivation to carry out such an attack varies from disgruntled employees to hacktivists trying to take down the system of an organization to make a statement or express disapproval.

2.2.7 Phishing

A phishing attack is characterized by fraudulent text messages, emails, websites and phone calls manufactured to trick the users to download a malicious set of code or a malware and sharing sensitive personal information like credit card numbers, login credentials, etc. [38]. They can further lead to ransomware attack, financial losses, data and identity theft of corporations and individuals. It is a type of social engineering attack in which the attacker tries to deceive, manipulate or pressurize the users to send sensitive information or assets to malicious sites, portals or people [39].

2.2.8 Ransomware

It is a type of malware. It is used by cybercriminals to steal information and hold it hostage for monetary purposes [40]. The information is released only if a ransom is paid. The organizations which have large stores of financial data, personal sensitive information of users and intellectual property-related information are vulnerable to a ransomware attack. They are the third most common type of attack used for cyber-attack [41]. Thereby making it responsible for about 10% of all the data breaches (Table 1).

2.3 Privacy and Security Requirements for Multimedia Data Within IoT Environment

The combination of images, texts, audio, video data and animation is referred to as multimedia. Some necessary security and privacy requirements needed for multimedia data are depicted in Fig. 2 and elaborated in the following sub-sections.

Table 1 Summary of cyber-attacks

S. no	Name	Brief explanation
1	Botnet [26, 27]	• A group of computers linked together to perform a specific task • Can be controlled by a bot herder • Can perform DDoS attacks
2	Man in the middle [28, 29]	• Criminals intercept the communication channel • Steal data • Include voice impersonation and spoofing
3	Data and integrity theft [30, 31]	• Unauthorized access to an organization's database • Misuse of personal identifiable information (PII) • Risk to financial security, reputation and future
4	Code injection [32, 33]	• Insertion of arbitrary code by a malicious user • Include SQL injection and XSS
5	DoS [34, 35]	• Malicious actors overload a device or computer, disrupting its normal functioning • Target machine is flooded with requests causing DoS for new users
6	DDoS [36, 37]	• Larger scale attack using multiple systems • Harder to combat • Carried out by disgruntled employees or hacktivists
7	Phishing [38, 39]	• Fraudulent communication to trick users into sharing sensitive information • Can lead to financial losses, ransomware attacks and identity theft
8	Ransomware [40, 41]	• A type of malware which steals and hold information as hostage • Third most common type of cyber-attack

2.3.1 Confidentiality

It refers to keeping the communication as secret and that it can be accessed only by the wanted communicating parties. This factor protects and controls private data from any unauthorized access. For the IoT scenarios, this aspect is desired and is critical for industrial and social applications [42]. As the major mode of information transfer in case of IoT applications is wireless hence this information is a lot more at risk as compared to the information that is transmitted through a wired media. Cryptographic systems can protect information from unauthorized access.

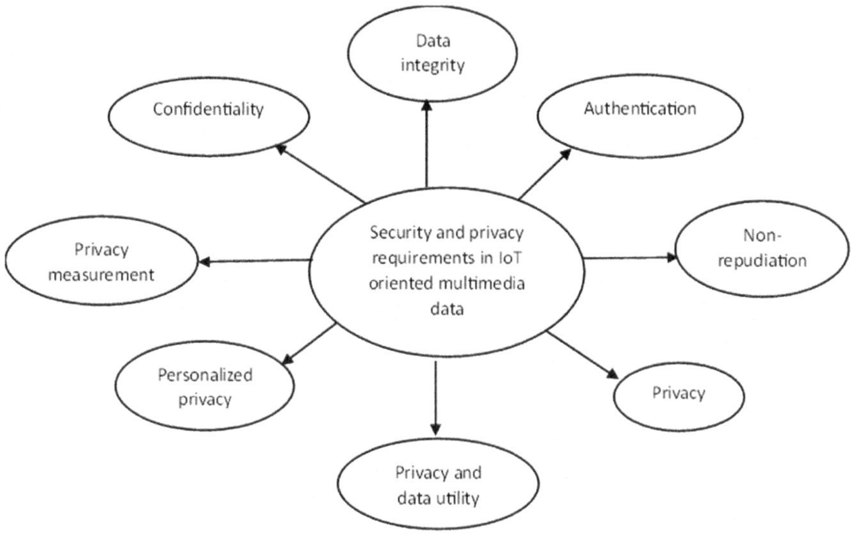

Fig. 2 Security and privacy requirements in IoT

2.3.2 Data Integrity

In order to maintain this aspect, it is to be verified by the communication recipients that the messages are not altered during the course of transmission. This ensures that the data cannot be manipulated. The quality of data is verified by integrity as it ensures the accuracy, reliability, relevance and the consistency of information in procedures like data storage, capture, updating and transfer of data. Modification of messages can be detected by all the parties involved in communication [43]. The tampering of the communicated data can be detected by using techniques like hash functions, watermarks, digital signatures (DS) and message authentication codes.

2.3.3 Authentication

It is the process which verifies a message's origins and content. It ensures the validity of data transmission, the message transmitted, its originator, and the individual authorized to receive that information [44]. The communication system should be able to verify the communicating parties. Some of the ways to authenticate are watermarking, biometrics/password logins and digital signatures.

2.3.4 Non-repudiation

It confirms a specific event or action. The transmission, creation and reception of a message can be considered as an event or action. In case of non-repudiation, the communicating parties will be informed when the information is delivered and the sender will have the delivery proof and the receiver will have the identity of the sender as a proof [45]. It can be ensured by DS, watermarks and authentication codes.

2.3.5 Privacy

Protecting individual's private and sensitive information from unauthorized access is the Privacy Aspect. To ensure privacy and proper utilization of multimedia data, policymakers, legal experts and researchers should collaborate to develop appropriate regulations and techniques [46]. Data privacy determines how much personal information about a person is shared and/or communicated with others. Like we exclude some people from our conversation similarly online users also want to control their personal data collection.

2.3.6 Trade-Off in Privacy and Data Utility

Users can present data in an anonymous manner to prevent privacy leakage and data degradation hence the focus here is to establish a balance between data utility and privacy. This trade-off further causes one out of the two factors to compromise on its function. When the data points are large, utility and privacy are chosen at each other's expense [47]. Perfect utility of data cannot be obtained if there exists complete privacy with no release of data. And utility cannot be achieved without the release of data thereby compromising on data privacy. Hence, a trade-off is bound to exist.

2.3.7 Personalized Privacy

All the categories of IoT-based multimedia data require constant privacy level. This becomes resource inefficient and impractical. Hence each category can be assigned with a different level of privacy thereby personalizing privacy [48]. This will further make our IoT solution more efficient and practical.

2.3.8 Privacy Measurement

Scientific work is based on measurement. But till date, the ways and methods to measure privacy are not well-defined. It is believed that measuring privacy for multimedia data in IoT should be a requirement in the future [49]. Relative measurements can be considered as an option as direct measurements may not be easily obtained for every scenario.

3 Multimedia Data Classification

Multimedia objects interacting with objects and outfitted with network links, in the IoT environments improve the daily lives of many people by bringing great opportunities. A wealth of information is contained in multimedia data as it contains both non-sensitive and normal data that does not require any protection, along with sensitive and private data like biometrics that needs protection so that it is not exposed or leaked [50]. This section classifies multimedia data on the basis of different application categories. It further mentions the levels of security for each type.

3.1 Type I: Biometric Data

This data includes a person's physical, behavioural and individual traits used in applications like fingerprint recognition and facial recognition. It is important to secure and maintain the integrity of biometrics as they are stronger than traditional recognition technologies like passwords [51]. The biometric data of an individual cannot be issued again if it is stolen, like stolen tokens or passwords. There should be de-identification of individual identifiers to safeguard the privacy of the user or there should be the reinforcement of protection in biometric multimedia data.

Joseph et al. [52] discussed the invariant feature matching in finger vein biometric framework along with its efficiency. Alyushin et al. [53] proposed an approach to create a meteorological database which is based on acquisition and storage of personal data that is of complex nature. Zhang et al. [54] introduced a large and diverse dataset, WiderPerson for dense pedestrian detection in the wild. Kant et al. [55] presented an approach based on watermarking to protect the templates used in multimodal biometric systems. Robinson et al. [56] proposed the first publicly available multi-task multimedia kinship dataset and developed the machinery for automatically collecting, annotating and preparing the data, which will require minimal human input and no financial cost. Kumar et al. [57] proposed a novel low-cost framework for the recognition of health monitoring of individual cattle on the basis of their sensed behavioural activity data and muzzle pattern image database by using incremental decision tree classification techniques and accelerometer-based activity monitoring method. Odya et al. [58] discussed an experiment in which 20 features that were extracted from eye movement data, related mainly to saccades and fixations, are used as a complementary biometric modality to authenticate clients at banking kiosks. Reis et al. [59] presented a deepfake detection method for detecting face-swapped videos of the person who is already known, with the help of a threshold classifier that uses similarity scores from deep convolutional neural network which is trained for facial recognition.

3.2 Type II: Healthcare Data

This form of multimedia data can be collected from and is available with hospitals, patients, medical insurers, pharmacies and medical professionals. This type of multimedia data is sensitive, often fragmented and very private and has personal identifiable information of the users hence, the sharing of this type of data is critical [60]. To share this data between different organizations is sometimes difficult. The healthcare multimedia data requires protection at each level of transmission and it should be ensured that the integrity of this data is intact.

Yuan et al. [61] presented a cloud-based framework to be used in healthcare monitoring systems that will allow fast indexing and data acquisition thereby ensuring strong privacy. Tawalbeh et al. [62] discussed networked healthcare with the role of big data analytics and mobile cloud computing in its enablement. Muhammed et al. [63] proposed a ubiquitous healthcare framework named UbeHealth that utilizes big data, deep learning, Internet of things, high performance computing and edge health to address challenges of bandwidth, reliability and network-latency. It also presented a cloud-computing infrastructure which is mobile and based on cloudlet to be used for applications dealing with big data in healthcare. Moniaga et al. [64] discussed an interactive multimedia application for integrating the data of stunting cases in Indonesia for increasing government interventions. Ren et al. [65] designed a strategy for task offloading with a low latency, centralized, reliable and secure decision-making algorithm which will have powerful emergency handling capacity and will facilitate resource-constrained edge devices in task offloading. Sutradhar et al. [66] proposed Dynamic stepwise encryption algorithm to encrypt data by using Fruit Fly Optimization thereby enhancing the Quality of Service of devices and systems based on Internet of Medical Things (IoMT). Siddiqui et al. [67] proposed a Decision Tree method which will acquire real-time patient data by using multiple biosensors. Almutairi et al. [68] outlined a diagnostic framework that is non-invasive and consists of four cascaded layers.

3.3 Type III: Agriculture Data

The agriculture multimedia data is collected with the help of IoT sensors and is used to develop and design an intelligent system which can be used to analyse the data with the help of images and the environmental information like Carbon dioxide levels, water levels, humidity and temperature, collected from the IoT sensors [69]. To enhance the crop quality and increase the production, the interpretation and analysis of this data is instrumental.

Bhanu et al. [70] proposed a multi-agent-based model for context-aware information gathering by using wireless multimedia sensor networks for agriculture monitoring. Liang et al. [71] proposed a novel solution by deploying an unmanned aerial vehicle, as a next-generation mobile network for achieving intermittent IoT device connections for data collection. Hussein et al. [72] presented a routing algorithm providing low packet loss and low end-to-end delay, referred to as a greedy and smart forwarding algorithm which is based on energy awareness and throughput. Chen et al. [73] presented a novel method for numerical environmental multimedia modelling for the assessment of the fate of emerging organic contaminants on the environment and the health risk they cause at a regional scale. Kaur et al. [74] developed a digital sensor network that can work across farmlands for monitoring large scale biodiversity. Zhou et al. [75] outlined a combination which is based on Savitzky–Golay (SG) filter, sparrow optimization algorithm and partial least squares, data dimensionality reduction and predictive models for improving the accuracy of humidity forecast. Quik et al. [76] presented a first of its kind multimedia model named SimpleBox4Plastic (SB4P) which addresses the full nano- and micro-plastics continuum, explored the validity of the model evaluated it on the basis of a microbeads case study and compared it with limited concentration data. Zhao et al. [77] presented graph convolution vector calibration network that is used for the recognition of fish group activity through an individual action relationship reasoning.

3.4 Type IV: Surveillance Data

The Internet-connected devices in the IoT networks are increasingly being used for streaming videos because of their ability to capture videos [78]. This class of data is derived from the applications of video surveillance like house monitors and traffic surveillance devices in which public safety, transport control, incident detection and human activities are monitored. This type of data is presented like a high-dimensional and unstructured, continuous-image series.

Blasch et al. [79] presented new developments in the domain of dynamic data-driven applications for context analysis and user support. Abdallah et al. [80] proposed an automated model for the multi-level security management for the processing of multimedia events. Shao et al. [81] presented a novel intelligent solution for processing and utilization for big surveillance video data that is based on alarming messages and event detection from smart cameras. Liu et al. [82] proposed a framework for progressive re-identification of vehicles through the use of deep neural networks. Yazici et al. [83] proposed a fusion-based wireless multimedia sensor network framework that uses intra-node processing and hence reduces the amount of transmitted data. Chen et al. [84] presented a novel framework for cluttered surveillance videos to find matching complex activities user-described queries. Aslam et al. [85] surveyed the state-of-the-art middleware methods based on the events and their suitability for each method of multimedia event processing. It also presented an object detection model evaluation by using the existing datasets. Cecchinato et al. [86] proposed an application-level security framework with a low-cost hardware and software implementation solution for encrypting and decrypting real-time multi-modal data for security applications in environments like monitoring, situational awareness, counter-UAV and espionage (Table 2).

Table 2 Literature review

Type	Reference	Year	Summary
Type I	JosephP et al. [52]	2017	Invariant feature matching in finger vein biometrics
	Alyushin et al. [53]	2018	Suggested a method for building a meteorological database with complex personal data
	Zhang et al. [54]	2020	WiderPerson, a comprehensive dataset for pedestrian detection
	Kant et al. [55]	2020	Watermarking technique to secure multimodal biometric system templates
	Robinson et al. [56]	2022	A cost-free, automated system for kinship data collection and annotation
	Kumar et al. [57]	2022	A cost-effective cattle health monitoring system using behaviour and muzzle patterns
	Odya et al. [58]	2023	Eye movement features for banking kiosk user authentication
	Reis et al. [59]	2023	A deepfake detection technique using a trained deep CNN
Type II	Yuan et al. [61]	2016	Cloud-based healthcare monitoring system with fast indexing and strong privacy
	Tawalbeh et al. [62]	2016	Networked healthcare via big data analytics and mobile cloud computing
	Muhammed et al. [63]	2018	UbeHealth: A comprehensive healthcare framework utilizing big data and IoT
	Moniaga et al. [64]	2019	Application to integrate and enhance interventions for stunting cases in Indonesia
	Ren et al. [65]	2022	Low-latency task offloading for edge devices
	Sutradhar et al. [66]	2023	Dynamic stepwise encryption algorithm for enhancing IoMT QoS
	Siddiqui et al. [67]	2023	Real-time patient data collection via Decision Tree method and bio-sensors
	Almutairi et al. [68]	2023	Four-layer framework for non-invasive COVID-19 detection
Type III	Bhanu et al. [70]	2018	Multi-agent-based model using sensor networks for agriculture monitoring
	Liang et al. [71]	2019	UAV deployment for intermittent IoT device connections and data collections
	Hussein et al. [72]	2021	Greedy and smart forwarding algorithm for packet loss and delay
	Chen et al. [73]	2022	Novel method for assessing organic contaminants and health risks
	Kaur et al. [74]	2023	Digital sensor network for large-scale biodiversity monitoring across farmlands

(continued)

Table 2 (continued)

Type	Reference	Year	Summary
	Zhou et al. [75]	2023	Combination of SG filter, sparrow optimization algorithm and partial least squares for humidity forecast accuracy
	Quik et al. [76]	2023	A multimedia model addressing nano- and micro-plastics continuum
	Zhao et al. [77]	2024	Graph convolutional vector calibration network for the recognition of fish group activity
Type IV	Blasch et al. [79]	2015	Dynamic data-driven applications for context analysis and user support
	Abdallah et al. [80]	2018	Automated model for multi-level security in multimedia events
	Shao et al. [81]	2018	Big surveillance video data processing through smart cameras
	Liu et al. [82]	2018	Progressive vehicle re-identification using neural networks
	Yazici et al. [83]	2019	Fusion-based wireless multimedia sensor network framework
	Chen et al. [84]	2019	Finding matching complex activities in cluttered surveillance videos
	Aslam et al. [85]	2020	Object detection model evaluation for multimedia event processing
	Cecchinato et al. [86]	2022	Low-cost security framework for real-time, multimodal encryption and decryption of data

4 Blockchain Background

The blockchain technology came in the year 2008 when it was used to track the transactions of the Bitcoin cryptocurrency [87]. It is an immutable ledger that is distributed and that stores data in blocks by using cryptographic mechanisms. It is laid on the foundation of distributed ledger database, peer-to-peer networks and encryption techniques. It is used to organize data inside a specific block structure and to chronologically arrange it in the form of a chain. The integrity and immutability of data is ensured through consensus mechanisms and encryption techniques [88].

The operation of blockchain is like an organized and systematic sequence which stores information similar to that of a distributed database. Every block contains specific content and a header. This header has the hash of current block and previous blocks. If the hash value of the next block is absent, the index can still be used to retrieve data which can further be added to the chain [89]. The blockchain architecture is presented in Fig. 3.

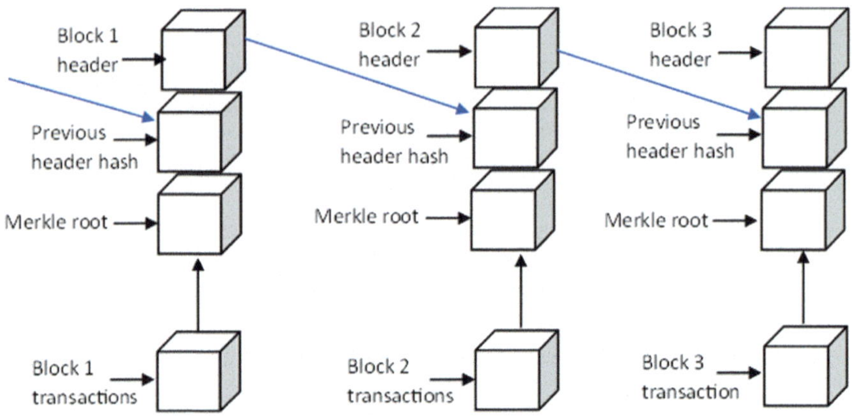

Fig. 3 Blockchain architecture

4.1 Architecture of Blockchain

This mechanism revolutionized information sharing as it uses a network of nodes and does not replicate data. The blockchain architecture utilizes two fundamental data structures, i.e. linked lists and pointers. Linked lists use data referencing with the help of pointers thereby forming a linked list-based structure [87]. Pointers are used to develop a connection between data points that is either pointing the location of referencing another piece [90]. The arrangement of blockchain is similar to that of a collection of pieces that are interconnected thereby leveraging linked lists and pointers. Each block has the data, the hash value of the current and previous block as it is depicted in Fig. 3.

Three types of Blockchain Architecture are, Public Blockchain which can be used by anyone and hence it allows everyone to participate [93], Private Blockchain which can only be used by authorized users that are granted access specifically, unlike public blockchain [94] and Consortium Blockchain which involves multiple organizations following defined rules and processes that are managed by specific users [94].

4.2 Blockchain Layers

For the development and design purposes of blockchain architectures, four layers are proposed, i.e. infrastructure, platform, distributed computing and application layer. Infrastructure layer: it consists of hardware components for running a blockchain [95].

4.2.1 Platform Layer

It facilitates the communication between the participants of a network through web Application Programming Interfaces (APIs), Representational State Transfer APIs and Remote Procedure Calls (RPC) [95].

4.2.2 Distributed Computing Layer

It ensures that the transaction data is authentic, secured, fault-tolerant, confidential, immutable and is accessible locally. Here immutability means that once the transaction records are updated in a ledger they cannot be modified [96]. The order of ledger updates, network transactions and miner selection for next block generation is maintained by the consensus protocol. To verify the user identity, an encryption technique is utilized and to maintain data privacy, hashing is utilized.

4.2.3 Application Layer

Clients access the applications that are built on blockchain network through this layer, as it consists of the business logic for smart contracts and transactions [96].

4.3 Technical Foundations of Blockchain

The foundation of blockchain is made up of cryptography, chaining-based data structures and peer-to-peer network. Hence, it gives rise to solutions that are secure when compared to centralized solutions. The following essential components are further discussed in the sub-sections below.

4.3.1 Cryptography

Secure systems are laid on the foundations of cryptography hence it is used by blockchain for cryptographic hashing [96]. One-way hash is the process of creating and verifying digital signatures by using public keys. This further ensures the authentication of users, accountability and non-repudiation.

4.3.2 Chaining-Based Data Structures

Data structures that are contained in blocks make up the blockchain. Each block, except the initial block also referred to as genesis block, consists of one parent block. This chain-based data structure is responsible for the name of blockchain. A block must be limited in size and consist of multiple transactions to ensure high-level throughput [97]. The detection of any block removal, alteration and reordering can be made simple by chaining the blocks together.

4.3.3 Peer-to-Peer Network

The blockchain works in a decentralized environment in a peer-to-peer network [90]. Hence, it is less prone to failure as it has no single point of failure (SPoF), compromised and corrupt nodes have few negative effects on the system thereby improving the reliability and robustness.

5 Security Solutions to Secure Multimedia Data in the IoT Environment

This section summarizes the approaches used in IoT security and the key findings. These are organized into four categories as depicted in Fig. 4.

5.1 Solutions Using Blockchain

The blockchain technology with its decentralized, tamper proof and secure platform has emerged as one of the promising solution to cater to the security issues in IoT devices. This subsection provides the overview of the proposed blockchain solutions to enhance IoT security.

Zulkifl et al. [98] proposed a novel framework, FBASHI that leverages fuzzy logic and blockchain technology to enhance the security of healthcare Internet of Things systems. This framework will help achieve the Authentication, Authorization and Audit services. This proposed system would be the most suitable for the IoT environments in healthcare as it uses Hyperledger blockchain platform that provides fast response and provides privacy. This framework proposes a behaviour-driven adaptive security framework for healthcare IoT environment and networks. This

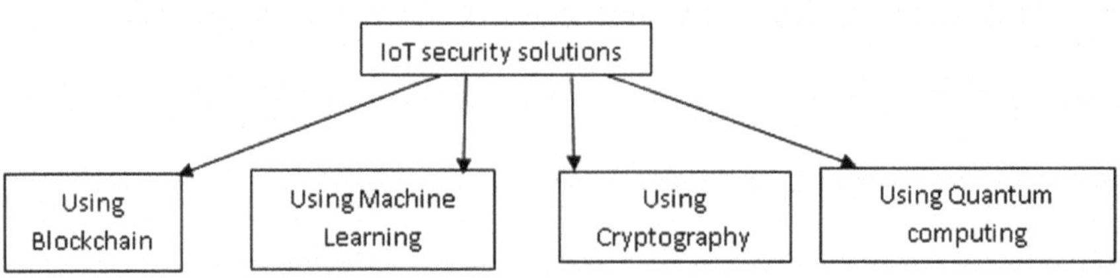

Fig. 4 IoT security solutions

mechanism uses fuzzy logic to provide a heuristic approach towards behaviour-based adaptive security, offering AAA services. Rathore et al. [99] proposed a security framework that leverages both blockchain and deep learning to work in intelligent 5 G-enabled IoT environments. The use of blockchain is meant for data security and deep learning is leverages for intelligent data analysis. Yang et al. [100] presented a framework that achieves effective authentication process by combining blockchain with modular square root algorithm. It also demonstrated the utility and security of the proposed framework with the help of a detailed experiment and security analysis. Zhang et al. [101] proposed a scheme for secure access management that can be used in IoT and is based on blockchain. This scheme uses decisional learning with the problem of error, to present a secure key encryption based on policy attribute which implements fine-grained access control. This scheme also uses blockchain in its second step for access management and authentication. Cai et al. [102] proposed a Graph neural network-based secure and smart Blockchain IoT architecture, GTxChain which obtains off-chain data and constructs the data structure for transaction used in blockchain, through the use of an intelligent prophecy machine and a blockchain-directed acyclic graph (DAG). Azbeg et al. [103] proposed a secure healthcare system that combines both blockchain and IoT, named BlockMedcare. It is used to monitor chronic diseases remotely. The proposed framework utilizes smart contracts for access control. A re-encryption proxy along with blockchain is used to ensure the security by storing hash data. Luo et al. [104] proposed an authentication scheme for IoT smart agriculture networks which is based on hybrid blockchain. The proposed scheme simultaneously mutually authenticates multiple devices which are from a number of service providers of data and from various agricultural systems. The proposed work also gives an algorithm for groupable batch verification which adjusts dynamically the batch size through group verification of a list of requests thereby improving the cross-domain flexibility of batch authentication. Khashan et al. [105] proposed a hybrid authentication framework for IoT systems which is centralized and based on blockchain. It used edge servers for providing centralized authentication for IoT devices associated with it. The proposed framework utilizes lightweight cryptographic methods for efficient resource implementation and authentication. Hennebelle et al. [106] proposed a system for risk-based diabetes prediction comprising IoT, edge computing, AI and blockchain. The proposed system ensures privacy and security of data by using blockchain and maintaining a collective view of the data from patients from different hospitals and their risk factors. Li et al. [107] proposed a framework for securing IoT by using the basic infrastructure as blockchain along with physically unclonable functions which are used to identify sensors uniquely.

5.1.1 Benefits of Blockchain in IoT Security

The decentralized architecture and distributed nature of blockchain allows the security solutions based on blockchain to be resilient against attacks like DDoS. It further eliminates single point of failure and offers efficient data immutability [98]. Without going through any intermediaries, the devices can communicate directly. The primary feature of blockchain that increases its security is its tamper-proof nature [99].

Blockchain offers integrity and authenticity with the help of cryptographic functions like hash functions and Elliptic Curve Cryptography (ECC). They also prove sender legitimacy by signing the transactions. Blockchain reduces the risk of any malicious activity and ensures trust in the IoT network by offering a transparent and validated record [101]. The smart contracts used in blockchain increase the trust, reduce human intervention and minimize delays by an automated execution of pre-defined rules.

5.1.2 Challenges in Securing IoT with Blockchain

IoT devices have limited power of processing and the high amount of processing time required for the addition and validation of blocks is not adequate for them. Therefore, to contribute and envision lightweight blockchain-based solutions is a challenge for the researchers [105]. IoT systems also generate excessive amount of data transactions thereby consuming high bandwidth and hence causing time latency for real-time applications. Consensus algorithms like Proof of Work require high energy and computational power.

Another challenge is with smart contracts, which if found with vulnerabilities in the source code can be applied to malicious purposes and hence risk the parties using that smart contract [103]. It is complicated to fix a smart contract bug in its source code. By using a blockchain solution for IoT security can make the IoT system susceptible to blockchain-related attacks like the 51% attack, where 51% of the network is in control of a single group or individual, thereby allowing them to manipulate the data transactions before its validation. Also, malware attacks can gain access to the private key of the user and further misuse it.

- Permission-less Blockchain

It can lead to privacy and security concerns for transactions containing sensitive information as it is open to all and the transactions on the network are visible to everyone [98].

- Permissioned Blockchain

Its access is restricted to selected users which makes it more trustworthy for the participants. However, the decentralization benefit is reduced. This type of blockchain requires a high level of oversight and governance thereby using more resources [99]. The literatures exploring blockchain-based federated learning in IoT systems often don't consider any necessity to go for a lightweight implementation. The synchronous transmission taking place in trained models is hindered by the heterogeneous nature of resources in IoT device thereby causing failures [101]. In case of consortium blockchain, the inclusion of any new consortium member for competition with the other powerless member, by the governing nodes' majority decision will also pose a risk to the sustainability of the system.

• Hybrid

The maintenance and implementation of this type of blockchain for the IoT environment is complex hence more costly [100].

5.2 Solutions Using Machine Learning

Data exchange and analysis among the sensors is done to provide efficient decision-making for the operation of the systems. AI, ML and DL forms the basis of these decisions. ML and DL algorithms play a significant role in cybersecurity by proving to be the alternate security solutions in IoT applications [109]. They are also used to design and implement security solutions for protecting the IoT devices from threats and unauthorized access.

Elsadig et al. [108] proposed an approach to detect DoS attacks in wireless sensor networks by using a lightweight and machine learning approach based on decision tree and Gini feature selection method. Kotenko et al. [109] discussed a novel framework for mobile Internet of Things and their security monitoring by the combined possibilities of machine learning and big data processing. Ekolle et al. [110] discussed the cybersecurity risks linked with healthcare IoT networks thereby providing a solution to safeguard medical devices and data by utilizing an agent-based approach. The proposed model Collabo, is based on a collaborative learning agents' network, which shares casual and mutual values of their actions when working on a common target. Shafiq et al. [111] proposed a novel method called CorrAUC. In the proposed model a feature selection algorithm that uses a wrapper technique to accurately filter and select effective features for the chosen machine learning algorithm is developed. This selection is done by using the AUC metric. To validate the selected features for malicious traffic identification, they apply an integrated TOPSIS and Shannon entropy based on a bijective soft set. The proposed approach is evaluated using the Bot-IoT dataset and four machine learning algorithms. Liu et al. [112] proposed a sparse learning algorithm and a machine learning-inspired probabilistic framework that can efficiently overcome the challenge of elimination of impulsive noise for wireless systems based on orthogonal frequency division multiplexing. Soto et al. [113] proposed a machine learning framework for resource-limited IoT environments that is decentralized and collaborative. It also proposed algorithmic approaches that can be used for prototype creation and prediction purposes.

Muñoz et al. [114] presented an anomaly detection system that works on unsupervised machine learning for the IoT architectures. Aloraini et al. [115] proposed a taxonomy of machine learning models and adversarial methods that can be exploited by the insider. It also discussed the applications of these methods in IoT environments working in real time. Gaber et al. [116] proposed a method to detect intrusions like injection attacks in applications of IoT like smart cities. The proposed method used two varieties of feature selection methods, recursive feature elimination and constant removal and is tested by a variety of ML classifiers like RF, DTs and SVMs.

Gaber et al. [117] proposed a deep learning-based intrusion detection system model. It is proposed for metaverse and IoT communication and utilizes two techniques: Convolutional neural network classification and Kernel PCA.

5.2.1 Benefits of ML in IoT Security

New threats and attacks are found to be continuously emerging with the evolution of IoT. Therefore, to address the emerging security issues in IoT the traditional methods of security prove to be insufficient. For such an IoT system that is under constant evolution, artificial intelligence techniques, machine learning and deep learning will prove to be beneficial in developing a security mechanism that is more advanced and up to date [109]. The AI can learn from data continuously thereby adapting to new and emerging threats and improving the accuracy with time.

The input data collection and analysis across the components of IoT system can be done by machine learning techniques thereby discerning interaction patterns and detecting malicious intent initially. Hence ML-based techniques are used to detect and identify threats and anomalies in IoT. A machine learning model learns to diagnose abnormal behaviours like a sudden network traffic increase which indicates a possible DDoS attack, by analysing different data sources [108]. A model of ML can also detect any changes in device performance or behaviour, thereby indicating malfunctioning and tempering. The tasks that need human intervention can be automated by using an ML model thereby saving time and cost along with enhancing security [111].

- Traditional ML techniques

Traditional ML models are utilized in some of the machine learning-based solutions. These models are computationally less intensive and simpler as compared to deep learning methods [109]. These are more suitable in decision-making for real-time or near real-time IoT security scenarios as they can be trained quickly since they utilize less amount of data for training [111].

- DL Techniques

Since IoT devices produce a large volume of data therefore deep learning techniques are more suitable for these environments as DL performs well with large datasets. The need for designing precise features and manually doing feature engineering is eliminated when DL is used as it learns features automatically from raw data [117]. This is useful in securing the IoT environments which has unstructured and complex data. DL-based models can adapt to new security threat, hence accordingly adjusting their responses.

- ML–DL Techniques

The strengths of both the approaches can be leveraged by combining both ML and DL techniques [111]. For example, ML can be used for decision-making and DL can extract features, or ML can be utilized for simpler scenarios and it can be switched to DL for intricate challenges [117].

5.2.2 Challenges in Securing IoT with ML

Data and learning methods are the two major and significant factors in machine learning systems utilized for IoT security. Data is collected, pre-processed and hence chosen for the learning algorithms as a dataset. The data must be of high quality for effective training. The sensors in IoT systems may collect incomplete, inaccurate and unnecessary data, which proves to be a challenging task. The heterogeneous data generated by IoT devices make it difficult for the ML models to pre-process the data [110]. Also, the machine learning models are difficult to interpret and complex therefore it is difficult to decipher how these models arrived at a particular decision. Some attacks target data by adding malicious data or manipulating the dataset, this further raises privacy and security concerns [115]. This malicious and irrelevant data leads to incorrect decision-making by the ML model as it does garbage processing. Hence, huge significance is given to the mechanisms that secure the dataset and analyse it deeply.

ML models that require a considerable number of resources in predictions, processing and training are not suitable for IoT devices as they are limited in their computational capacity, i.e. power, energy and memory. Therefore, the implementation of a sophisticated and a lightweight ML model becomes a challenge [109]. IoT security includes rapid detection and response to security threats. Therefore, real-time predictions need to be considered.

- **Traditional ML Models**

Issues like true negatives and false positives hamper the reliability and accuracy of the model as it effects the precision of the prediction [109]. Hence, proper guidance and amendments are required to form a precise prediction.

- **DL Models**

These models require significant hardware resources to train and deploy thereby being more intensive than machine learning models. The duration of training can extend significantly as the size of the training dataset extends [117]. A combination of ML–DL models also requires high computational resources [111].

5.3 Cryptography-Based Solutions

The process used to encrypt information for maintaining the integrity and security when it is communicated between objects are all included in cryptography. Classic cryptography cannot be implemented in IoT because of its limited capacities. The designing of cryptographic mechanisms is essential to maintain the confidentiality, integrity and authentication of the data [119]. Cryptography is hence used to build solutions optimized for resource-constrained IoT devices and hence maintaining their level of protection.

Thabit et al. [118] proposed a novel cryptographic algorithm that is effective, lightweight and homomorphic with two encryption layers. The first layer of the proposed model utilizes the lightweight cryptographic algorithms that are more effective and the other layer consists of multiplicative homomorphic schemes which improve data security in cloud computing. Devi et al. [119] presented an approach to maintain data transmission security among IoT devices and designed a malware detection model. The proposed model for malware detection uses deep learning and Deep LSTM classifier. Achary et al. [120] discussed ECC lightweight cryptographic algorithm for securing IoT devices which are resource-constrained and performs an evaluation by comparing it to the Rivest Shamir Algorithm (RSA) cryptosystem. Cacep et al. [121] discussed the characteristics and comparisons of development boards currently used in IoT systems and cryptographic applications. The basic requirements discussed were a security mechanism based out of cryptography and which is also low-cost. Dhar et al. [122] proposed a novel method for IoT device security using Quantum key distribution (QKD) and blockchain. QKD offers secure key exchange, while blockchain enhances the overall security. This combination aims to protect multimedia data from IoT devices against malicious attacks. Wang et al. [123] presented an asymmetric computing cryptosystem and an asymmetric key exchange protocol named ACKE by combining Subset Product Problem and Diffie–Hellman key exchange protocol. Awan et al. [124] proposed a framework to resolve the privacy and security issues in IoT ecosystems by augmenting the efficacy and privacy of big data analytics with the help of federated learning. Wang et al. [125] discussed the need to enhance security in smart cities using physical layer approaches in IoT and Mobile edge computing. It proposed physical layer security technologies to mitigate threats that arise from integrating cloud computing capabilities within the radio access network. Samaila et al. [126] proposed a security framework named IoT Hardware Platform Security Advisor (IoT-HarPSecA) for providing support to IoT producers. It includes the functions of security requirement elicitation, secure development and best practice guidelines, lightweight cryptographic algorithm suggesting feature. Atiewi et al. [127] proposed an IoT environment that is cloud-enabled and supports multifactor authentication and protects Big Data systems by the use of lightweight cryptographic encryption.

5.3.1 Benefits of Cryptography in IoT Security

For maintaining the security and integrity of the system cryptographic techniques have been used for centuries and they are beneficial for IoT security too. Cryptographic methods play a crucial role in meeting security requirements like authentication, integrity and confidentiality [118]. Over the years these safety measures have undergone rigorous testing. For instance, hash functions generate unique hash for the input data thereby maintaining data integrity. Among these, SHA-256 is regarded high in security than MD5. Cryptography can be supported and standardized by a large variety of devices and systems because it is interoperable [120]. Therefore, cryptography makes it easier to maintain the security of IoT devices that are heterogeneous in nature and have different hardware and software specifications.

- Lightweight

The cryptographic solutions used for IoT security are designed to be lightweight. Because of their adaptability and suitability to IoT devices which are usually constrained with computational power, battery life and memory, lightweight protocols tend to be more efficient than emerging solutions [118]. The purpose of a lightweight protocol is to minimize the energy consumption, memory complexity and computational time. Researchers continue to explore and develop security solutions for IoT based on cryptography [126].

- Symmetric Cryptography

When compared with asymmetric cryptographic algorithms, symmetric cryptography is more efficient and fast for encryption and decryption of large amount of data [118]. They can provide better security if combined together [123]. A type of Symmetric cryptographic algorithm used in IoT security is AES.

- Asymmetric cryptography

A prior secret sharing is not required in this as it ensures key establishment and secure exchange of data even without it. The authenticity is verified by the use of digital signatures. The IoT solutions based on cryptography largely use asymmetric cryptographic methods like RSA and ECC [120, 123].

5.3.2 Challenges in Securing IoT with Cryptography

Traditional cryptographic protocols, which do not account for lightweight implementations are not practical for IoT devices as they have limited capabilities. Attacks like physical attacks and side channel attacks used to steal sensitive parameters utilized in cryptographic methods cannot be prevented by only classical forms of encryption [122]. It is still a challenge to keep pace and adapt to the ever-evolving attacks even after continuous research.

Classical computers find it difficult to break efficient and classical cryptography-based protocols. But these mathematical problems will be solved in no time once quantum computers become a reality. Therefore, the cryptographic solutions would then be broken in a short span of time. Thereby making them vulnerable to Quantum computing attacks [126].

- Lightweight

To employ a cryptographic technique that has the properties of both being secure and lightweight is a challenge for the researchers as the lightness of the protocol is a significant factor [126].

- Symmetric

The key sharing is a complex task in case of systems like IoT which are distributed in nature [118]. The past and future communication that was encrypted by using that key will be compromised if the key is exposed. With increase in the number of parties in communication, there is an exponential increase in the need for keys thereby causing a problem of scalability. As compared with asymmetric ciphers, Symmetric ciphers however are more prone to attacks as they are less complex and use a reduced key length [123].

- Asymmetric

As compared to symmetric algorithms they are more intensive computationally [126]. They can cause a slowdown in performance when it uses approaches like RSA as it involves using a key length that is longer than symmetric algorithms.

5.4 Solutions Using Quantum Computing

The IoT security is one of the fields that would be revolutionized by Quantum Computing. Hence, quantum computing-based privacy and security solutions are proposed by the researchers.

Rehman et al. [128] presented an image encryption scheme that can be used for colour images and uses a combination of schemes of encryption like controlled Rubik's cube transformations, random and alternate quantum walks and hill cipher-based Elliptic curve cryptosystem. Bey et al. [129] proposed a novel method for service placement that is quantum-inspired and based on particle swarm optimization. The proposed algorithm is for edge computing environment. Chen et al. [130] proposed a novel method for IoT environments for the purpose of user perceptual task offloading and it is based on the optimizing strategy of particle population's quantum behaviour. Visalaxi et al. [131] proposed the employment of basic distributed computational techniques in a natural and open setting for providing membrane systems as a framework which provides unique security for cloud environment. Mangla et al. [132] discussed the existing quantum solutions for the issues of 5G security, their mitigation methods and suggested the security of 6G quantum world. Mohanty et al. [133] proposed a novel protocol for quantum computing-based threshold private set intersection which is directly used for privacy-preserving ride sharing. Latif et al. [134] proposed a novel image encryption scheme that is lightweight and is based on quantum walks for securely transferring data in the IoT platforms and wireless networking along with edge computing.

Prajapat et al. [135] presented a quantum designated verifier signature scheme based on identity. Latif et al. [136] presented a framework for fog cloud IoT for securing information. It also proposed a novel protocol for quantum steganography which is based on quantum entangled states and hash function. Shi et al. [137] presented a quantum secret permutating protocol that is feasible and is used with single photons. It also presented a quantum key distribution-based grouping strategy for complexity reduction and improving the efficiency of the two-part computation solution.

5.4.1 Benefits of Quantum Computing in IoT Security

Quantum Computing plays a crucial role in enhancing the security of Internet of Things through its application in quantum cryptography. This technology enables the generation of encryption keys that are considered unbreakable [131]. The reason behind this is the superior computational speed of quantum computers, which allows them to decipher traditional cryptographic keys more efficiently. Quantum cryptography operates on the principles of quantum mechanics, ensuring secure data transmission [134]. It offers a level of security where any attempt to read or interfere with the transmitted data becomes noticeable thereby providing a robust defence against potential attackers.

- Quantum Cryptography

Techniques that are based on quantum computing and can ensure IoT security are quantum walks, continuous wave technology and quantum circuit, etc. [128].

- Quantum key distribution (QKD)

It can help to detect attacks like eavesdropping as compared to classical algorithms of cryptography [137].

- Post-quantum cryptography

It offers the advantage of creating cryptographic methods that can withstand attacks from quantum computing, a feat that traditional cryptographic methods cannot achieve. One of the most potential approaches within post-quantum cryptography is lattice-based cryptography [138]. This technique is grounded in complex mathematical problems, providing a strong defence against attacks on both classical and quantum computers. Furthermore, lattice-based methods present encryption and decryption algorithms that are well-suited for Internet of Things (IoT) devices, which often operate under resource constraints [137].

- Random number generation

Random numbers to calculate classical cryptographic key can be generated by using Quantum cryptography. Conventional methods for generating random numbers rely on pre-set algorithms, which makes them predictable and potentially vulnerable to attackers [139]. On the other hand, QC, which is grounded in the principles of quantum mechanics, has the ability to produce vast quantities of random numbers that are inherently unpredictable. This unpredictability enhances the security and robustness of QC-generated random numbers.

5.4.2 Challenges in Securing IoT with Quantum Computing

To ensure IoT security with the help of Quantum cryptography, quantum hardware is required as it generates and transmits quantum states and to ensure the presence of such hardware is a challenge. Therefore, there is research on developing lightweight and effective solutions for the implementation of quantum cryptography in IoT [134]. The drawback of QKD is that the secret key it establishes is only between two entities. Therefore, to apply that for greater than two devices wanting to share a key becomes a challenge [137].

While some post-quantum cryptographic algorithms are engineered to be resource-efficient, their implementation can still pose difficulties due to their higher resource consumption compared to traditional algorithms [138]. Additionally, these algorithms demand larger key sizes. This necessitates a comprehensive evaluation of factors such as key size, security levels, network performance and scalability when incorporating these cryptographic techniques into IoT networks [130]. Transitioning to post-quantum cryptography also entails substantial modifications to current protocols and systems, which could be time-consuming and costly.

6 Conclusion

The IoT environment which is ever-evolving and is integrated into our daily lives in the form of devices like sensors and cameras have immense potential and can cater to large volumes of data thereby making it an indispensable part of our daily routines. This makes the IoT environment an area that requires privacy and security enhancement solutions and mitigations. Therefore, this paper dives deep into the realm of multimedia security in the IoT environment by first categorizing the multimedia data into four categories, i.e. biometric, healthcare, agriculture and surveillance. It then discusses the different solutions that can be applied to multimedia security within the IoT environment by using the techniques like Blockchain, Machine Learning, Cryptography and Quantum Computing with special emphasis on the blockchain technology. The future directions for research can be to increase the scalability of the research solutions in order to handle larger IoT networks, to develop models that can work in resource constrained IoT environment and to explore emerging technologies like Quantum key distribution and Blockchain for securing the evolving IoT communications. Some of the open research directions are to develop hybrid models that can utilize the strengths of each approach, need for lightweight and efficient cryptographic algorithms for the resource-constrained IoT set-ups and to develop newer quantum solutions for the emerging 6G technology.

References

1. Hwang, J., Nkenyereye, L., Sung, N., Kim, J., Song, J.: IOT service slicing and task offloading for edge computing. IEEE Internet Things J. **8**(14), 11526–11547 (2021). https://doi.org/10.1109/jiot.2021.3052498
2. Swamy, S.N., Kota, S.R.: An empirical study on system level aspects of internet of things (IOT). IEEE Access **8**, 188082–188134 (2020). https://doi.org/10.1109/access.2020.3029847
3. Chettri, L., Bera, R.: A comprehensive survey on internet of things (IOT) toward 5G wireless systems. IEEE Internet Things J. **7**(1), 16–32 (2020). https://doi.org/10.1109/jiot.2019.2948888
4. Frustaci, M., Pace, P., Aloi, G., Fortino, G.: Evaluating critical security issues of the IOT World: present and future challenges. IEEE Internet Things J. **5**(4), 2483–2495 (2018). https://doi.org/10.1109/jiot.2017.2767291
5. Bedi, G., Venayagamoorthy, G.K., Singh, R., Brooks, R.R., Wang, K.-C.: Review of internet of things (IOT) in electric power and energy systems. IEEE Internet Things J. **5**(2), 847–870 (2018). https://doi.org/10.1109/jiot.2018.2802704
6. Fujii, Y., Shan, J., Han, Y., Nishihara, Y.: Sentiment analysis of user reviews transition in multimedia franchise. Proced. Comput. Sci. **225**, 1533–1541 (2023). https://doi.org/10.1016/j.procs.2023.10.142
7. Wang, Y., Zhu, Z., Wang, L., Sun, G., Liang, R.: Visualization and visual analysis of multimedia data in manufacturing: a survey. Vis. Inf. **6**(4), 12–21 (2022). https://doi.org/10.1016/j.visinf.2022.09.001
8. Shabaz, M., Sharma, A., Ajrawi, S.A., Estrela, V.V.: Multimedia-based emerging technologies and data analytics for neuroscience as a service (NAAS). Neurosci. Inf. **2**(3), 100067 (2022). https://doi.org/10.1016/j.neuri.2022.100067
9. Ma, S., Zhang, T., Wu, A., Zhao, X.: Lightweight and privacy-preserving data aggregation for mobile multimedia security. IEEE Access **7**, 114131–114140 (2019). https://doi.org/10.1109/access.2019.2935513
10. Yassine, A., Shirehjini, A.A., Shirmohammadi, S.: Bandwidth on-demand for multimedia big data transfer across geo-distributed cloud data centers. IEEE Trans. Cloud Comput. **8**(4), 1189–1198 (2020). https://doi.org/10.1109/tcc.2016.2617369
11. Ma, S., Wang, S., Tsai, W.-T.: Delay analysis of consensus communication for Blockchain-based applications using network calculus. IEEE Wirel. Commun. Lett. **11**(9), 1825–1829 (2022). https://doi.org/10.1109/lwc.2022.3183197
12. Wang, S., Ouyang, L., Yuan, Y., Ni, X., Han, X., Wang, F.-Y.: Blockchain-enabled smart contracts: architecture, applications, and future trends. IEEE Trans. Syst. Man Cybern. Syst. **49**(11), 2266–2277 (2019). https://doi.org/10.1109/tsmc.2019.2895123
13. Kharche, A., Badholia, S., Upadhyay, R.K.: Implementation of blockchain technology in integrated IOT networks for constructing scalable its systems in India. Blockchain Res. Appl. 100188 (2024). https://doi.org/10.1016/j.bcra.2024.100188
14. Liu, S., Zheng, Q.: A study of a blockchain-based judicial evidence preservation scheme. Blockchain Res. Appl. 100192 (2024). https://doi.org/10.1016/j.bcra.2024.100192
15. Zhaisanova, D., Mansurova, M.: Blockchain concept for the educational purposes: bibliometric analysis and conceptual structure. Proced. Comput. Sci. **231**, 753–758 (2024). https://doi.org/10.1016/j.procs.2023.12.142
16. Neshenko, N., Bou-Harb, E., Crichigno, J., Kaddoum, G., Ghani, N.: Demystifying IOT security: an exhaustive survey on IOT vulnerabilities and a first empirical look on internet-scale IOT exploitations. IEEE Commun. Surv. Tutor. **21**(3), 2702–2733 (2019). https://doi.org/10.1109/comst.2019.2910750
17. Metzger, F., Hobfeld, T., Bauer, A., Kounev, S., Heegaard, P.E.: Modeling of aggregated IOT traffic and its application to an IOT cloud. Proc. IEEE **107**(4), 679–694 (2019). https://doi.org/10.1109/jproc.2019.2901578

18. Langer, G., Schulze, H., Kühl, S.: From intentions to adoption: investigating the attitudinal and emotional factors that drive IOT sensor use among dairy farmers. Smart Agric. Technol. **7**, 100404 (2024). https://doi.org/10.1016/j.atech.2024.100404
19. Saharkhizan, M., Azmoodeh, A., Dehghantanha, A., Choo, K.-K.R., Parizi, R.M.: An ensemble of deep recurrent neural networks for detecting IOT cyber attacks using network traffic. IEEE Internet Things J. **7**(9), 8852–8859 (2020). https://doi.org/10.1109/jiot.2020.2996425
20. Yu, P., Li, Y., Zhang, M., Xiong, A., Li, W., Qiu, X., Meng, L.: Self-organized and distributed green resource allocation for space–air–ground IOT Networks. IEEE Internet Things J. **10**(11), 9385–9397 (2023). https://doi.org/10.1109/jiot.2022.3222238
21. Shahraki, A., Taherkordi, A., Haugen, O., Eliassen, F.: A survey and future directions on clustering: from WSNS to IOT and modern networking paradigms. IEEE Trans. Netw. Serv. Manage. **18**(2), 2242–2274 (2021). https://doi.org/10.1109/tnsm.2020.3035315
22. Zhang, Z., Wei, X., Lung, C.-H., Zhao, Y.: ICache: an intelligent caching scheme for dynamic network environments in ICN-based IOT Networks. IEEE Internet Things J. **10**(2), 1787–1799 (2023). https://doi.org/10.1109/jiot.2022.3209256
23. Lei, T., Xue, J., Wang, Y., Baker, T., Niu, Z.: An empirical study of problems and evaluation of IOT malware classification label sources. J. King Saud Univ. Comput. Inf. Sci. **36**(1), 101898 (2024). https://doi.org/10.1016/j.jksuci.2023.101898
24. Akinbi, A., MacDermott, Á., Ismael, A.M.: A systematic literature review of blockchain-based internet of things (IOT) forensic investigation process models. Forensic Sci. Int. Digital Investig. **42–43**, 301470 (2022). https://doi.org/10.1016/j.fsidi.2022.301470
25. Schiller, E., Aidoo, A., Fuhrer, J., Stahl, J., Ziörjen, M., Stiller, B.: Landscape of IOT security. Comput. Sci. Rev. **44**, 100467 (2022). https://doi.org/10.1016/j.cosrev.2022.100467
26. Arshad, A., Jabeen, M., Ubaid, S., Raza, A., Abualigah, L., Aldiabat, K., Jia, H.: A novel ensemble method for enhancing internet of things device security against botnet attacks. Decis. Anal. J. **8**, 100307 (2023). https://doi.org/10.1016/j.dajour.2023.100307
27. Hussain, F., Abbas, S.G., Pires, I.M., Tanveer, S., Fayyaz, U.U., Garcia, N.M., Shah, G.A., Shahzad, F.: A two-fold machine learning approach to prevent and detect IOT botnet attacks. IEEE Access **9**, 163412–163430 (2021). https://doi.org/10.1109/access.2021.3131014
28. Salem, O., Alsubhi, K., Shaafi, A., Gheryani, M., Mehaoua, A., Boutaba, R.: Man-in-the-middle attack mitigation in internet of medical things. IEEE Trans. Industr. Inf. **18**(3), 2053–2062 (2022). https://doi.org/10.1109/tii.2021.3089462
29. Zhang, X.-G., Yang, G.-H., Wasly, S.: Man-in-the-middle attack against cyber-physical systems under random access protocol. Inf. Sci. **576**, 708–724 (2021). https://doi.org/10.1016/j.ins.2021.07.083
30. Burnes, D., DeLiema, M., Langton, L.: Risk and protective factors of identity theft victimization in the United States. Prev. Med. Rep. **17**, 101058 (2020). https://doi.org/10.1016/j.pmedr.2020.101058
31. Wang, C., Zhu, H., Yang, B.: Composite behavioralmodeling for identity theft detection in online social networks. IEEE Trans. Comput. Soc. Syst. **9**(2), 428–439 (2022). https://doi.org/10.1109/tcss.2021.3092007
32. Alnabulsi, H., Islam, R., Talukder, M.: GMSA: gathering multiple signatures approach to defend against code injection attacks. IEEE Access **6**, 77829–77840 (2018). https://doi.org/10.1109/access.2018.2884201
33. George, T.K., Jacob, K.P., James, R.K.: Token based detection and neural network based reconstruction framework against code injection vulnerabilities. J. Inf. Secur. Appl. **41**, 75–91 (2018). https://doi.org/10.1016/j.jisa.2018.05.005
34. Sakthibalan, P., Devarajan, K.: DFMS: differential flow management scheme for denial of service impact mitigation in 5G communications. J. King Saud Univ. Comput. Inf. Sci. **34**(8), 5366–5374 (2022). https://doi.org/10.1016/j.jksuci.2020.12.005
35. Alzahrani, S., Hong, L.: Detection of distributed denial of service (ddos) attacks using artificial intelligence on cloud. 2018 IEEE World Congress on Services (SERVICES) (2018). https://doi.org/10.1109/services.2018.00031

36. Dong, S., Sarem, M.: DDoS attack detection method based on improved KNN with the degree of ddos attack in software-defined networks. IEEE Access **8**, 5039–5048 (2020). https://doi.org/10.1109/access.2019.2963077
37. López Bernal, S., Perez Martins, D., Huertas Celdrán, A.: Towards the mitigation of distributed denial-of-service cyberbioattacks in bacteria-based biosensing systems. Digital Signal Process. **118**, 103241 (2021). https://doi.org/10.1016/j.dsp.2021.103241
38. Marshall, N., Sturman, D., Auton, J.C.: Exploring the evidence for email phishing training: a scoping review. Comput. Secur. **139**, 103695 (2024). https://doi.org/10.1016/j.cose.2023.103695
39. Pham, C., Nguyen, L.A., Tran, N.H., Huh, E.-N., Hong, C.S.: Phishing-aware: a neuro-fuzzy approach for anti-phishing on fog networks. IEEE Trans. Netw. Serv. Manage. **15**(3), 1076–1089 (2018). https://doi.org/10.1109/tnsm.2018.2831197
40. Min, D., Ko, Y., Walker, R., Lee, J., Kim, Y.: A content-based ransomware detection and backup solid-state drive for ransomware defense. IEEE Trans. Comput. Aided Des. Integr. Circuits Syst. **41**(7), 2038–2051 (2022). https://doi.org/10.1109/tcad.2021.3099084
41. Benmalek, M.: Ransomware on cyber-physical systems: taxonomies, case studies, security gaps, and open challenges. Internet of Things Cyber-Phys. Syst. **4**, 186–202 (2024). https://doi.org/10.1016/j.iotcps.2023.12.001
42. Goplakrishnan, S., Hussain, M.Z., Ashraf, M., Senthilkumar, G., Murthy, M.Y.: Sensitive product feature integrity and confidentiality using blockchain-based internet of things (IOT) architecture. Meas. Sens. **27**, 100798 (2023). https://doi.org/10.1016/j.measen.2023.100798
43. Zhang, Q.: Artistic expression and data protection: balancing aesthetics with data privacy in IOT. Heliyon **9**(9) (2023). https://doi.org/10.1016/j.heliyon.2023.e19380
44. Masud, M., Gaba, G.S., Kumar, P., Gurtov, A.: A user-centric privacy-preserving authentication protocol for IOT-ami environments. Comput. Commun. **196**, 45–54 (2022). https://doi.org/10.1016/j.comcom.2022.09.021
45. Akkad, A., Wills, G., Rezazadeh, A.: An information security model for an IOT-enabled smart grid in the Saudi energy sector. Comput. Electr. Eng. **105**, 108491 (2023). https://doi.org/10.1016/j.compeleceng.2022.108491
46. Popoola, O., Rodrigues, M., Marchang, J., Shenfield, A., Ikpehia, A., Popoola, J.: A critical literature review of security and privacy in smart home healthcare schemes adopting IOT & blockchain: problems, challenges and solutions. Blockchain Res. Appl. 100178 (2023). https://doi.org/10.1016/j.bcra.2023.100178
47. Wu, Q., Tang, J., Dang, S., Chen, G.: Data privacy and utility trade-off based on mutual information neural estimator. Expert Syst. Appl. **207**, 118012 (2022). https://doi.org/10.1016/j.eswa.2022.118012
48. Lavado-Nalvaiz, N., Lucia-Palacios, L., Pérez-López, R.: The role of the humanisation of smart home speakers in the personalisation–privacy paradox. Electron. Commer. Res. Appl. **53**, 101146 (2022). https://doi.org/10.1016/j.elerap.2022.101146
49. Bi, R., Zhao, M., Ying, Z., Tian, Y., Xiong, J.: Achieving dynamic privacy measurement and protection based on reinforcement learning for mobile edge crowdsensing of IOT. Digital Commun. Netw. (2022). https://doi.org/10.1016/j.dcan.2022.07.013
50. Iftikhar, A., Qureshi, K.N., Shiraz, M., Albahli, S.: Security, trust and privacy risks, responses, and solutions for high-speed smart cities networks: a systematic literature review. J. King Saud Univ. Comput. Inf. Sci. **35**(9), 101788 (2023). https://doi.org/10.1016/j.jksuci.2023.101788
51. Fenu, G., Marras, M.: Demographic fairness in multimodal biometrics: a comparative analysis on audio-visual speaker recognition systems. Proced. Comput. Sci. **198**, 249–254 (2022). https://doi.org/10.1016/j.procs.2021.12.236
52. JosephP, R.B., Ezhilmaran, D.: A smart computing algorithm for finger vein matching with affine invariant features using fuzzy image retrieval. Proced. Comput. Sci. **125**, 172–178 (2018). https://doi.org/10.1016/j.procs.2017.12.024
53. Alyushin, M.V., Kolobashkina, L.V.: Development of a metrological database with images of a human face in the infrared range to evaluate the effectiveness of biometric algorithms. Proced. Comput. Sci. **123**, 7–11 (2018). https://doi.org/10.1016/j.procs.2018.01.002

54. Zhang, S., Xie, Y., Wan, J., Xia, H., Li, S.Z., Guo, G.: Widerperson: a diverse dataset for dense pedestrian detection in the wild. IEEE Trans. Multimed. **22**(2), 380–393 (2020). https://doi.org/10.1109/tmm.2019.2929005
55. Kant, C., Chaudhary, S.: A watermarking based approach for protection of templates in multimodal biometric system. Proced. Comput. Sci. **167**, 932–941 (2020). https://doi.org/10.1016/j.procs.2020.03.392
56. Robinson, J.P., Khan, Z., Yin, Y., Shao, M., Fu, Y.: Families in wild multimedia: a multimodal database for recognizing kinship. IEEE Trans. Multimed. **24**, 3582–3594 (2022). https://doi.org/10.1109/tmm.2021.3103074
57. Kumar, S., Chaube, M.K., Kumar, S.: Secure and sustainable framework for cattle recognition using wireless multimedia networks and machine learning techniques. IEEE Trans. Sustain. Comput. **7**(3), 696–708 (2022). https://doi.org/10.1109/tsusc.2021.3123496
58. Odya, P., Górski, F., Czyżewski, A.: User authentication by eye movement features employing SVM and XGBoost classifiers. IEEE Access **11**, 93341–93353 (2023). https://doi.org/10.1109/access.2023.3309000
59. Reis, P.M., Ribeiro, R.O.: A forensic evaluation method for deepfake detection using DCNN-based facial similarity scores. Forensic Sci. Int. 111747 (2023).https://doi.org/10.1016/j.forsciint.2023.111747
60. Mehmood, I., Lv, Z., Zhang, Y.-D., Ota, K., Sajjad, M., Singh, A.K.: Mobile cloud-assisted paradigms for management of multimedia big data in healthcare systems: research challenges and opportunities. Int. J. Inf. Manage. **45**, 246–249 (2019). https://doi.org/10.1016/j.ijinfomgt.2018.10.020
61. Yuan, X., Wang, X., Wang, C., Weng, J., Ren, K.: Enabling secure and fast indexing for privacy-assured healthcare monitoring via compressive sensing. IEEE Trans. Multimed. **18**(10), 2002–2014 (2016). https://doi.org/10.1109/tmm.2016.2602758
62. Tawalbeh, L.A., Mehmood, R., Benkhlifa, E., Song, H.: Mobile cloud computing model and big data analysis for healthcare applications. IEEE Access **4**, 6171–6180 (2016). https://doi.org/10.1109/access.2016.2613278
63. Muhammed, T., Mehmood, R., Albeshri, A., Katib, I.: UbeHealth: a personalized ubiquitous cloud and edge-enabled networked healthcare system for smart cities. IEEE Access **6**, 32258–32285 (2018). https://doi.org/10.1109/access.2018.2846609
64. Moniaga, J.V., Ohyver, M., Siregar, J., Yauwito, P.H.: Map-type modelling and analysis of children stunting case data in Indonesia with interactive multimedia method. Proced. Comput. Sci. **157**, 530–536 (2019). https://doi.org/10.1016/j.procs.2019.09.010
65. Ren, J., Li, J., Liu, H., Qin, T. (2022). Task offloading strategy with emergency handling and blockchain security in SDN-empowered and fog-assisted healthcare IOT. Tsinghua Science and Technology, 27(4), 760–776. https://doi.org/10.26599/tst.2021.9010046
66. Sutradhar, S., Karforma, S., Bose, R., Roy, S.: A dynamic step-wise tiny encryption algorithm with fruit fly optimization for quality of service improvement in healthcare. Healthc. Anal. **3**, 100177 (2023). https://doi.org/10.1016/j.health.2023.100177
67. Siddiqui, E.F., Ahmed, T., Nayak, S.K.: A decision tree approach for enhancing real-time response in exigent healthcare unit using edge computing. Meas. Sens **32**, 100979 (2024). https://doi.org/10.1016/j.measen.2023.100979
68. Almutairi, S.A.: A multimodal AI-based non-invasive COVID-19 grading framework powered by deep learning, Manta Ray, and fuzzy inference system from multimedia vital signs. Heliyon **9**(6) (2023). https://doi.org/10.1016/j.heliyon.2023.e16552
69. Kour, V.P., Arora, S.: Recent developments of the internet of things in agriculture: a survey. IEEE Access **8**, 129924–129957 (2020). https://doi.org/10.1109/access.2020.3009298
70. Bhanu, K.N., Reddy, T.B., Hanumanthappa, M.: Multi-agent based context aware information gathering for agriculture using wireless multimedia sensor networks. Egypt. Inf. J. **20**(1), 33–44 (2019). https://doi.org/10.1016/j.eij.2018.07.001
71. Liang, H., Gao, W., Nguyen, J.H., Orpilla, M.F., Yu, W.: Internet of things data collection using unmanned aerial vehicles in infrastructure free environments. IEEE Access **8**, 3932–3944 (2020). https://doi.org/10.1109/access.2019.2962323

72. Hussein, W.A., Ali, B.M., Rasid, M., Hashim, F.: Smart geographical routing protocol achieving high QoS and energy efficiency based for wireless multimedia sensor networks. Egypt. Inf. J. **23**(2), 225–238 (2022). https://doi.org/10.1016/j.eij.2021.12.005
73. Chen, Z., Dong, J., Asif, Z.: A regional numerical environmental multimedia modeling approach to assess spatial eco-environmental exposure risk of perfluorooctane sulfonate (PFOS) in the Pearl River Basin. Environ. Int. **161**, 107101 (2022). https://doi.org/10.1016/j.envint.2022.107101
74. Kaur, M., Singh, K., Kumar, S.: Biodiversity sensor: a customized and power efficient solution for biodiversity surveillance. IEEE Sens. J. **23**(24), 31159–31170 (2023). https://doi.org/10.1109/jsen.2023.3328067
75. Zhou, W., Xu, L., Yang, L., Liu, S., He, M., Sheng, Q., Liu, T., Guo, J., Feng, D., Hassan, S.G., Cao, L.: A hybrid model for predicting the environment humidity of pigeon sheds. IEEE Access **11**, 92258–92272 (2023). https://doi.org/10.1109/access.2023.3298649
76. Quik, J.T.K., Meesters, J.A.J., Koelmans, A.A.: A multimedia model to estimate the environmental fate of microplastic particles. Sci. Total Environ. **882**, 163437 (2023). https://doi.org/10.1016/j.scitotenv.2023.163437
77. Zhao, Z., Yang, X., Liu, J., Zhou, C., Zhao, C.: GCVC: graph convolution vector distribution calibration for fish group activity recognition. IEEE Trans. Multimed. **26**, 1776–1789 (2024). https://doi.org/10.1109/tmm.2023.3287339
78. AL-Dhief, F.T., Sabri, N., Fouad, S., Latiff, N.M.A., Albader, M.A.: A review of forest fire surveillance technologies: mobile AD-HOC network routing protocols perspective. J. King Saud Univ. Comput. Inf. Sci. 31(2):135–146.https://doi.org/10.1016/j.jksuci.2017.12.005
79. Blasch, E.P., Aved, A.J.: Dynamic data-driven application system (DDDAS) for video surveillance user support. Proced. Comput. Sci. **51**, 2503–2517 (2015). https://doi.org/10.1016/j.procs.2015.05.359
80. Abdallah, H.B., Abdellatif, T., Chekir, F.: AMSEP: automated multi-level security management for multimedia event processing. Proced. Comput. Sci. **134**, 452–457 (2018). https://doi.org/10.1016/j.procs.2018.07.184
81. Shao, Z., Cai, J., Wang, Z.: Smart monitoring cameras driven intelligent processing to big surveillance video data. IEEE Trans. Big Data **4**(1), 105–116 (2018). https://doi.org/10.1109/tbdata.2017.2715815
82. Liu, X., Liu, W., Mei, T., Ma, H.: Provid: progressive and multimodal vehicle reidentification for large-scale urban surveillance. IEEE Trans. Multimed. **20**(3), 645–658 (2018). https://doi.org/10.1109/tmm.2017.2751966
83. Yazici, A., Koyuncu, M., Sert, S.A., Yilmaz, T.: A fusion-based framework for wireless multimedia sensor networks in surveillance applications. IEEE Access **7**, 88418–88434 (2019). https://doi.org/10.1109/access.2019.2926206
84. Chen, Y., Wang, J., Bai, Y., Castanon, G., Saligrama, V.: Probabilistic semantic retrieval for surveillance videos with activity graphs. IEEE Trans. Multimed. **21**(3), 704–716 (2019). https://doi.org/10.1109/tmm.2018.2865860
85. Aslam, A., Curry, E.: A survey on object detection for the internet of multimedia things (IOMT) using deep learning and event-based middleware: approaches, challenges, and future directions. Image Vis. Comput. **106**, 104095 (2021). https://doi.org/10.1016/j.imavis.2020.104095
86. Cecchinato, N., Toma, A., Drioli, C., Oliva, G., Sechi, G., Foresti, G.L.: A secure real-time multimedia streaming through robust and lightweight AES encryption in UAV networks for operational scenarios in military domain. Proced. Comput. Sci. **205**, 50–57 (2022). https://doi.org/10.1016/j.procs.2022.09.006
87. Soori, M., Dastres, R., Arezoo, B.: AI-powered blockchain technology in industry 4.0, a review. J. Econ. Technol. **1**, 222–241 (2023). https://doi.org/10.1016/j.ject.2024.01.001
88. Islam, S., Islam, M.J., Hossain, M., Noor, S., Kwak, K.-S., Islam, S.M.: A survey on consensus algorithms in blockchain-based applications: architecture, taxonomy, and operational issues. IEEE Access **11**, 39066–39082 (2023). https://doi.org/10.1109/access.2023.3267047

89. Wang, Y., Cai, S., Lin, C., Chen, Z., Wang, T., Gao, Z., Zhou, C.: Study of blockchains's consensus mechanism based on credit. IEEE Access **7**, 10224–10231 (2019). https://doi.org/10.1109/access.2019.2891065
90. Xiao, Y., Zhang, N., Lou, W., Hou, Y.T.: A survey of distributed consensus protocols for blockchain networks. IEEE Commun. Surv. Tutor. **22**(2), 1432–1465 (2020). https://doi.org/10.1109/comst.2020.2969706
91. Luo, H., Yu, H., Luo, J.: Praft and RPBFT: a class of blockchain consensus algorithm and their applications in electric vehicles charging scenarios for V2G Networks. Internet of Things Cyber-Phys. Syst. **3**, 61–70 (2023). https://doi.org/10.1016/j.iotcps.2023.02.003
92. Bandara, E., Liang, X., Foytik, P., Shetty, S., Mukkamala, R., Rahman, A., Ranasinghe, N., De Zoysa, K., Ng, W.K.: Lightweight, geo-scalable deterministic blockchain design for 5G networks sliced applications with hierarchical CFT/BFT consensus groups, ipfs and Novel Hardware Design. Internet of Things **25**, 101077 (2024). https://doi.org/10.1016/j.iot.2024.101077
93. Salimitari, M., Chatterjee, M., Fallah, Y.P.: A survey on consensus methods in blockchain for resource-constrained IOT Networks. Internet of Things **11**, 100212 (2020). https://doi.org/10.1016/j.iot.2020.100212
94. Shrimali, B., Patel, H.B.: Blockchain state-of-the-art: architecture, use cases, consensus, challenges and opportunities. J. King Saud Univ. Comput. Inf. Sci. **34**(9), 6793–6807 (2022). https://doi.org/10.1016/j.jksuci.2021.08.005
95. Hasan, A.S.M.T., Sabah, S., Daria, A., Haque, R.U.: A peer-to-peer blockchain-based architecture for trusted and reliable agricultural product traceability. Decis. Anal. J. **9**, 100363 (2023). https://doi.org/10.1016/j.dajour.2023.100363
96. Tabatabaei, M.H., Vitenberg, R., Veeraragavan, N.R.: Understanding blockchain: definitions, architecture, design, and system comparison. Comput. Sci. Rev. **50**, 100575 (2023). https://doi.org/10.1016/j.cosrev.2023.100575
97. Andrew, J., Isravel, D.P., Sagayam, K.M., Bhushan, B., Sei, Y., Eunice, J.: Blockchain for healthcare systems: architecture, security challenges, trends and future directions. J. Netw. Comput. Appl. **215**, 103633 (2023). https://doi.org/10.1016/j.jnca.2023.103633
98. Zulkifl, Z., Khan, F., Tahir, S., Afzal, M., Iqbal, W., Rehman, A., Saeed, S., Almuhaideb, A.M.: FBASHI: fuzzy and blockchain-based adaptive security for healthcare IoTs. IEEE Access **10**, 15644–15656 (2022). https://doi.org/10.1109/access.2022.3149046
99. Rathore, S., Park, J.H., Chang, H.: Deep learning and blockchain-empowered security framework for intelligent 5G-enabled IOT. IEEE Access **9**, 90075–90083 (2021). https://doi.org/10.1109/access.2021.3077069
100. Yang, X., Yang, X., Yi, X., Khalil, I., Zhou, X., He, D., Huang, X., Nepal, S.: Blockchain-based secure and lightweight authentication for internet of things. IEEE Internet Things J. **9**(5), 3321–3332 (2022). https://doi.org/10.1109/jiot.2021.3098007
101. Zhang, J., Xin, Y., Gao, Y., Lei, X., Yang, Y.: Secure abe scheme for access management in blockchain-based IOT. IEEE Access **9**, 54840–54849 (2021). https://doi.org/10.1109/access.2021.3071031
102. Cai, J., Liang, W., Li, X., Li, K., Gui, Z., Khan, M.K. GTxChain: a secure IOT smart blockchain architecture based on graph neural network. IEEE Internet Things J. 1–1.https://doi.org/10.1109/jiot.2023.3296469
103. Azbeg, K., Ouchetto, O., Jai Andaloussi, S.: BlockMedCare: a healthcare system based on IOT, blockchain and ipfs for data management security. Egypt. Inf. J. **23**(2), 329–343 (2022). https://doi.org/10.1016/j.eij.2022.02.004
104. Luo, F., Huang, R., Xie, Y.: Hybrid blockchain-based many-to-many cross-domain authentication scheme for smart agriculture IOT networks. J. King Saud Univ. Comput. Inf. Sci. **101946**.https://doi.org/10.1016/j.jksuci.2024.101946
105. Khashan, O.A., Khafajah, N.M.: Efficient hybrid centralized and blockchain-based authentication architecture for heterogeneous IOT systems. J. King Saud Univ. Comput. Inf. Sci. **35**(2), 726–739 (2023). https://doi.org/10.1016/j.jksuci.2023.01.011

106. Hennebelle, A., Ismail, L., Materwala, H., Al Kaabi, J., Ranjan, P., Janardhanan, R.: Secure and privacy-preserving automated machine learning operations into end-to-end integrated IOT-edge-artificial intelligence-blockchain monitoring system for diabetes mellitus prediction. Comput. Struct. Biotechnol. J. **23**, 212–233 (2024). https://doi.org/10.1016/j.csbj.2023.11.038
107. Li, Z., Chu, Y., Liu, X., Zhang, Y., Feng, J., Xiang, X.: Physical unclonable function based identity management for IOT with blockchain. Proced. Comput. Sci. **198**, 454–459 (2022). https://doi.org/10.1016/j.procs.2021.12.269
108. Elsadig, M.A.: Detection of denial-of-service attack in wireless sensor networks: a lightweight machine learning approach. IEEE Access **11**, 83537–83552 (2023). https://doi.org/10.1109/access.2023.3303113
109. Kotenko, I., Saenko, I., Branitskiy, A.: Framework for mobile internet of things security monitoring based on big data processing and machine learning. IEEE Access **6**, 72714–72723 (2018). https://doi.org/10.1109/access.2018.2881998
110. Ekolle, Z.E., Ochiai, H., Kohno, R.: Collabo: a collaborative machine learning model and its application to the security of heterogeneous medical data in an IOT Network. IEEE Access **11**, 142663–142675 (2023). https://doi.org/10.1109/access.2023.3341837
111. Shafiq, M., Tian, Z., Bashir, A.K., Du, X., Guizani, M.: CORRAUC: a malicious bot-IOT traffic detection method in IOT network using machine-learning techniques. IEEE Internet Things J. **8**(5), 3242–3254 (2021). https://doi.org/10.1109/jiot.2020.3002255
112. Liu, S., Xiao, L., Huang, L., Wang, X.: Impulsive noise recovery and elimination: a sparse machine learning based approach. IEEE Trans. Veh. Technol. **68**(3), 2306–2315 (2019). https://doi.org/10.1109/tvt.2019.2891617
113. González-Soto, M., Díaz-Redondo, R.P., Fernández-Veiga, M., Fernández-Castro, B., Fernández-Vilas, A.: Decentralized and collaborative machine learning framework for IOT. Comput. Netw. **239**, 110137 (2024). https://doi.org/10.1016/j.comnet.2023.110137
114. Arnau Muñoz, L., Berná Martínez, J.V., Maciá Pérez, F., Lorenzo Fonseca, I.: Anomaly detection system for data quality assurance in IOT infrastructures based on machine learning. Internet Things **25**, 101095 (2024). https://doi.org/10.1016/j.iot.2024.101095
115. Aloraini, F., Javed, A., Rana, O., Burnap, P.: Adversarial machine learning in IOT from an insider point of view. J. Inf. Secur. Appl. **70**, 103341 (2022). https://doi.org/10.1016/j.jisa.2022.103341
116. Gaber, T., El-Ghamry, A., Hassanien, A.E.: Injection attack detection using machine learning for smart IOT applications. Phys. Commun. **52**, 101685 (2022). https://doi.org/10.1016/j.phycom.2022.101685
117. Gaber, T., Awotunde, J.B., Torky, M., Ajagbe, S.A., Hammoudeh, M., Li, W.: Metaverse-ids: deep learning-based intrusion detection system for metaverse-IOT networks. Internet Things **24**, 100977 (2023). https://doi.org/10.1016/j.iot.2023.100977
118. Thabit, F., Can, O., Alhomdy, S., Al-Gaphari, G.H., Jagtap, S.: A novel effective lightweight homomorphic cryptographic algorithm for data security in cloud computing. Int. J. Intell. Netw. **3**, 16–30 (2022). https://doi.org/10.1016/j.ijin.2022.04.001
119. Aiyshwariya Devi, R., Arunachalam, A.R.: Enhancement of IOT device security using an improved elliptic curve cryptography algorithm and malware detection utilizing DEEP LSTM. High-Confid. Comput. **3**(2), 100117 (2023). https://doi.org/10.1016/j.hcc.2023.100117
120. Achary, R., Shelke, C.J., Marx, K., Rajesh, A.: Security implementation on IOT using CoAP and elliptical curve cryptography. Proced. Comput. Sci. **230**, 493–502 (2023). https://doi.org/10.1016/j.procs.2023.12.105
121. Caraveo-Cacep, M.A., Vázquez-Medina, R., Hernández Zavala, A.: A survey on low-cost development boards for applying cryptography in IOT systems. Internet Things **22**, 100743 (2023). https://doi.org/10.1016/j.iot.2023.100743
122. Dhar, S., Khare, A., Dwivedi, A.D., Singh, R.: Securing IOT devices: a novel approach using blockchain and quantum cryptography. Internet Things **25**, 101019 (2024). https://doi.org/10.1016/j.iot.2023.101019
123. Wang, H., Wen, J., Liu, J., Zhang, H.: Acke: asymmetric computing key exchange protocol for IOT environments. IEEE Internet Things J. **10**(20), 18273–18281 (2023). https://doi.org/10.1109/jiot.2023.3279283

124. Awan, K.A., Din, I.U., Almogren, A., Rodrigues, J.J.: Privacy-preserving big data security for IOT with federated learning and cryptography. IEEE Access **11**, 120918–120934 (2023). https://doi.org/10.1109/access.2023.3328310
125. Wang, D., Bai, B., Lei, K., Zhao, W., Yang, Y., Han, Z.: Enhancing information security via physical layer approaches in heterogeneous IOT with multiple access mobile edge computing in smart city. IEEE Access **7**, 54508–54521 (2019). https://doi.org/10.1109/access.2019.2913438
126. Samaila, M.G., Sequeiros, J.B., Simoes, T., Freire, M.M., Inacio, P.R.: IOT-harpseca: a framework and roadmap for secure design and development of devices and applications in the IOT space. IEEE Access **8**, 16462–16494 (2020). https://doi.org/10.1109/access.2020.2965925
127. Atiewi, S., Al-Rahayfeh, A., Almiani, M., Yussof, S., Alfandi, O., Abugabah, A., Jararweh, Y.: Scalable and secure big data IOT system based on multifactor authentication and lightweight cryptography. IEEE Access **8**, 113498–113511 (2020). https://doi.org/10.1109/access.2020.3002815
128. Rehman, M.U., Shafique, A., Usman, A.B.: Securing medical information transmission between IOT devices: an innovative hybrid encryption scheme based on quantum Walk, DNA encoding, and chaos. Internet Things **24**, 100891 (2023). https://doi.org/10.1016/j.iot.2023.100891
129. Bey, M., Kuila, P., Naik, B.B., Ghosh, S.: Quantum-inspired particle swarm optimization for efficient IOT service placement in edge computing systems. Expert Syst. Appl. **236**, 121270 (2024). https://doi.org/10.1016/j.eswa.2023.121270
130. Chen, L., Zhang, D.-G., Zhang, J., Zhang, T., Wang, W.-J., Cao, Y.-H.: A novel offloading approach of IOT user perception task based on quantum behavior particle swarm optimization. Futur. Gener. Comput. Syst. **141**, 577–594 (2023). https://doi.org/10.1016/j.future.2022.12.016
131. Visalaxi, G., Muthukumaravel, A.: IOT monitoring membrane computing based on quantum inspiration to enhance security in cloud network. Meas. Sens. **27**, 100755 (2023). https://doi.org/10.1016/j.measen.2023.100755
132. Mangla, C., Rani, S., Faseeh Qureshi, N.M., Singh, A.: Mitigating 5G security challenges for next-gen industry using quantum computing. J. King Saud Univ. Comput. Inf. Sci. **35**(6), 101334 (2023). https://doi.org/10.1016/j.jksuci.2022.07.009
133. Mohanty, T., Srivastava, V., Debnath, S.K., Das, A.K., Sikdar, B.: Quantum secure threshold private set intersection protocol for IOT-enabled privacy-preserving ride-sharing application. IEEE Internet Things J. **11**(1), 1761–1772 (2024). https://doi.org/10.1109/jiot.2023.3291132
134. El-Latif, A.A., Abd-El-Atty, B., Venegas-Andraca, S.E., Elwahsh, H., Piran, Md.J., Bashir, A.K., Song, O.-Y., Mazurczyk, W.: Providing end-to-end security using quantum walks in IOT networks. IEEE Access **8**, 92687–92696 (2020). https://doi.org/10.1109/access.2020.2992820
135. Prajapat, S., Kumar, P., Kumar, S., Das, A.K., Shetty, S., Hossain, M.S.: Designing high-performance identity-based quantum signature protocol with strong security. IEEE Access **12**, 14647–14658 (2024). https://doi.org/10.1109/access.2024.3355196
136. El-Latif, A.A., Abd-El-Atty, B., Hossain, M.S., Elmougy, S., Ghoneim, A.: Secure quantum steganography protocol for fog cloud internet of things. IEEE Access **6**, 10332–10340 (2018). https://doi.org/10.1109/access.2018.2799879
137. Shi, R.-H., Yu, H.: Privacy-preserving range query quantum scheme with single photons in edge-based internet of things. IEEE Trans. Netw. Serv. Manage. **20**(4), 4923–4936 (2023). https://doi.org/10.1109/tnsm.2023.3253858
138. Bandara, H., Herath, Y., Weerasundara, T., Alawatugoda, J.: On advances of lattice-based cryptographic schemes and their implementations. Cryptography **6**(4), 56 (2022). https://doi.org/10.3390/cryptography6040056
139. Baldanzi, L., Crocetti, L., Falaschi, F., Bertolucci, M., Belli, J., Fanucci, L., Saponara, S.: Cryptographically secure pseudo-random number generator IP-core based on SHA2 algorithm. Sensors **20**(7), 1869 (2020). https://doi.org/10.3390/s20071869

Multimedia Data Processing and Analysis in IoT

Ganga Rama Koteswara Rao, D. Rajendra Prasad, Suneel Miriyala, Ravindra Janga, and F. Ajesh

Abstract The integration of Multimedia Data Processing and Analysis within the IoT framework is a pivotal domain shaping the landscape of intelligent and context-aware systems. This chapter explores the synergy between multimedia data and IoT, focusing on the intricacies of processing and analyzing multimedia content in IoT environments. The study delves into methodologies for efficient multimedia data handling, encompassing visual, auditory, and sensor-generated information. Emphasis is placed on the challenges and opportunities associated with real-time processing, storage, and extraction of meaningful insights from multimedia data in the context of IoT applications. Through a comprehensive examination, this paper seeks to enhance our understanding of the multifaceted interactions between Multimedia Data Processing and Analysis within the dynamic and evolving IoT ecosystem.

Keywords Multimedia data · IoT · Smart City · Data analysis

G. R. K. Rao (✉)
Department of CSE, Koneru Lakshmaiah Education Foundation, Vaddeswaram, Guntur, Andhra Pradesh, India
e-mail: grkraoganga@gmail.com

D. Rajendra Prasad
Department of ECE, St. Ann's College of Engineering and Technology, Chirala, Bapatla, Andhra Pradesh, India

S. Miriyala
Department of ECE, Bapatla Engineering College, Bapatla, Andhra Pradsh, India

R. Janga
Department of EEE, Bapatla Engineering College, Bapatla, Andhra Pradsh, India

F. Ajesh
Department of Computer Science and Engineering, TKM Institute of Technology, Kollam, Kerala, India

R. Sharma et al. (eds.), *Multimedia Technologies in the Internet of Things Environment, Volume 4*, Studies in Big Data 173, https://doi.org/10.1007/978-981-96-4356-1_4

1 Overview of Multimedia Data in IoT

The Internet of Things (IoT) is transforming the way devices and systems communicate and interact with the environment. One of the most significant types of data in IoT applications is multimedia data, which includes images, video, audio, and sensor data. These types of data enable systems to gather richer, more detailed information about the environment, leading to more intelligent and responsive applications. The integration of multimedia data with IoT opens up possibilities in various domains such as surveillance, healthcare, smart cities, and industrial automation. In this section, we will explore the types of multimedia data in IoT and their sources.

1.1 Types of Multimedia Data

Image and Video Data: Visual data captured by IoT devices such as cameras, drones, and other visual sensors plays a critical role in many IoT applications. Image and video data are essential in surveillance and security systems, providing real-time visual monitoring of environments. For example, security cameras are used to monitor public spaces, while drones capture aerial footage for agricultural or environmental monitoring. Additionally, video streams are key in object detection, where algorithms can identify and track objects or individuals in real time, enabling smart monitoring systems. Video and image data also help in healthcare applications, such as telemedicine, where doctors rely on high-quality images for diagnosis.

Audio Data: Audio data in IoT systems is commonly captured through microphones and used in various applications like voice recognition, environmental sound analysis, and safety monitoring [1]. For instance, in smart homes, voice assistants rely on audio data to respond to user commands and perform actions like adjusting the temperature or controlling smart appliances. Audio data is also integral in monitoring sound events, such as detecting glass breaking or emergency alarms in surveillance systems. Additionally, it can be used in health-monitoring systems where sound analysis helps detect conditions like sleep apnea or respiratory issues.

Sensor Data: IoT systems also generate and process various sensor data, including temperature, humidity, motion, proximity, and environmental conditions. These sensor readings provide contextual information that is often used alongside multimedia data [2]. For example, a camera in a smart home can capture video footage, while environmental sensors can detect temperature changes or motion within the space. By combining video or audio data with sensor data, IoT systems can offer more accurate analyses and enable intelligent responses. In smart cities, sensor data can provide real-time information on traffic, air quality, and public safety, complementing video footage to improve decision-making (Fig. 1).

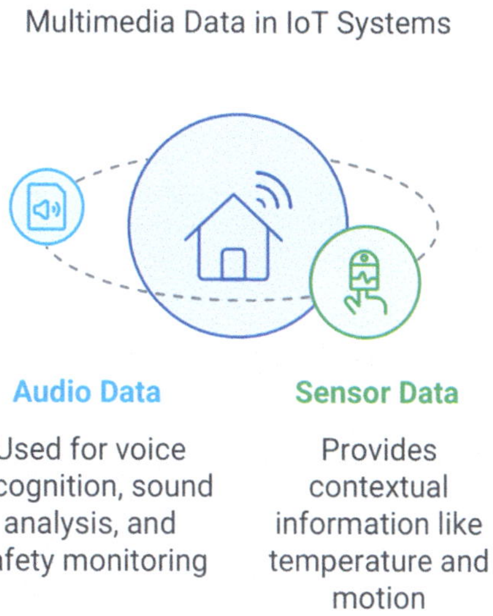

Fig. 1 Multimedia data in IoT systems

1.2 Multimedia Data Sources in IoT

Smart Cameras: Smart cameras are widely used in IoT systems for applications such as security, surveillance, and traffic monitoring. These cameras are equipped with advanced image processing and analysis capabilities, enabling real-time object detection, facial recognition, and movement tracking. In smart cities, cameras mounted on streetlights or traffic poles monitor traffic flow, detect accidents, and improve public safety. In industrial settings, cameras are used to monitor production lines, identify defects in products, and ensure worker safety.

Wearable Devices: Wearable devices, such as smartwatches, fitness trackers, and health-monitoring devices, are another crucial source of multimedia data in IoT. These devices capture both video and audio for health monitoring and user interaction. For example, a smartwatch can record heart rate, temperature, and movement data, while a fitness tracker may include video or audio capabilities for guided workouts or health assessments. Wearable devices can also capture environmental sounds, enabling applications such as hearing aids or noise level monitoring in the workplace.

Environmental Sensors: Environmental sensors are often integrated with multimedia data sources to enhance the contextual information provided by IoT systems. These sensors can measure variables like temperature, humidity, air quality, and noise levels, which are essential for monitoring environmental conditions [3]. For instance, in a smart building, a camera may monitor a room, while environmental sensors measure the room's temperature and air quality. This integration allows for more intelligent responses, such as adjusting HVAC systems based on temperature readings or activating ventilation systems when air quality deteriorates. Similarly, environmental sensors are used in agriculture to monitor soil moisture and temperature, combined with drone-captured video to assess crop health (Fig. 2).

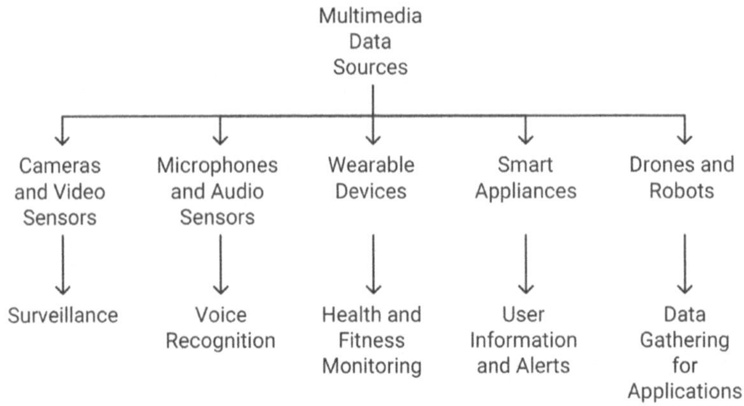

Fig. 2 Multimedia data sources in IoT

2 Key Technologies in Multimedia Data Processing

The initial phase of multimedia data processing is data acquisition, which involves gathering real-time images, video streams, audio signals, or sensor readings from IoT sensors embedded in various connected devices. These IoT sensors play a crucial role in collecting environmental data in real time [4]. However, due to the limited computational power of IoT devices, edge computing becomes essential for preprocessing this multimedia data before it is transmitted to central servers. This preprocessing step reduces the load on the network and improves efficiency by enabling initial data analysis closer to the data source.

Once acquired, multimedia data—typically large in volume—requires efficient storage and transmission methods to optimize bandwidth usage. Compression techniques, such as video compression standards H.264 and H.265, as well as audio compression formats like MP3 and AAC, are commonly employed to reduce the size of multimedia files. For long-term storage, cloud storage solutions are widely used to handle these substantial datasets. In addition, fog and edge computing further facilitate local processing and decrease latency, making them valuable for real-time applications [5]. When transmitting real-time video and audio data, protocols like RTP (Real-time Transport Protocol) ensure efficient streaming and maintain the quality of service required for seamless multimedia experiences.

2.1 Data Acquisition

The first stage in multimedia data processing is the acquisition of data through various IoT sensors. Data Acquisition is a fundamental stage in multimedia data processing that involves gathering data from a variety of Internet of Things (IoT) sensors. This stage encompasses the collection of images, video streams, audio signals, and sensor readings in real-time from connected devices. Here's an expanded look at how data acquisition works and the components involved, along with a diagram to illustrate the process.

Key Components of Data Acquisition in Multimedia Processing

IoT Sensors: IoT sensors are crucial in capturing real-time data from various sources, including environmental conditions, motion, sound, and video. These sensors collect data continuously or at specified intervals and can be installed in devices like surveillance cameras, smart home devices, wearable health monitors, or industrial machinery [6].

Image and Video Capture: Specialized sensors, such as cameras, capture high-resolution images and video streams. These sensors are often integrated into devices to monitor and record visual information, which can be processed for various purposes, such as surveillance, traffic monitoring, and facial recognition.

Audio Capture: Microphones or audio sensors collect sound waves and convert them into digital signals. This captured audio data is often used in voice recognition, environmental monitoring, and various multimedia applications where sound is relevant.

Environmental and Motion Sensors: These sensors measure physical conditions, such as temperature, humidity, pressure, or movement. For instance, accelerometers can detect motion, which is crucial in automotive and mobile applications, while other sensors monitor environmental conditions in smart homes or industrial settings.

Edge Computing for Preprocessing: Due to the limited computational capacity of many IoT devices, it is often necessary to perform initial data processing at the edge, close to where data is captured. Edge devices can filter, preprocess, and analyze data to reduce the volume of raw information transmitted to central servers. This helps reduce latency, conserve bandwidth, and increase the efficiency of multimedia data processing.

Data Transmission: Once processed, the data is transmitted to cloud servers or data centers for further processing, storage, and analysis. Transmission often relies on Wi-Fi, cellular networks, or dedicated communication protocols optimized for IoT devices (Fig. 3).

3 Data Storage and Transmission

In multimedia data processing, data storage and transmission play a critical role, as multimedia files are often large and need to be optimized for efficient transmission across IoT networks. Due to their size, multimedia files typically require compression techniques to reduce bandwidth requirements and storage space. Common video compression standards, such as H.264 and H.265, and audio compression formats, like MP3 and AAC, are widely used to minimize file sizes without significantly compromising quality.

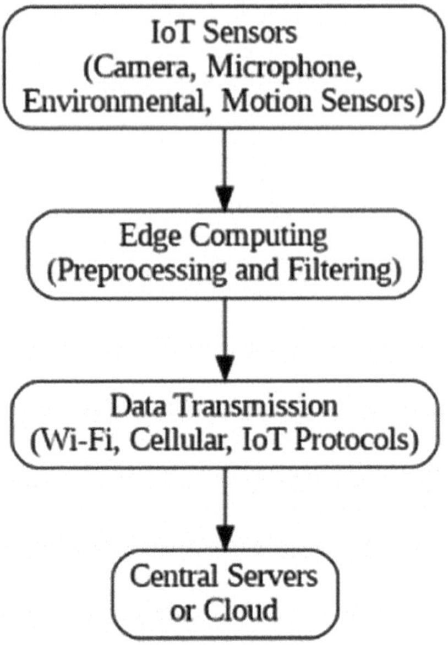

Fig. 3 Data acquisition flow

To handle these large datasets, cloud storage is frequently utilized, offering scalable storage solutions for multimedia data. Cloud storage enables efficient handling and retrieval of vast amounts of data, supporting both storage demands and data accessibility. Additionally, fog and edge computing are used to bring processing closer to the data source, reducing latency and ensuring quicker response times in real-time applications. By enabling local processing, edge, and fog computing help alleviate network congestion and optimize resource use, which is especially useful for applications that require immediate feedback.

Data transmission of multimedia data relies on protocols designed for real-time streaming, such as the Real-time Transport Protocol (RTP). RTP ensures that video and audio data are transmitted smoothly and in a timely manner, which is crucial for applications like video conferencing, live streaming, and other real-time communication services. These protocols help maintain the quality of service needed for uninterrupted multimedia experiences, even over networked environments.

3.1 Cloud Storage for Multimedia Data

Cloud storage offers a centralized, scalable, and highly accessible solution for storing massive multimedia datasets, such as high-resolution images, long video streams, and large audio files. Traditional on-premises storage systems may struggle to handle such large volumes of data and typically lack the flexibility needed to scale on demand [7]. Cloud storage, by contrast, offers virtually unlimited capacity and scalability,

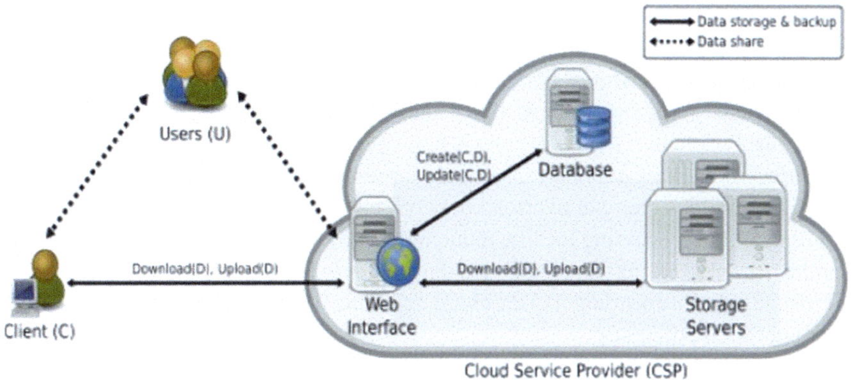

Fig. 4 Architecture of cloud data storage

which is crucial for multimedia applications that generate terabytes or even petabytes of data.

Scalability: As multimedia data volumes grow, cloud storage can expand seamlessly without the need for physical infrastructure updates. Organizations can increase or decrease storage resources as needed, only paying for what they use, making it cost-effective.

Data Accessibility: Cloud storage allows users to access multimedia data from anywhere with an internet connection, facilitating remote collaboration and global access. This is particularly useful for distributed teams and applications where data needs to be available for analysis across multiple locations (Fig. 4).

Data Redundancy and Backup: Cloud storage providers often offer data redundancy by replicating data across multiple servers or locations. This ensures that data is preserved even in cases of hardware failure, providing robust disaster recovery options that are crucial for long-term multimedia data storage [8].

Cost Efficiency: Cloud storage eliminates the need for maintaining extensive on-premises hardware, which can reduce costs related to hardware, maintenance, and upgrades. Many cloud providers offer tiered storage options, where "cold" storage is used for infrequently accessed data at a lower cost.

3.2 Fog and Edge Computing for Real-Time Multimedia Applications

While cloud storage is ideal for storing large datasets, some multimedia applications require fast, real-time data processing. Fog and edge computing bring computation closer to the data source, which is especially valuable for applications like video surveillance, virtual reality, and real-time analytics in smart cities.

Reduced Latency: Fog and edge computing reduce the time it takes to process and respond to data by performing computations near the data source. This is critical for real-time applications where latency (delay in data transmission) must be minimized, such as in autonomous vehicles, live video streaming, and IoT-based healthcare monitoring systems.

Bandwidth Optimization: Transmitting large amounts of multimedia data to a centralized cloud server can overload network bandwidth. Edge computing handles preliminary data processing locally, reducing the volume of data that needs to be sent over the network. For example, an edge device might filter, compress, or summarize video footage before sending it to the cloud, reducing bandwidth usage and improving overall network efficiency.

Privacy and Security: Keeping data close to its source can improve privacy and security, as sensitive information doesn't need to traverse long network paths. Local processing at the edge can ensure that raw data is not exposed to potential network vulnerabilities, which is especially important in industries like healthcare and finance.

Support for Distributed Architecture: Fog and edge computing work well with distributed networks of devices, such as in smart cities or industrial IoT. This architecture allows for efficient data sharing and processing among connected devices without relying heavily on centralized cloud resources.

4 Challenges in Multimedia Data Processing for IoT

Processing multimedia data in Internet of Things (IoT) environments involves unique challenges due to the complexity and high resource requirements of multimedia data. IoT devices generate vast amounts of images, video, and audio, often in real time, which poses technical and logistical hurdles in handling, transmitting, and analyzing this data efficiently. Here are some key challenges in multimedia data processing for IoT (Fig. 5).

4.1 High Data Volume

Multimedia data from IoT devices, such as surveillance cameras or audio sensors, generates high volumes of data that are difficult to store, transmit, and process efficiently. The sheer size of images, videos, and audio files means that data centers and processing units can quickly become overwhelmed, especially in applications like smart cities or healthcare. This challenge necessitates effective data compression and efficient storage solutions, such as cloud and edge storage.

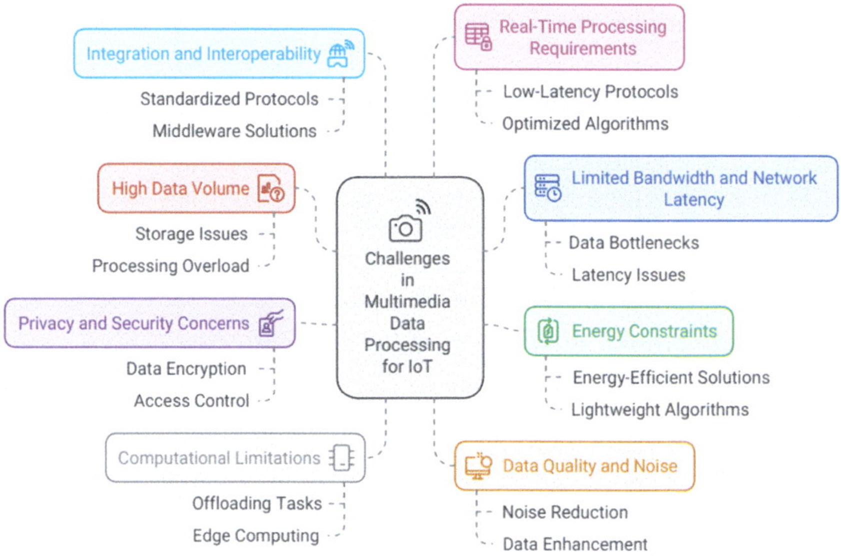

Fig. 5 Challenges in multimedia data processing for IoT

4.2 *Limited Bandwidth and Network Latency*

Transmitting large multimedia files over IoT networks can be bandwidth-intensive and prone to latency issues. Many IoT applications require real-time data processing, where even minor delays can impact performance and user experience. However, transmitting large multimedia data over constrained networks leads to bottlenecks and delays. Techniques like data compression, edge computing, and prioritizing data transmissions are essential to mitigate these issues.

4.3 *Energy Constraints on IoT Devices*

Most IoT devices, especially in remote or mobile environments, are powered by limited battery resources. Processing, storing, or transmitting multimedia data can quickly deplete these resources. IoT devices need energy-efficient solutions, such as lightweight processing algorithms, edge computing to handle data locally, and protocols designed for low power consumption to extend the life of these devices.

4.4 Privacy and Security Concerns

Multimedia data often contains sensitive information, such as images of individuals or voice recordings, raising privacy concerns. In IoT environments, ensuring data security during transmission, storage, and processing is crucial, as multimedia data can be vulnerable to cyber-attacks. Implementing robust encryption, access control, and secure protocols is necessary to protect sensitive data from unauthorized access.

4.5 Computational Limitations of IoT Devices

Many IoT devices have limited processing power and memory, which makes it challenging to handle complex multimedia processing tasks like object detection or facial recognition. These devices may struggle to process data locally, especially for applications that demand real-time analysis. Offloading tasks to edge devices or cloud servers can help, but it also requires careful management to balance data transfer and processing efficiency.

4.6 Data Quality and Noise

Multimedia data can often be noisy or of low quality, especially when captured in challenging conditions (e.g., poor lighting, background noise). Low-quality data affects the accuracy of analysis, as algorithms may struggle to extract useful information. Techniques such as noise reduction, data enhancement, and sensor calibration are necessary to improve data quality before processing.

4.7 Integration and Interoperability

IoT ecosystems are diverse, with devices from various manufacturers using different standards and protocols [9]. Ensuring that multimedia data from different devices is compatible and can be processed together is challenging. Achieving interoperability among devices requires standardized data formats and protocols, as well as middleware solutions to handle data integration seamlessly.

4.8 Real-Time Processing Requirements

Some IoT applications, such as autonomous vehicles or surveillance systems, require real-time multimedia processing to make immediate decisions based on the data. Achieving real-time performance can be difficult due to the data volume, computational requirements, and latency constraints. Leveraging edge computing, low-latency transmission protocols, and optimized processing algorithms are essential to meet these real-time demands.

5 Applications of Multimedia Data Processing in IoT

Multimedia data processing plays a crucial role in enhancing various IoT applications across multiple domains. Below is a detailed description of five key application areas.

5.1 Smart Home and Smart City Applications

In smart homes, multimedia data such as video, audio, and sensor information is processed to create a connected living environment.

Applications include:

Security Systems: Real-time video surveillance with face recognition helps in identifying intruders. *Voice-Activated Assistants*: Devices like Amazon Alexa and Google Home use audio data to process user commands. Energy Management: Smart meters analyze energy consumption data to optimize usage. In smart cities, traffic management systems use video feeds to monitor congestion, detect accidents, and manage traffic lights dynamically. Public safety is enhanced with CCTV systems using real-time video analytics for crime detection.

5.2 Health Monitoring and Telemedicine

Multimedia data is crucial in healthcare IoT systems, providing real-time monitoring and remote diagnostics: *Wearable Devices*: Wearables like smartwatches capture physiological signals (heart rate, ECG) and use audio/video feeds for fall detection. Telemedicine: Video conferencing allows remote consultations between patients and doctors, with image processing used for analyzing medical scans (e.g., X-rays, MRIs).

Health Data Analytics: Processing audio and video feeds can help monitor patient activities and detect symptoms like coughing or irregular breathing.

5.3 Autonomous Vehicles and Smart Transportation

Autonomous vehicles rely heavily on multimedia data processing to navigate safely and make real-time decisions: *Computer Vision:* Cameras and sensors capture video data for lane detection, obstacle recognition, and traffic sign detection. *Sensor Fusion:* Combining data from LiDAR, cameras, and radar improves object detection accuracy. *Traffic Monitoring*: Smart transportation systems use video feeds from road cameras to analyze traffic flow and detect incidents, reducing congestion and accidents.

5.4 Industrial IoT and Predictive Maintenance

In industrial settings, multimedia data processing is employed to monitor equipment and predict failures: *Machine Vision*: Cameras inspect product quality on assembly lines, detecting defects using image processing techniques. *Vibration Analysis*: Audio and vibration data from sensors are analyzed to detect anomalies in machinery, indicating potential issues before breakdowns. *Remote Monitoring*: Video feeds from industrial sites enable real-time supervision, reducing the need for on-site personnel.

5.5 Environmental Monitoring and Agriculture

Multimedia data helps in monitoring environmental conditions and enhancing agricultural practices: *Drones*: Equipped with cameras, drones capture high-resolution images of crops, which are analyzed to assess health, detect pests, and optimize irrigation. Weather Monitoring: Sensors collect audio data (e.g., rain impact sound) and video feeds to assess weather conditions in real-time. *Wildlife Surveillance*: Cameras and microphones placed in natural habitats monitor wildlife activities, aiding in conservation efforts.

6 Tools and Frameworks for Multimedia Data Processing in IoT

Tools and Frameworks for Multimedia Data Processing in IoT are essential for efficiently handling, analyzing, and managing the vast amounts of multimedia data generated by IoT devices. Popular tools like TensorFlow and PyTorch provide robust deep learning capabilities for tasks such as image recognition, object detection, and audio processing, enabling developers to build advanced AI models for real-time multimedia analysis. OpenCV is widely used for computer vision tasks, offering a comprehensive suite of functions for image and video processing, which is crucial in applications like smart surveillance and autonomous vehicles. For real-time data streaming and processing, frameworks such as Apache Kafka and Apache Spark Streaming are commonly employed. Apache Kafka allows efficient handling of data streams, while Apache Spark Streaming provides the ability to process and analyze large-scale multimedia data in real-time.

IoT-specific platforms like AWS IoT, Azure IoT Hub, and Google Cloud IoT offer integrated services that facilitate multimedia data processing at scale. These platforms provide edge computing capabilities, reducing latency and enabling real-time processing close to the source of data generation. For instance, AWS Greengrass allows local processing of data on edge devices, reducing the need to send all data to the cloud, which is particularly beneficial for bandwidth-limited environments. Additionally, multimedia data processing libraries such as FFmpeg are used for video and audio processing tasks, enabling encoding, decoding, and streaming of multimedia content efficiently. Together, these tools and frameworks form a powerful ecosystem for developing, deploying, and managing multimedia data processing solutions in IoT, catering to diverse use cases across smart cities, healthcare, industrial automation, and more.

7 Future Trends and Research Directions in Multimedia Data Processing for IoT

As the Internet of Things (IoT) continues to evolve, the integration of multimedia data processing capabilities is shaping the future of various applications across industries. Several emerging trends and research directions are driving advancements in this area.

7.1 Role of 5G and Beyond in Enhancing IoT Multimedia Capabilities

The advent of 5G and upcoming 6G networks are revolutionizing multimedia data processing in IoT by providing ultra-high-speed, low-latency communication. With faster data transmission rates and reduced latency, 5G enables seamless real-time processing of high-resolution video and audio data, which is critical for applications like autonomous vehicles, smart city surveillance, and telemedicine. The high bandwidth offered by 5G networks allows for more connected devices and higher-quality multimedia streaming, paving the way for advanced IoT applications such as augmented reality (AR) and virtual reality (VR). Future research is expected to focus on optimizing 5G networks for better Quality of Service (QoS) in multimedia data transmission, as well as exploring the potential of 6G for enhanced multimedia analytics in IoT environments.

7.2 Integration of AI and IoT for Enhanced Multimedia Analytics

The convergence of Artificial Intelligence (AI) and IoT, often referred to as AIoT, is significantly enhancing the ability to process and analyze multimedia data. By leveraging machine learning and deep learning algorithms, AIoT systems can perform complex tasks such as image recognition, voice analysis, and video analytics in real-time. For example, AI-powered cameras in smart cities can detect suspicious activities, and voice-activated assistants can interpret and respond to user commands using natural language processing (NLP). The integration of AI models into IoT devices, coupled with real-time multimedia data processing, is expected to drive further innovation in areas such as predictive maintenance, personalized healthcare, and smart home automation. Ongoing research aims to improve the efficiency and accuracy of AI models, making them more suitable for resource-constrained IoT environments.

7.3 Advances in Edge AI and TinyML

Edge AI and TinyML are transforming the landscape of multimedia data processing in IoT by enabling machine learning directly on edge devices. Edge AI refers to deploying AI algorithms on local devices (e.g., cameras, sensors, smartphones) rather than relying on cloud servers. This approach reduces latency, enhances data privacy, and lowers bandwidth requirements, making it ideal for time-sensitive applications such as autonomous driving and industrial automation. TinyML extends this concept by focusing on implementing lightweight machine learning models that can run on microcontrollers and other low-power IoT devices. This allows for on-device processing of multimedia data, such as real-time image and audio analysis, even in environments with limited computational resources. Future research will likely focus on optimizing TinyML models for better performance and power efficiency, expanding the scope of edge AI in multimedia data processing.

7.4 Emerging Technologies (Quantum Computing, Federated Learning)

Emerging technologies like Quantum Computing and Federated Learning are poised to play a significant role in the future of multimedia data processing in IoT. Quantum computing has the potential to solve complex optimization problems and accelerate data analysis tasks that are currently beyond the capabilities of classical computers. In multimedia processing, quantum algorithms could significantly speed up tasks like video compression and real-time image recognition. Federated Learning, on the other hand, offers a privacy-preserving approach to training machine learning models by keeping the data localized on devices while only sharing model updates. This technique is particularly useful for multimedia applications in IoT, where sensitive data (e.g., video feeds from smart home cameras) needs to be protected. Research in these areas is exploring new algorithms and architectures to make these technologies more practical and scalable for real-world IoT applications.

8 Open Research Challenges

Despite the rapid advancements, several open research challenges remain in multimedia data processing for IoT:

Scalability: Handling the exponential growth of multimedia data generated by billions of IoT devices requires scalable processing frameworks and efficient data storage solutions.

Energy Efficiency: Many IoT devices operate on limited battery power, making it challenging to perform intensive multimedia processing tasks. Research is needed to develop energy-efficient algorithms and hardware accelerators.

Data Privacy and Security: Ensuring the security and privacy of sensitive multimedia data remains a significant concern. New techniques, such as homomorphic encryption and privacy-preserving machine learning, are being explored to address these issues.

Interoperability: The diversity of IoT devices and data formats presents a challenge for seamless integration and interoperability. Standardized protocols and data formats are essential to enable effective multimedia data processing across different IoT systems.

Real-Time Processing: Achieving real-time processing of high-resolution multimedia data, especially in dynamic and resource-constrained environments, requires innovative solutions in hardware and software optimization.

9 Case Studies

The use of multimedia data processing in IoT has led to significant advancements across various industries, enabling real-time decision-making and improving system efficiencies. The following case studies illustrate practical applications of these technologies:

9.1 Case Study 1: Real-Time Video Surveillance in Smart Cities

Background: In modern smart cities, ensuring public safety and efficient traffic management is a top priority. Real-time video surveillance systems play a crucial role in monitoring urban environments, providing valuable insights to city administrators and law enforcement agencies.

Implementation: A city deployed an IoT-based video surveillance system with cameras placed at strategic locations such as traffic intersections, public squares, and key infrastructure points. These cameras capture high-definition video streams that are processed in real-time using advanced computer vision techniques, such as object detection and facial recognition. The processing is performed using edge computing nodes to reduce latency, enabling immediate detection of incidents like traffic violations, accidents, and unauthorized access to restricted areas.

Impact: The system has significantly improved traffic management by providing real-time alerts about congestion and accidents, allowing for timely intervention by traffic authorities. It also enhances public safety by enabling quick responses to criminal activities and suspicious behavior. Moreover, the use of AI-powered analytics helps in identifying patterns and predicting potential issues, leading to a proactive approach in city management.

Challenges: The primary challenges include ensuring data privacy, as video feeds can contain sensitive personal information, and managing the vast amounts of data generated by numerous cameras across the city.

9.2 Case Study 2: Audio Data Analysis for Health Monitoring

Background: In the healthcare sector, continuous monitoring of patients' health is crucial, especially for the elderly and those with chronic conditions. Audio data analysis has emerged as an effective tool for non-invasive health monitoring, enabling early detection of health issues.

Implementation: An IoT-based health-monitoring system was developed using wearable devices and smart speakers equipped with microphones. These devices continuously capture audio data, such as breathing patterns, coughing sounds, and voice characteristics. Machine learning algorithms analyze the audio signals to detect anomalies, such as irregular breathing, prolonged coughing, or changes in vocal tone, which may indicate respiratory issues or other health conditions.

Impact: The system allows for early detection of potential health problems, enabling timely medical intervention and reducing hospital readmissions. It has been particularly beneficial for remote patient monitoring, allowing healthcare providers to assess patients' conditions without the need for in-person visits, which is especially valuable in rural or underserved areas.

Challenges: Ensuring the privacy and security of the captured audio data is a major concern, as it may contain sensitive information about the patient's health. Additionally, achieving high accuracy in detecting health conditions based on audio signals requires extensive training data and robust machine learning models.

9.3 Case Study 3: Image Processing for Precision Agriculture

Background: Precision agriculture aims to optimize farming practices by using data-driven insights to enhance crop yield and reduce resource usage. Image processing techniques are widely used to monitor crop health, detect pests, and assess soil conditions.

Implementation: In this case study, drones equipped with high-resolution cameras and multispectral sensors are used to capture images of agricultural fields. The captured images are processed using computer vision algorithms to analyze various parameters, such as leaf color, plant height, and canopy coverage. By examining the spectral signatures of plants, the system can detect signs of nutrient deficiencies, pest infestations, and water stress.

Impact: The use of image processing in precision agriculture has led to improved crop management by providing farmers with detailed insights into the health of their crops. It allows for targeted interventions, such as applying fertilizers or pesticides only where needed, thus reducing costs and minimizing environmental impact. Additionally, the early detection of issues helps prevent crop losses and boosts overall productivity.

Challenges: The main challenges include handling the large volume of image data, requiring efficient data processing and storage solutions. Environmental factors like lighting conditions and weather variations can also affect the accuracy of image analysis.

9.4 Case Study 4: Multimedia Data Processing in Autonomous Vehicles

Background: Autonomous vehicles rely heavily on multimedia data processing to navigate safely and make real-time decisions. The integration of cameras, LiDAR, radar, and ultrasonic sensors allows these vehicles to perceive their surroundings accurately.

Implementation: In this case study, an autonomous vehicle is equipped with a comprehensive array of sensors and cameras that continuously capture video, image, and other sensory data. The onboard processing unit uses machine learning algorithms to analyze the multimedia data, identifying objects such as pedestrians, vehicles, traffic signs, and road markings. Techniques like image segmentation and object recognition are used to understand the environment, while sensor fusion combines data from different sensors to enhance the vehicle's perception capabilities.

Impact: The ability to process multimedia data in real-time allows autonomous vehicles to make informed decisions, such as braking to avoid collisions, changing lanes, and navigating complex traffic scenarios. This has the potential to significantly reduce road accidents, improve traffic flow, and enhance the overall safety of transportation systems.

Challenges: Real-time multimedia data processing in autonomous vehicles requires high computational power and low latency to ensure timely and accurate decision-making. Managing the massive volume of data generated by sensors and ensuring reliable operation under different weather and lighting conditions are also critical challenges.

10 Conclusion

The integration of Multimedia Data Processing and Analysis within the IoT framework represents a significant advancement in developing intelligent, context-aware systems capable of transforming vast amounts of visual, auditory, and sensor-generated data into actionable insights. This chapter has explored the various methodologies and technologies that enable efficient handling of multimedia data, highlighting both the challenges and the potential of real-time processing in dynamic IoT environments. The synergy between multimedia data and IoT opens up numerous opportunities across various domains, such as smart cities, healthcare, agriculture, and autonomous vehicles, by enhancing decision-making capabilities and delivering enriched user experiences.

Despite the promising advancements, several challenges remain, particularly in areas such as data privacy, security, interoperability, and real-time processing under resource constraints. The rapid evolution of enabling technologies like 5G, edge computing, AI, and emerging paradigms such as federated learning and quantum computing, are paving the way for more scalable, secure, and efficient multimedia data processing solutions. These innovations hold the potential to overcome existing limitations and enable more robust, scalable, and intelligent IoT systems. As the IoT ecosystem continues to evolve, future research and development must focus on creating holistic frameworks that address the intricacies of multimedia data processing while maintaining a balance between performance, privacy, and security. By doing so, the full potential of IoT systems can be realized, fostering innovation across industries and driving the adoption of smart, interconnected solutions that significantly improve quality of life and operational efficiencies.

References

1. Singh, M., Verma, P.: Federated learning for privacy preservation in IoT. ACM Trans. Multimed. (2023)
2. Yadav, S.: Privacy-preserving multimedia data sharing in IoT networks. IEEE Trans. Multimed. (2024)
3. Yang, X.: Edge computing solutions for high-volume IoT data. Springer J. Real-Time Process. (2024)
4. Lee, J., Kim, H.: Blockchain-based multimedia data integrity in IoT. J. Internet Serv. (2023)
5. Wang, Y.: Deep learning for real-time video analysis in IoT systems. IEEE Access (2023)
6. Nguyen, T.: Security frameworks for IoT multimedia data transmission. IEEE Trans. Inf. Forensics (2024)
7. Singh, V.: Federated learning approaches in autonomous vehicle IoT systems. IEEE Trans. Veh. Technol. (2024)
8. Zhou, L.: Hybrid cloud-edge architecture for multimedia data processing in IoT. Springer Multimed. Tools (2024)
9. Kumar, N.: 5G-enhanced IoT for low latency video streaming. IEEE Trans. Mob. Comput. (2024)

Embedded and Cloud Computing for Ingesting Big Multimedia Data in IOT Sensor Networks Applications

Suneeta Mohanty, Chinmay Jain, Rudransh Bharadwaj, and Prasant Kumar Pattnaik

Abstract The rapid expansion of the Internet of Things (IoT) has led to the generation of vast volumes of multimedia data across diverse sectors. This surge in data necessitates efficient processing mechanisms to extract meaningful insights in real time. This research paper examines the pivotal role played by cloud computing and embedded computing in optimizing multimedia data processing within IoT ecosystems. Cloud computing offers scalable and resource-rich environments ideal for offloading intensive multimedia processing tasks from IoT devices. By harnessing the computational power of remote servers, cloud-based solutions alleviate the burden on constrained IoT devices, enabling them to focus on essential functions while outsourcing resource-demanding computations. Simultaneously, embedded computing techniques empower IoT devices with local processing capabilities, facilitating quick decision-making and reducing reliance on external resources. Through efficient utilization of onboard processing units, embedded computing enhances system responsiveness and reduces latency, critical factors in real-time multimedia applications. This paper presents insights gleaned from empirical studies highlighting the transformative impact of cloud and embedded computing on multimedia data processing in IoT environments. By leveraging these computing paradigms, IoT systems can achieve enhanced scalability, adaptability, and efficiency, paving the way for innovative applications across domains such as smart cities, healthcare, and industrial automation. The findings underscore the importance of integrating cloud

S. Mohanty, C. Jain, R. Bharadwaj and P. K. Pattnaik are contributed equally to this work

S. Mohanty (✉) · C. Jain · R. Bharadwaj · P. K. Pattnaik
School of Computer Engineering, Kalinga Institute of Industrial Technology Deemed to be University, Bhubaneswar, Odisha, India
e-mail: smohantyfcs@kiit.ac.in

C. Jain
e-mail: 21051392@kiit.ac.in

R. Bharadwaj
e-mail: 2105481@kiit.ac.in

P. K. Pattnaik
e-mail: patnaikprasantfcs@kiit.ac.in

R. Sharma et al. (eds.), *Multimedia Technologies in the Internet of Things Environment, Volume 4*, Studies in Big Data 173, https://doi.org/10.1007/978-981-96-4356-1_5

and embedded computing strategies to unlock the full potential of IoT technology in handling multimedia data effectively and sustainably.

Keywords Internet of Things (IoT) · Embedded computing · Cloud computing · Multimedia data

1 Introduction

The Internet of Things (IoT) connects various objects to the Internet, enabling applications such as smart homes, cities, and healthcare. Multimedia IoT (MIoT) combines sensor networks with limited resources to collect real-time data. However, IoT networks face challenges as they grow in size and include cameras, vehicles, and sensors. Wireless communication is flexible, but it exposes security risks. In MIoT, huge amounts of multimedia data are sent to the cloud, but current solutions are inadequate due to resource and security constraints. To overcome these challenges, a proposed model uses intelligent edges and security features, improving the multimedia industry.

This model reduces data traffic over wireless channels, making the best use of MIoT nodes. It also addresses security issues, preventing malicious attacks and preserving privacy and integrity. Moreover, an optimization model with Software Defined Networking (SDN) architecture and Artificial Intelligence of Things (AIoT) is presented. This model enhances the Quality of Service (QoS) for multimedia sensors, providing dependable services with high performance. AIoT, along with mobile edges, supports multihop routing services, lowering communication overhead. SDN controllers control network resources centrally, ensuring efficient flow management and information security. The model not only offers more bandwidth but also protects against unauthorized actions.

On the other hand, multimedia data usage has increased significantly due to technological progress and easy access. Platforms like YouTube, Twitter, and Instagram contribute largely to the enormous amount of multimedia data shared on the internet. Statistics show a remarkable growth, with global data production reaching 40ZB in 2020. People spend a lot of their time online, sharing multimedia content in different forms, such as sound, visuals, written material, and images. Multimedia big data is characterized by its large scale, diverse variety, and fast speed. Conventional data processing tools fail to handle its unstructured nature, requiring advanced solutions.

Multimedia applications and services offer opportunities for processing multimedia big data. By 2020, an estimated $4 * 10^{24}$ bytes may be produced, with IoT devices playing a key role. This creates challenges in terms of mobility, data collection, and User Experience (UX). The chapter provides a thorough investigation of challenges, impacts, features, and possible research directions in multimedia big data in the context of IoT.

The relation between embedded and cloud computing is that they both work together to enable MIoT applications. Embedded computing refers to the use of microcontrollers or microprocessors in IoT devices, such as sensors and cameras, to perform specific tasks. Cloud computing refers to the use of remote servers on the Internet to store, process, and access data. Embedded computing provides local processing and intelligence to IoT devices, while cloud computing provides global storage and analysis to IoT data. Embedded and cloud computing complement each other, as embedded computing reduces the amount of data sent to the cloud, and cloud computing provides scalability and reliability to embedded computing [1].

2 What Are Embedded Systems?

Embedded systems are combinations of hardware and software designed in complex ways for specific tasks. Software dictates their operation, while hardware ensures optimal performance. Real-time operating systems (RTOS) play a central role in ensuring these tasks run smoothly. The main components of an embedded system include embedded hardware, RTOS, device drivers, communication stack, and embedded applications. The overarching goal is to create cost-effective, interoperable devices that meet stringent standards for safety, security, reliability, and availability without exceeding budget limits. Embedded systems often serve as controllers, overseeing specific functions within a device. Typically designed for a singular task, in more complex setups, they might even govern the entire operating system. Although some advanced embedded systems can handle multiple tasks, these operations are generally straightforward and do not require substantial computing power. What distinguishes embedded systems is their nonprogrammable nature. Once configured for a specific purpose, they operate reliably without the need for intervention. While certain embedded devices allow software modifications for planned enhancements, the core functionality remains steadfast. Renowned for their reliability and minimal maintenance needs, embedded systems seamlessly integrate into devices. Their purpose-built design makes them highly dependable electronic components. Unlike many other systems, they exhibit resistance to failure and seldom require reprogramming, making them invaluable for devices that demand autonomous operation, such as home appliances.

3 What is Cloud Computing?

Cloud computing is garnering significant attention in modern technology, introducing noteworthy attributes such as self-service on demand, wide network accessibility, resource pooling, rapid scalability, and metered service. This paradigm shift differs from traditional models by providing flexible and scalable access to computing resources via the Internet, consolidating technical elements in remote servers within data centers. Industries spanning various sectors are increasingly adopting the trend of cloud computing, making sophisticated technologies more accessible and fostering the introduction of innovative concepts. When scrutinizing its collaboration with embedded devices to manage expanded multimedia data in IoT sensor networks, it is imperative to recognize that cloud computing serves as the linchpin of these endeavors. A comprehensive understanding of this synergy is essential for navigating the intricate landscape of modern computing.

Cloud computing does a lot to boost what embedded systems and IoT devices can do. One big plus is that it lets these devices send resource heavy tasks to cloud servers. This means embedded devices get to tap into the huge computing power of the cloud, making sure they run efficiently.

Not only that, cloud computing gives a powerful and scalable way to store the massive amounts of multimedia data that IoT devices churn out. This scalability is super useful for embedded systems with not a lot of storage space, letting them easily use cloud storage to handle the growing data stream in different IoT applications.

On top of that, the cloud is great for managing data remotely for embedded devices in IoT networks. With a cloud-based system, organizations can keep an eye on, control, and manage embedded devices. This makes sure everything runs smoothly and responses happen in real time, boosting the overall efficiency and reliability of IoT applications.

In the world of data analytics, bringing cloud computing into the mix lets us do real-time processing and analysis of data collected by onboard devices. The cloud manages this data at a big scale, giving us important insights that are crucial for making decisions. This is especially key in areas like smart cities, healthcare, and industrial IoT, where we need quick and informed actions.

Cloud computing is like a bridge that improves how embedded systems connect. By using cloud services, embedded devices can talk to each other and central servers without a hitch. This helps create a more complex and collaborative IoT ecosystem. This whole connectivity thing shows how much cloud computing can shake up the data processing and management abilities of embedded systems in the ever-changing world of IoT.

4 Cloud Computing Architecture

The structure of cloud computing consists of two essential components: the backend and the frontend. Functioning as the gateway, the frontend acts as the initial point of contact for users, initiating communication with the backend through network or internet connections [2]. Within the framework of cloud computing, the frontend facing the client is outwardly observable, whereas the backend remains shielded from external interactions, yet possesses the capability to directly engage with its client through a predetermined protocol. The backend of the cloud architecture plays a critical role in protecting vital information from the demands of client facing technologies [3]. It efficiently handles requests for data and provides appropriate responses. As an integral part of the overall computer system, the backend represents a significant portion of the overarching concept of cloud computing.

In a corporate setting, it is crucial to ensure seamless compatibility between the software and hardware components that constitute the cloud environment. A wise starting point involves using readily available components. Versatile servers are essential for managing the increasing needs for data storage and processing power [4]. Additionally, a comprehensive set of reliable applications tailored to the specific requirements and financial considerations of the business is essential. While there are many options for transitioning to cloud architecture, choosing the most suitable one can be a daunting task. The optimal cloud architecture for a business involves a combination of software and hardware elements that work together effectively. Inadequate planning for cloud computing can result in higher operational costs and limited scalability, leading to a reduced return on investment. Operating within the framework of the most appropriate computing architecture is crucial, as it not only saves time and financial resources but also ensures readiness for future demands, thus avoiding concerns about data becoming obsolete during situations requiring increased computational power.

4.1 Frontend

The frontend architecture in cloud computing connects with end users via diverse elements, encompassing internal networks, online browsers, and software platforms. It facilitates seamless interactions without requiring local software installation on devices such as smartphones, desktops, or tablets. Key elements include the user interface, client-side software, and hardware like input devices. In cloud computing, the client-side device doesn't need significant computational power as the cloud infrastructure handles processing efficiently.

4.2 Backend

The backend architecture of cloud technology empowers the frontend structure, comprising hardware and storage components managed by the cloud service provider. It's engineered for resilience and longevity, supporting the cloud-based system. Key components include the application layer providing backend services, diverse service components for task execution, storage frameworks utilizing various mediums like SSDs and HDDs, infrastructure encompassing technologies like CPUs and GPUs, and management mechanisms ensuring resource allocation efficiency.

4.3 Cloud Delivery

Cloud solutions offer a range of digital resources accessible through your web browser on any internet-connected device. These resources encompass computing power, tools, and operating systems, eliminating the need for local installation or hardware management. Tech companies bundle products into user-friendly packages and deliver them through various methods.

Software as a Service (SaaS): Software is distributed through cloud-based systems, eliminating the need for additional IT infrastructure. Examples include Google Workspace, Dropbox, and Salesforce.

Platform as a Service (PaaS): Organizations host platforms for developing and running applications in the cloud, with third-party support. Examples include Google App Engine and Heroku.

Infrastructure as a Service (IaaS): Basic infrastructure is provided for enterprise operations, including servers and storage. Examples include AWS and Microsoft Azure. Deployment models include Public, Private, and Hybrid Clouds [5].

5 Fog Computing Architecture

Fog computing, also known as edge computing, extends cloud computing to the network's periphery to address challenges arising from the growing data generated by IoT devices and widespread internet usage. It introduces a three-layer architecture comprising devices, an intermediary fog layer, and the cloud layer. The fog layer reduces latency by processing data directly through fog nodes, bridging the gap between data-generating devices and the cloud.

5.1 Components of Fog Computing Architecture

(1) **IoT Devices**: These devices connect to the IoT network using various wired and wireless technologies, continuously generating data.
(2) **Fog Nodes**: Essential building blocks equipped with computational, storage, and network capabilities, deployed across diverse applications for high-sensitivity data processing.
(3) **Fog Aggregate Nodes**: Analyze data within seconds to minutes, with varying data storage capacities, spread over broader territories to handle moderate sensitivity data.
(4) **Remote Cloud**: Central hub connecting all aggregate fog nodes, responsible for processing, analysis, and storage of less time-sensitive data.

(5) **Local Servers and Clouds**: May include dedicated private servers or clouds to enhance data security and privacy, safeguarding organizational data.

5.2 *Layered Fog Architecture*

The layered fog architecture comprises six tiers:

(1) **Physical and Virtualization Stratum**: Includes physical and virtual nodes for data capture, with sensors gathering data transmitted via gateways.
(2) **Monitoring Stratum**: Focuses on monitoring operational aspects of nodes such as uptime and application performance.
(3) **Preprocessing Stratum**: Manipulates and cleans data, extracting meaningful information before analysis.
(4) **Ephemeral Storage**: Transiently distributes and replicates data, eventually migrating it to the cloud for permanent storage.
(5) **Security Stratum**: Ensures data privacy, integrity, and cryptographic protocols, addressing data-centric privacy and location-based privacy.
(6) **Transport Stratum**: Securely transfers partially processed data to the cloud for permanent storage, optimizing resource usage through efficient data collation and transmission.

6 What is Big Multimedia Data?

Multimedia big data represents an intricate theoretical paradigm, sharply diverging from conventional big data due to its inherent attributes, including heterogeneity, a human-centric focus, diverse media forms, and substantial size. This conceptual framework introduces a myriad of salient features that distinctly demarcate it from traditional big data, heralding a new era in data analytics and comprehension.

Comprising an extensive and varied array of data types, multimedia big data transcends the limitations of its machine-centric counterpart, rendering it more comprehensible to human cognition. The intricacy is not solely rooted in the sheer volume but also in the diverse data, encompassing various audio and video formats such as 3D videos and social videos. This intricate nature necessitates sophisticated processing methods, demanding a nuanced approach to derive meaningful insights that go beyond conventional analytical methodologies.

Modeling and characterizing multimedia big data pose formidable challenges, stemming from its origins across diverse and heterogeneous sources. Smartphones, sensor-equipped gadgets, electronic games, immersive environments, and social platforms collectively contribute to the richness and diversity of the data landscape, complicating the establishment of a unified model. This diversity accentuates the intricacies involved in understanding and representing the multifaceted dimensions inherent in multimedia big data.

Analyzing the substance and backdrop of multimedia big data becomes an intellectually stimulating endeavor, given its dynamic nature, subject to continuous change over time and space. The inherent variability in multimedia content mandates adaptive analytical approaches capable of capturing evolving patterns within the data. This dynamism challenges traditional data analytics paradigms, necessitating an innovative and flexible approach to uncover meaningful insights.

The security dimension of vast data sets adds a layer of intricacy, primarily attributed to the rapid surge in sensitive video data during communication. Safeguarding abundant data involves addressing potential vulnerabilities and threats, a task exacerbated by the sheer volume and diverse nature of the information. Advanced encryption methods, robust access control mechanisms, and secure communication protocols become imperative in securing the integrity and confidentiality of copious data.

Furthermore, the imperative for rapid and uninterrupted processing of enormous data becomes increasingly vital, especially to align with the speed of data transmission networks and fulfill real-time computing requirements. The dynamic nature of sizable applications necessitates swift and efficient processing to maintain responsiveness and exceed user expectations. Advanced processing techniques, such as parallel computing and distributed computing, emerge as essential components to tackle the intricacies posed by the sheer magnitude and complexity of substantial data.

In the realm of scientific voluminous data, fundamental challenges abound, encompassing the complexity in cognitive understanding, intricate data analysis, management of distributed data security, assurance of quality of experience and service, meticulous attention to detailed requirements, and overcoming performance restrictions. This multifaceted spectrum of challenges extends across processing, storing, transmitting, and analyzing substantial data, paving the way for continuous research and innovation.

Voluminous data originates from diverse sources, and the term"big data" is aptly employed to describe datasets that transcend the capacities of traditional processing tools due to their significant size and intricate complexity. The challenges associated with managing such datasets are multifaceted, involving querying, sharing, moving, revising, collecting, retaining, displaying, and analyzing. In addition, critical considerations for security and privacy further compound these challenges. The structured nature of voluminous data, which encompasses images, sound files, visual content, online content, and a variety of multimedia materials, significantly deviates from fixed database formats. This deviation amplifies the complexity, necessitating comprehensive and adaptable approaches in the analysis and utilization of substantial data.

User-generated data, proliferating on platforms like Facebook, Twitter, and YouTube, further contributes to the dynamic nature of varied datasets within the realm of voluminous data. In contrast, organized data types boast a static size and an organized format, rendering them easily manageable and storable in databases. The juxtaposition of structured and voluminous data accentuates the multifaceted nature of substantial data, underscoring the necessity for innovative methodologies in its analysis and utilization.

The advent of Internet services has precipitated a rapid surge in indexed and queried content, posing formidable challenges for search engine companies tasked with processing the burgeoning volumes of voluminous data. In handling the increasing volumes of diverse data generated by applications like video surveillance, medical imaging, entertainment, social media, and sound and visual, real time, streaming, distribution, parallel, and online analysis methods become crucial for effective management. Traditional data processing systems prove inadequate, necessitating the implementation of efficient management strategies for large volumes of voluminous data.

Largescale voluminous data, often distributed across different owners maintaining various parts, introduces variability and noise due to differences in data quality across sources. This distribution poses challenges for analysis, as user requirements for the same data may differ significantly. The complexities arising from the distributed nature of large-scale voluminous data necessitate innovative solutions to ensure coherence and reliability in analysis outcomes.

Human-centered analysis methods emerge as pivotal, as the understanding of user experience remains a formidable challenge despite the rapid advancements in data analysis technology. Creating efficient techniques for representing photos and videos to enhance the comprehension of expressive user experiences becomes of utmost importance. Social behavioral networks should be considered as models of user intentions in large-scale voluminous data analysis, introducing a human-centric dimension to the analytical process.

Security remains an ongoing concern in the realm of voluminous data, with unprecedentedly large amounts stored in a distributed manner. Different analysts necessitate varying permissions, justifiable by data owners, demanding the implementation of sophisticated security mechanisms. These mechanisms encompass effective encryption methods, meticulous access control, robust security controls, and secure communication protocols. Considerations for the nature of attacks, loopholes, and viruses add further complexity to the design of security mechanisms for large volumes of voluminous data.

In conclusion, voluminous data transcends the boundaries of traditional data processing paradigms, ushering in a new era marked by complexity, diversity, and dynamicity. The challenges inherent in its analysis, storage, and utilization necessitate advanced methodologies and innovative solutions. The multifaceted nature of voluminous data, originating from diverse sources and spanning various formats, demands a holistic and adaptable approach to unravel its profound insights and potential. As technology continues to evolve, research and innovation in the field of voluminous data remain imperative to harness its transformative power across diverse domains [6].

7 Ingesting Data

In the ever-evolving landscape of cloud computing and embedded systems within IoT networks, data ingestion serves as the foundational gateway, marking the initiation of a sequence that encompasses the entire data lifecycle. This pivotal process involves the aggregation, transmission, and reception of multimedia information, setting the stage for subsequent stages like data processing, analysis, and informed decision-making.

Embedded devices, equipped with an array of sensors, operate as the frontline data collectors in the IoT setup. These sensors, ranging from temperature and humidity gauges to cameras and motion detectors, continually gather data from their surroundings. This diversity in sensor capabilities allows for the comprehensive capture of environmental variables, enabling a nuanced understanding of the context in which IoT devices operate.

Wireless communication acts as the conduit through which collected data is transmitted. Utilizing protocols like WIFI, Zigbee, or Bluetooth, embedded devices establish seamless communication with the cloud, facilitating the efficient transfer of data for further processing and analysis. This wireless connectivity is crucial for the flexibility and adaptability of IoT systems.

Gateways emerge as essential intermediaries in the data ingestion process. They not only aggregate data from various embedded devices but may also perform preliminary tasks on the data before forwarding it to the cloud. This intermediary role enhances the efficiency of data flow, ensuring optimal communication between devices and the cloud infrastructure.

Cloud computing platforms contribute significantly to data ingestion by offering scalable setups. This scalability proves invaluable in handling the dynamic and often unpredictable data volumes generated by IoT sensor networks. It ensures that the system can effectively manage fluctuations in data loads, thereby maintaining robust performance.

A complementary concept to cloud-based data ingestion is edge computing, where data undergoes initial processing closer to the source of the embedded devices before being transmitted to the cloud. This approach reduces latency, optimizes bandwidth usage, and enhances the overall efficiency of the system, particularly in real-time applications.

Within this orchestrated process, data streams and queues play pivotal roles. Data streams enable the real-time flow of data from diverse sources, while queues ensure orderly data processing, mitigating the risk of information loss during transit [5]. This dynamic interplay ensures a synchronized and efficient handling of incoming data. However, amid these advancements, challenges loom large, particularly in the domains of security and privacy. Robust measures such as encryption, secure communication protocols, and stringent access controls are imperative to safeguard sensitive information and respect user privacy.

Ensuring that ingested data follows standardized formats is another critical consideration. This standardization streamlines subsequent processing within cloud platforms, necessitating attention to data formatting at the source or through intermediary gateways.

Equally crucial is the validation and cleaning of ingested data. Implementing quality assurance measures helps identify and rectify inconsistencies, inaccuracies, or anomalies in the data, ensuring its reliability for subsequent stages of analysis and decision-making.

In summary, data ingestion within the domain of cloud computing and embedded systems in IoT networks emerges as a meticulous and multifaceted process. This integral step involves the seamless collection, transmission, and reception of multimedia data, facilitated by scalable cloud setups, wireless communication, and edge computing. The effective addressing of challenges such as security, standardization, and quality control is imperative to uphold the integrity and reliability of the ingested data. It ensures that the data serves as a robust foundation for shaping subsequent analytical insights and fostering informed decision-making processes. Multimedia big data demands sophisticated strategies for the ingestion of data, encompassing the methodical accumulation, importation, and storage of extensive multimedia datasets from diverse sources. This pivotal phase establishes the foundational framework for subsequent analytical endeavors and the derivation of meaningful insights. Among the diverse methodologies applied in the realm of multimedia big data ingestion, streaming processing emerges as a notable approach, adept at managing data in real-time or near real-time scenarios. This involves the ingestion of live video feeds and other data streams as they are generated, facilitated by platforms such as Apache Kafka or Apache Flink, enabling instantaneous processing.

Data lakes assume a critical role within multimedia big data ecosystems, serving as centralized repositories capable of accommodating both structured and unstructured data at scale. In this context, multimedia data finds a fitting repository, stored within data lakes utilizing technologies like the Hadoop Distributed File System (HDFS) or cloud-based solutions such as Amazon S3 and Azure Data Lake Storage.

Distributed file systems play a pivotal role in ensuring scalable and fault-tolerant storage and retrieval of multimedia data. Technologies such as the Hadoop Distributed File System (HDFS) are employed to distribute multimedia data across multiple nodes, thereby facilitating parallel processing.

Cloud-based solutions provide a comprehensive framework for multimedia big data storage and processing, offering scalability and flexibility. Platforms like Amazon S3, Google Cloud Storage, and Azure Blob Storage play a critical role in the ingestion, storage, and processing of multimedia data within a cloud environment.

Content Delivery Networks (CDNs) introduce an additional dimension to the landscape, streamlining the distribution of multimedia content closer to end users. CDNs, leveraging efficient caching mechanisms, significantly contribute to reducing latency and enhancing the overall speed of content delivery. This proves particularly advantageous in scenarios where the widespread distribution of multimedia content is paramount.

These methodologies are not mutually exclusive but can be strategically combined or tailored to suit specific use cases and requirements. Consideration of factors such as data volume, velocity, variety, and the imperative for real-time processing assumes a pivotal role in determining the most apt approach within the multimedia big data processing landscape [7].

8 Case Studies: How IOT and Big Data Can Be Connected

In the swift progression of the Internet of Things (IoT), an expansive range of sensors has been seamlessly integrated into various devices, spanning from individual gadgets to industrial machinery, all intricately connected to the Internet. These embedded sensors diligently collect data from diverse origins, including household gadgets, environmental conditions, scientific data, geographical information, transportation metrics, medical records, individual human data, mobile device particulars, communal records, and cosmic insights. The multimedia voluminous data amassed from IoT devices exhibit distinctive attributes compared to conventional extensive data, attributed to the diverse nature of sources, including heterogeneity, various data formats, unstructured features, and noise.

Anticipating a substantial influx of IoT-generated data by 2030, with projections surpassing 125 billion interconnected IoT devices, as highlighted in an IHS Markit report. The inadequacy of existing technological capabilities for handling multimedia big data is evident in meeting the challenges expected in the future era. The imperative to embrace IoT applications for advancing multimedia big data is acknowledged by many operators in the IoT space, recognizing its importance and evolution in the IoT landscape. Opportunities for the growth of multimedia big data arise from the considerable volume of multimedia data and the rapid expansion of IoT. This acknowledgment of significance and evolution in the IoT landscape intensifies the impetus for operators to adopt IoT applications. The collaboration between these two interdependent technological developments offers additional avenues for research in the realm of IoT.

Abdellah et al. [8] conducted an extensive survey exploring the integration of IoT and big data analytics in the realm of intelligent residences. Their investigation delved into the transformative influence of the digital evolution on the construction industry, unveiling emerging patterns facilitated by data processing. Through establishing a connection between the internet and physical objects, IoT forms a repository ecosystem for the copious data generated by interconnected devices. The examination of this voluminous data is crucial for actualizing the notion of intelligent living spaces, optimizing construction management, addressing security breaches, and fostering a safe and comfortable residential environment.

In a concurrent initiative, Rathore et al. [9] introduced and executed a model for intelligent urban management rooted in IoT and extensive data analysis. Their approach involved implementing diverse sensors across various domains, including intelligent dwellings, parking spaces, meteorological stations, water reservoirs, automotive networks, and surveillance tools. The model underwent a systematic progression, encompassing data acquisition, formatting, filtering, amalgamation, categorization, preprocessing, and decision-making. Extensive data processing employed Spark within the Hadoop ecosystem. Although the system aimed to contribute to advanced urban development, its real-world implementation is pending, necessitating further assessments of precision.

Vuppalapati et al. [10] explored big data analytics within the smart medical management system, emphasizing the substantial volumes of vital health-related data produced by on-body or in-body sensors. Their proposed solution involved correlating patient fitness activities with healthcare standards, utilizing supervised machine learning models for cluster analysis, and disseminating notifications to relevant stakeholders. Additionally, they advocated for a sensor assimilation scheme with Electronic Health Records (EHR) and recommended a scalable cloud architecture. Apache Kafka and Spark were employed for processing the vast amount of data and presenting it visually to doctors in real time.

Within the scope of futuristic buildings, Yassine et al. [11] provided a knowledge driven solution that harnesses big data analytics through fog and cloud computing. Their study concentrated on cognitive and anticipatory analysis of energy consumption habits in residences. Highlighting applications adept at recognizing consumption patterns and forecasting device maintenance schedules, this approach prevents costly servicing and ensures efficient service operations from an energy consumption standpoint.

Bashir and Gill [12] presented a strategic solution to address the hurdles linked to the real-time acquisition, processing, and analysis of data from intelligent residences. Their proposed analytical framework integrates IoT, big data management, and data analytics, aiming to autonomously uphold consistent oxygen levels, identify perilous gases or smoke, and manage luminous conditions. The implementation of this framework on the Cloudera Hadoop distribution platform, utilizing PySpark for extensive data analysis, demonstrated encouraging outcomes for the enhancement of smart building management through big data analytics.

Highlighting its capacity to revolutionize traditional systems into contemporary digitized alternatives, Mourtzis et al. [13] demonstrated the integration of IoT in the manufacturing industry. Their approach was put to practical examination in an authentic case study conducted at a mold making facility.

Directing attention to the power sector, Ramakrishnan et al. [14] delved into the amplification of electrical power in India, pinpointing potential advantages facilitated by the synergy of cloud computing and data analytics. Their scrutiny encompassed variables such as present measurements, power efficiency, drawn current, unauthorized usage, impairments, prepaid invoicing, and more. The authors contended that the utilization of data analytics could provide policymakers with enhanced capabilities, elevate energy conservation efforts, and curtail operational costs.

Yen and colleagues [15] delved into the functions of state-of-the-art technologies in elevating the sophistication of intelligent environments. Their examination of innovations such as artificial intelligence and big data sought to deduce causation and execute subsequent operations within a clever physical setting, encompassing individuals, tangible entities, and virtual entities. They introduced a collaborative sourcing strategy grounded in gaming activity to emulate human capabilities, emphasizing the potency of service exploration and composition methodologies to tackle real-world inquiries derived from data generated by the Internet of Things (IoT).

Jara and collaborators [16] carried out a thorough examination of prevailing technology solutions for extensive data, concentrating particularly on the hurdles presented by Cyber-Physical Systems within cloud settings. Their investigation zoomed in on critical elements of privacy and security within cloud frameworks and the intricate synthesis of information from various origins. The research also emphasized the imperative need to construct sophisticated data exploration systems and mechanisms tailored for analytics in real time.

Ahmed and colleagues [17] explored the symbiotic interplay between IoT and Big Data, underscoring the potential opportunities stemming from their intersection. The research delved into requisites, prospects, and hurdles within the realm of data refinement and analytics in intelligent IoT systems. It scrutinized human conduct within interactive social IoT environments through the lens of big data and analytics, providing perspectives on the conceivable synergies amid these technological realms.

In conclusion, these diverse studies collectively underscore the burgeoning intersection of IoT and big data across various domains. From smart homes and cities to healthcare, manufacturing, and power sectors, these technological advancements hold immense promise. The challenges associated with data collection, processing, and analysis are acknowledged, paving the way for ongoing research and innovation to unlock the full potential of IoT and big data integration. The continuous collaboration and exploration in this field are expected to reshape industries and enhance the efficiency and intelligence of diverse applications.

9 Applications

In the contemporary landscape, the surge in multimedia big data surpasses that of traditional data. Addressing the substantial volume of data in real-time scenarios, multimedia IoT big data analytics furnishes intricately designed tools for nuanced decision-making. The fusion of multimedia big data with the Internet of Things (IoT) introduces a spectrum of fresh challenges and promising avenues for the development of intelligent environments.

The practice of Social Media Analytics involves the systematic collection of data from platforms like Facebook, Twitter, Google Plus, blogs, Wikipedia, etc. The methodical analysis and statistical interpretation of this data aim to accumulate comprehensive knowledge. Numerous E-commerce vendors strategically employ social media analytics to extract discernible business value, amplify sales and profits, ensure sustained customer satisfaction, foster a robust company reputation, and instill widespread brand awareness among the public.

In the context of Intelligent Urban Centers, significant initiatives are underway, fueled by the evolution of multimedia big data and IoT technologies. The amalgamation of IoT and multimedia big data constitutes a burgeoning research area, presenting captivating challenges and opportunities in the trajectory toward future sophisticated urban ecosystems. IoT emerges as a pivotal source for accumulating substantial volumes of multimedia big data, necessitating swift and efficient processing, analysis, and transmission. A cutting-edge security notification system for intelligent residences, as envisioned by Tanwar and colleagues, utilizes pyroelectric infrared sensors and Raspberry Pi modules.

Within the Medical field, big data emerges as a transformative force with immense potential. Intelligent medical devices generate a wealth of information, including ECG readings, climate monitoring, sugar levels, and more. These devices enable real-time monitoring of patient well-being information, contributing significantly to the reduction of overall costs related to illness prevention and management. Through the meticulous analysis of medical data, healthcare professionals can identify and uncover ailments at their nascent stages. The widespread availability of high-speed internet has prompted a growing number of individuals to utilize mobile applications for the effective management of their well-being concerns. This integration of mobile applications with intelligent devices creates a cohesive and integrated Medical Internet of Things (MIoT) system.

In summary, the current era witnesses an unprecedented growth in multimedia big data, offering profound implications across diverse domains. From the fusion of multimedia big data with IoT to the strategic use of social media analytics in Ecommerce and the transformative impact of big data in the medical field, these trends shape the landscape of technology and data analytics. As research and innovation progress, the intricate interplay between these technological advancements promises to redefine the future landscape of intelligent environments and data driven decision-making [1].

10 Conclusion

In conclusion, this chapter has illuminated the intricate relationship between cloud computing, IoT, and embedded computing, showcasing their collective impact on handling multimedia big data in IoT ecosystems. Through a synthesis of theoretical frameworks and practical case studies, we've underscored the critical role of these technologies in addressing the challenges associated with managing dynamic, high-volume multimedia traffic.

The integration of cloud computing offers scalable and resource-rich environments, enabling efficient processing and storage of multimedia data. This is complemented by embedded computing, which empowers IoT devices with local processing capabilities, facilitating quick decision-making and reducing reliance on external resources. By leveraging both cloud and embedded computing paradigms, IoT systems can achieve enhanced scalability, adaptability, and efficiency.

Moreover, the chapter has provided insights into the architecture of cloud and fog computing, highlighting their respective roles in optimizing data processing and communication in IoT environments. Through a series of case studies, we've illustrated how IoT devices are seamlessly integrated with embedded computing systems to enable diverse applications across smart homes, agriculture, health care, and more.

Furthermore, we've discussed the challenges associated with handling multimedia big data and the importance of efficient data ingestion mechanisms in IoT systems. By addressing these challenges and leveraging the synergies between cloud computing, IoT, and embedded computing, we can unlock the full potential of IoT technology to revolutionize various domains, from smart cities to industrial automation.

In summary, this chapter serves as a comprehensive exploration of the intersection between cloud computing, IoT, and embedded computing, offering valuable insights and practical guidance for researchers, practitioners, and stakeholders in the field.

References

1. Sharmila, K., et al.: Introduction to multimedia big data computing for IoT. In: Multimedia big data computing for IoT applications, vol. 163, Intelligent Systems Reference Library. Springer, Singapore
2. Mohanty, S., et al.: Confidentiality preserving auditing for cloud computing environment. In: 2018 International Conference on Research in Intelligent and Computing in Engineering (RICE). IEEE (2018)
3. Mohanty, S., et al.: Privacy preserving auction based virtual machine instances allocation scheme for cloud computing environment. Int. J. Electr. Comput. Eng. (IJECE) **7**(5), 26452650 (2017)
4. Mohanty, S., Pattnaik, P.K., Mund, G.B.: Framework for auditing in cloud computing environment. J. Theor. Appl. Inf. Technol. **65**(1) (2014)
5. Mohanty, S., et al.: QoS aware groupbased workload scheduling in cloud environment. In: Data Engineering and Communication Technology: Proceedings of 3rd ICDECT2K19, pp. 953–960. Springer Singapore, Singapore (2020).
6. Pouyanfar, S., Yang, Y., Chen, S.C., Shyu, M.L., Iyengar, S.S.: Multimedia big data analytics: a survey. ACM Comput. Surv. **51**:134 (2018). https://doi.org/10.1145/3150226
7. XenonStack.: Data Ingestion Pipeline Architecture. XenonStack (2019). https://www.xenonstack.com/blog/dataingestionpipelinearchitecture
8. Daissaoui, A., Boulmakoul, A., Karim, L., Lbath, A.: IoT and big data analytics for smart buildings: a survey. Procedia Comput. Sci. **167**, 22442253 (2020). https://doi.org/10.1016/j.procs.2020.03.021
9. Rathore, M.M., Ahmad, A., Paul, A.: IoT based smart city development using big data analytical approach. In: 2016 International Conference on Advances in Computing, Communications and Informatics (ICACCI), Jaipur, India, pp. 2438–2442 (2016). https://doi.org/10.1109/ICACCA.2016.7778510.
10. Vuppalapati, C., Ilapakurti, A., Kedari, S.: The role of big data in creating sense EHR, an integrated approach to create next generation mobile sensor and wearable data driven Electronic Health Record (EHR). In: 2016 IEEE Second International Conference on Big Data Computing Service and Applications (BigDataService), Ox ford, UK, pp. 116–121 (2016). https://doi.org/10.1109/BigDataService.2016.18

11. Yassine, A., Singh, S., Hossain, M.S., Muhammad, G.: IoT big data analytics for smart homes with fog and cloud computing. Futur. Gener. Comput. Syst. **91**, 563573 (2019). https://doi.org/10.1016/j.future.2018.08.040
12. Bashir, M.R., Gill, A.Q.: Towards an IoT big data analytics framework: smart buildings systems. In: 2016 IEEE 18th International Conference on High Performance Computing and Communications; IEEE 14th International Conference on Smart City; IEEE 2nd International Conference on Data Science and Systems (HPCC/SmartCity/DSS), Sydney, Australia, pp. 1325–1332 (2016). https://doi.org/10.1109/HPCCSmartCityDSS.2016.0188
13. Mourtzis, D., Vlachou, E., Milas, N.: Industrial big data as a result of IoT adoption in manufacturing. Procedia CIRP **55**, 290295 (2016). https://doi.org/10.1016/j.procir.2016.07.038
14. Ramakrishnan, R., Gaur, L.: Smart electricity distribution in residential areas: Internet of Things (IoT) based advanced metering infrastructure and cloud analytics. In: 2016 IEEE International Conference on Internet of Things (iThings) and IEEE Green Computing and Communications (GreenCom) and IEEE Cyber, Physical and Social Computing (CPSCom) and IEEE Smart Data (SmartData), Chengdu, China, pp. 104–109 (2016). https://doi.org/10.1109/IOTA.2016.7562693
15. Yen, I.L., Zhou, G., Zhu, W., Bastani, F., Hwang, S.Y.: A smart physical world based on service technologies, big data, and game based crowd sourcing. In: 2015 IEEE International Conference on Web Services, pp. 789–796, New York, NY, USA (2015). https://doi.org/10.1109/ICWS.2015.111
16. Jara, A.J., Genoud, D., Bocchi, Y.: Big data for cyber physical systems an analysis of challenges, solutions and opportunities. In: 2014 Eighth International Conference on Innovative Mobile and Internet Services in Ubiquitous Computing, pp. 546–551, Birmingham, UK (2014). https://doi.org/10.1109/IMIS.2014.139
17. Ahmad, A., Rathore, M.M., Paul, A., Rho, S.: Defining human behaviors using big data analytics in social internet of things. In: 2016 IEEE 30th International Conference on Advanced Information Networking and Applications (AINA), pp. 994–997, CransMontana, Switzerland (2016). https://doi.org/10.1109/AINA.2016.104

IoT Security Measures and Social Network Content-Sharing Behavior: A Social Cognitive Theory Perspective

Anh Duc Do, Pham Chien Thang, and Ngo Ngoc Linh

Abstract This study examines the interplay between Internet of Things (IoT) security awareness, social cognitive factors, and social sharing behaviors. The study applies social cognitive theory (SCT) to examine the cognitive determinants of online security behavior in the context of IoT. Using a survey-based approach, data were collected from 393 social media users. Structural equation modeling was used to analyze the relationships between IoT security awareness, self-efficacy, outcome expectations, behavioral ability, and social sharing behavior. The results show that IoT security awareness does not have a significant impact on content-sharing behavior. Meanwhile, self-efficacy, outcome expectations, and behavioral ability had significant positive effects on sharing behavior. These results suggest that individuals with higher self-confidence in their security skills, a stronger belief in the positive outcomes of safe behavior, and greater practical security knowledge are more likely to share safe content on social networks. This study contributes to the theoretical understanding of online safety behavior by extending the applicability of SCT to the digital domain. This highlights the importance of self-efficacy, outcome expectations, and behavioral skills in shaping online safety behavior. This study, therefore, highlights the need for tailored educational and technological interventions to promote safe online practices. However, limitations such as the cross-sectional design and regional focus suggest opportunities for future research, including longitudinal studies and the examination of additional variables that influence digital behavior in the evolving IoT and social networking landscape.

Keywords Internet of things security · Social cognitive · Self-efficacy · Content-sharing behavior · Structural equation modeling

A. D. Do
VNU University of Social Sciences and Humanities, Hanoi, Vietnam

P. C. Thang (✉) · N. N. Linh
TNU-University of Sciences, Thai Nguyen, Vietnam
e-mail: thangpc@tnus.edu.vn

R. Sharma et al. (eds.), *Multimedia Technologies in the Internet of Things Environment, Volume 4*, Studies in Big Data 173, https://doi.org/10.1007/978-981-96-4356-1_6

1 Introduction

In the age of digitalization, advances in science and technology have brought numerous conveniences to human life. The Internet of Things (IoT) is a modern technology used in numerous areas including infrastructure management, healthcare, construction and automation, transportation, communications, and more. Modern science has actively explored various facets of IoT and confirmed its potential to improve users' convenience, efficiency, and level of automation [1]. However, the emergence of IoT also brings with it ethical considerations such as cybercrime and security concerns. Strengthening security, privacy, authentication, and recovery capabilities in IoT becomes imperative. Rani et al. [2] identify 12 latent risks associated with IoT, particularly related to social ethics, including unauthorized access to confidential information, data breaches, misuse of private data, or identity theft. Despite built-in features such as firewalls, antivirus software, and Address Space Layout Randomization (ASLR) in some devices, securing the IoT remains a challenge today. Threat sources and incidents of IoT attacks are identified as the biggest challenges in IoT security [3].

The rapid expansion of IoT and the ubiquity of social networks have created complex challenges and opportunities in the digital landscape [4, 5]. The integration of IoT devices into daily life and the proliferation of social media platforms have raised critical questions about security measures and content-sharing behaviors [6, 7]. Although technological advances in IoT provide unprecedented convenience and connectivity, they also pose unique vulnerabilities and security risks [8]. At the same time, the dynamics of content sharing on social networks have changed the way information is disseminated and consumed, influencing both individual behaviors and broader societal trends [9, 10].

Despite the extensive study of IoT security and social media content sharing in isolation, there is a significant research gap in understanding the convergence of these two areas from a theoretical perspective [11, 12]. In particular, the application of Social Cognitive Theory (SCT) has been little researched in this context [13, 14]. While studies separately address the technological and behavioral aspects of IoT security and social network interactions, there is a lack of comprehensive research examining these elements within the SCT framework [22, 38]. This gap requires an integrated approach that takes into account not only the technological foundations but also the cognitive and social factors that influence user behavior in IoT environments and on social media platforms.

The current study aims to fill this gap by examining IoT security measures and social sharing behavior from the perspective of SCT [21, 24]. This research seeks to understand how SCT principles such as observational learning, self-efficacy, and self-regulation influence users' interactions and behaviors in the context of IoT security and social media content sharing [15, 16]. By integrating the theoretical constructs of SCT with empirical findings in these areas, this research provides deeper insights into the cognitive and social determinants that influence user behavior in digital environments [17, 18]. This research is crucial for developing effective strategies to improve IoT security measures and promote responsible content-sharing behavior, thereby addressing the diverse challenges arising from the convergence of IoT and social networks.

2 Literature Review

2.1 IoT Security Measures

Contemporary academic discourse in the area of IoT security measures is increasingly focused on a triad of critical aspects: evolving challenges and threats, innovative solutions, and user perceptions in this rapidly evolving technology area. Overall, these works highlight the dynamic and complex interplay between technological advances, user awareness, and the ever-changing threat landscape in IoT security. At the heart of this discussion is the concept that a holistic understanding of IoT security is essential. Studies such as those by Roman et al. [4] have emphasized the need to fully understand the IoT ecosystem, critically assess the current state of Internet security, and identify possible solutions for a more secure IoT framework [4]. This fundamental approach has set the stage for an in-depth study of specific security threats and challenges prevalent in the IoT space and highlights the need for robust security measures and strategies to mitigate these risks [6].

A significant portion of the literature is devoted to the development of IoT security measures, characterized by efforts to refine and adapt existing frameworks in response to new threats. This includes a review of current security standards, particularly in smart environment contexts, where gaps have been identified and the development of more comprehensive solutions has been suggested [8, 19]. Complementary to these frameworks, ethical hacking practices have been integrated into existing security measures, advocating a systematic and proactive approach to strengthening IoT security [20]. Application-specific aspects of IoT security have also received attention, with research examining privacy protection algorithms, needs to meet security and privacy requirements in cloud-based IoT environments, and the potential of blockchain technology in secure IoT applications [21–23]. Another focus is on lightweight architectures for IoT end devices that maintain efficiency without compromising security [11].

Human factors in IoT security have emerged as important issues. Studies have highlighted the influence of users' attitudes, risk perceptions, and security awareness on the overall security posture of IoT systems [15, 24]. These findings highlight the need to increase user awareness to effectively mitigate vulnerabilities. Furthermore, research into digital inequalities and the complexity of users' perceptions highlights the need to understand users' knowledge and attitudes towards IoT technologies, especially in the context of smart homes [16–18, 25, 26].

When it comes to users' awareness of IoT security, many scholars have considered IoT data which can extract sensitive information about privacy, such as the number of residents, personal habits, and daily routines, from smart devices within homes, such as smartwatches [27]. In response to these risks, researchers have assessed users' awareness and concern regarding IoT security.

A qualitative study, utilizing in-depth interviews with 11 owners of smart devices in their homes, was conducted by Serena Zheng and colleagues [28]. The study revealed that device owners tend to trust IoT device manufacturers to protect their privacy without verifying the implementation of these protective measures. Consequently, users are often unaware of privacy risks associated with inference algorithms operating on non-audio/visual data. Nonetheless, users deem security and privacy highly necessary for an IoT device [29]. However, when asked to prioritize factors in decision of purchasing an IoT device, research results indicate that cost is a more significant concern for users. Accordingly, there is no statistically significant difference between technical knowledge and experience concerning the detection of virus-infected devices without any noticeable abnormalities.

On the other hand, awareness of IoT can positively influence users' knowledge of privacy and security [30]. This knowledge also affects users' trust in IoT and the intention to continue using IoT-integrated products. Zlatolas's exploratory research offers another perspective on users' awareness of IoT, suggesting that data breaches, ransomware attacks, violations of personal information access rights, and device vulnerabilities significantly impact the importance of IoT security [26]. Despite this awareness, users tend to habitually feel secure when using IoT devices and do not engage in any security configuration checks.

Up to now, research on users' awareness of IoT security remains a relatively new and underexplored topic. However, the glimpses provided here underscore the importance of this research area and the need for further exploration in the future. A number of researchers have asserted that current users have a low awareness of the necessity and proactive security measures for IoT [26, 28, 30]. Users self-assess that they possess knowledge and experience in the technology field but find it challenging to recognize virus attacks on IoT devices [29].

The literature provides a comprehensive picture of the challenges, solutions, and evolving user perceptions in the context of IoT security. It advocates a holistic approach that integrates technological advances, user awareness, and adaptive security measures to effectively address the dynamic and complex nature of IoT security.

2.2 Social Network Content-Sharing Behavior

The current discourse on social sharing behavior has brought together numerous studies to provide a comprehensive overview of the diverse influences and dynamics underlying this phenomenon. This synthesis of academic research navigates various aspects, including identity salience, business models, influence propagation, influencer marketing, user engagement, user-generated content, social activism, academic skills, and privacy concerns, and weaves them into a coherent narrative.

Zhang et al. [5] mark a crucial understanding of the role of employee identity salience in content sharing within social networks. This study underscores the significant influence of employee-created content on sharing behavior, thereby illuminating the interplay between professional identity and social media engagement. Complementing this perspective, Cortimiglia et al. [7] delved into the critical success factors pertinent to business models in content sharing and social network services, particularly highlighting revenue models and delivery channels in the Spanish-speaking market. Their insights are pivotal for demystifying the commercial aspects of content-sharing behaviors.

The models for the spread of influence through social networks, as studied by Kempe et al. [9], offer a vital understanding of the mechanisms through which ideas and influences are disseminated within these digital ecosystems. This research contributes significantly to our understanding of the intricate processes governing content sharing.

In the context of information explosion, social media has become the primary channel for the public to access and share information [31]. While social media platforms share common purposes with traditional media, there are certain notable differences in user-content interactions [32]. Studies also indicate that interaction and information sharing on social media have become a prominent and significant phenomenon, transcending social, economic, and political boundaries [33].

User news-sharing behavior on social media has been studied based on the Elaboration Likelihood Model and Social Influence Theory [32]. The authors conducted a survey with 513 active social media users to identify factors influencing news-sharing behavior on social media platforms. The study results underscore that users' awareness of online civic engagement, attitudes toward news-sharing, perception of influence on others, and social influence are crucial predictors of their news-sharing behavior on social media. Furthermore, the research discovers that the online news quality and the reliability of news sources reinforce the relationship between users' awareness of online civic engagement and their news-sharing behavior. Regarding the experience of sharing news on social media, some studies find that sharing low-quality news may damage an individual's reputation [34]. Thus, it is crucial to assess the quality of news from social media platforms.

In the realm of educational environment, some scholars argue that trust and perceived reciprocity benefits are strong predictors of knowledge sharing on Facebook [35]. This study involved 170 subjects, university students in Malaysia, and emphasized that students with higher levels of gratitude are more open to sharing knowledge unconditionally compared to their counterparts.

Another study conducted in Germany tried to explain "Why do users share news?" [36]. The survey involved 333 participants and revealed that news-sharing attitudes and subjective norms positively impact the intention to share news, consequently affecting actual news-sharing behavior. On the other hand, the influence of socialization suggests that individuals not only share news for dissemination and information gathering but also for interacting with others, potentially placing less emphasis on the actual content shared. The study also found that political interest positively influences news-sharing behavior. Conducted in Nigeria amid the COVID-19 pandemic, a study by Apuke and Omar [37] demonstrated that motives for information sharing, social integration, information seeking, and leisure predicted the dissemination of misinformation about COVID-19. In contrast, there was no relationship between entertainment motives and information-sharing behavior, supporting previous research findings [36].

In the realm of marketing, Berjozkina and Garanti [10] discussed the burgeoning role of social media influencers, especially in promoting travel. Their work sheds light on the evolving marketing strategies employed by companies and destinations, indicating a shift towards leveraging influencer marketing in content dissemination.

The relationship between public engagement and organizational content on social networking sites analyzed by Smith and Gallicano [38] further deepens our understanding of user interaction dynamics. Similarly, Tang et al. [39] provide valuable insight into the impact of user-generated content on product sales, highlighting the sheer volume and dynamism of content shared on platforms such as Facebook, Twitter, and YouTube.

In the context of social activism, Haq et al. [12] underscored the significant role of platforms such as Instagram in garnering higher visibility and engagement for social causes. Complementary to this, Aoun and Jerbi [40] explored the transformative potential of social media networks in content sharing and knowledge exchange, particularly focusing on the impact on students' academic skills.

Privacy and security concerns on social networks have emerged as critical factors influencing sharing behavior. There are a lot of previous studies that collectively underscore the complex relationship between privacy concerns, security issues, and sharing behavior [41–52]. These studies span various contexts, from general social network usage to specific domains, such as healthcare and SMEs, revealing the nuanced ways in which privacy concerns manifest and influence content-sharing behaviors.

For example, Gross and Acquisti's study [41] illustrated that Facebook users exhibited a lack of concern regarding privacy risks when sharing personal information on online social platforms. "The biggest lie on the Internet" is when individuals are compelled to accept and click on the "agree" button for privacy and security terms without reading and often without understanding how it will impact their personal experience in the online environment. These risks associated with information disclosure that users may face include physical and network-related threats, particularly the ease with which third parties can create digital profiles of their behavior.

According to Obar and Oeldorf-Hirsch [53], up to 74% of social media users overlook reading and understanding privacy-related issues and terms of service when participating in these platforms. Another research by Yuniar [54] also revealed that young users possess sufficient fundamental knowledge regarding crucial points of security policies and privacy rights, however, they tend to overlook them. Social media users view privacy policies and terms of service as a nuisance, disregarding them to pursue production goals and digital engagement. Subsequent studies validate these results, discovering that users rarely protect their privacy and tend to delete cookies and browser history or reject cookies to safeguard their online privacy [54, 55].

There is a significant relationship between various aspects of privacy, including awareness and protective behavior, with trust and activity. Furthermore, user trust and activity significantly impact the intention to share photos on Facebook [56]. The study's findings are intriguing, indicating that as users' awareness and protective behavior increase, their motivation to share images on the social media platform also increases. However, this research approach has limitations, stopping at users' self-assessment of their capabilities without employing additional authenticating methods. Factors influencing users' social media sharing behavior regarding privacy include users' awareness of the importance of information privacy and the effectiveness of information privacy [57]. Security-related factors not only make the public feel secure sharing information but also stimulate purchasing intentions, especially as current social media platforms integrate e-commerce functionalities [58].

In terms of gender studies, some researchers investigate gender differences in sharing information on social media platforms. A study by Lin and Wang [59] revealed significant gender differences in their perception of sharing information on social media platforms. Accordingly, women tend to be more concerned about privacy-related issues and are more committed to security than men, despite no significant differences in their digital literacy.

Although privacy and security policy terms are often overlooked by users, they still show significant concern for issues related to personal security and privacy. Some studies suggest that privacy concerns, such as security and trust, have a positive impact on sharing information on social media platforms [60, 61]. Thus, existing research hints at the relationship between privacy protection awareness and information-sharing behavior on social media platforms.

In summary, the literature on Social Network Content-Sharing Behavior presents a rich body of research exploring various facets of this phenomenon. Based on the influence of employee identity, business models, and influencer marketing on the pivotal role of privacy and security concerns, these studies collectively enhance our understanding of the dynamics shaping content-sharing behaviors in the digital age.

2.3 Theoretical Framework and Hypothesis Development

Social Cognitive Theory (SCT) is a psychological model that emphasizes the role of observational learning, social experience, and reciprocal determinism in personality development. According to Bandura [13], the key concepts of SCT include observational learning, self-efficacy, and self-regulation. Observational learning refers to the process of acquiring new behaviors and knowledge by observing others, whereas self-efficacy pertains to an individual's belief in their ability to execute specific actions to achieve desired outcomes. Self-regulation involves managing thoughts, behaviors, and emotions to attain personal goals [14].

SCT has been widely applied in different fields, including health promotion, education, and online behavioral studies. In the context of online behavior studies, SCT has been utilized to understand and predict user behaviors on online platforms.

For instance, the theory has been employed to analyze user participation in online political discussions, with a focus on how communication mechanisms enforce gender norms [62]. Additionally, this model has been used to investigate users' continuous use behavior of online leisure and entertainment platforms, demonstrating its wide applicability in understanding online user behavior [30].

Moreover, SCT has been instrumental in examining the determinants of users' participation in online social question-and-answer communities. Some researchers found that the theory provided a robust framework for analyzing how user-generated content emerges and influences users' future behaviors in online communities [63]. Furthermore, SCT can serve studies into knowledge-seeking behavior in online communities, highlighting its relevance in understanding information-sharing and dissemination on online platforms [64].

In education, SCT has been used to improve students' mathematical thinking skills through social cognitive learning in online environments. The application of SCT highlights its importance in shaping learning experiences and outcomes in digital environments [65]. SCT can also validate research on knowledge sharing in cyberspace, particularly in the context of collaborative business innovation, and highlight its role in facilitating knowledge sharing and collaboration in online environments [66].

In addition, SCT was found to be useful in examining the influence of social cognitive determinants on diet and physical activity among web health users participating in online interventions. This study highlights the relevance of SCT for understanding the psychosocial characteristics and health behaviors of online users, thereby providing information for the development of theory-based online behavior change interventions [67].

In the area of IoT security awareness, SCT can be used to improve users' understanding of security risks and encourage proactive security behavior. The concept of self-efficacy is particularly relevant in this context because it can influence individuals' confidence in their ability to adopt safe practices and mitigate potential threats in IoT environments [68]. Additionally, integrating social cognitive theories into SETA (Security Education, Training, and Awareness) programs can help reduce moral disengagement tendencies and improve employee security awareness, thereby contributing to a more secure IoT ecosystem [69]. SCT can provide a better understanding of content-sharing behavior in the context of IoT. The emphasis on reciprocal interactions between behavior, cognition, and social environmental influences is consistent with the dynamics of content exchange in IoT environments [70]. By considering the social-cognitive determinants of content sharing, such as self-regulation and social trust, organizations can develop strategies to promote safe and responsible content-sharing practices within IoT networks [71].

Last but not least, the application of SCT in IoT security awareness and content sharing aligns with the growing emphasis on context-aware security solutions in IoT environments. Context-aware security models informed by SCT propositions can enhance situation prediction, improve network security, and mitigate social engineering attacks in IoT ecosystems [72–74]. Additionally, the integration of SCT with edge intelligence and blockchain-based authorization management can contribute to the development of comprehensive security frameworks tailored to the unique challenges of IoT devices and networks [72, 75, 76].

In summary, Social Cognitive Theory (SCT) propositions, particularly those related to self-efficacy, self-regulation, and social cognitive determinants, hold significant relevance for enhancing IoT security awareness and promoting responsible content-sharing practices. Based on SCT principles, we hypothesize the following.

Hypothesis 1 (H1): IoT security awareness has a positive effect on content-sharing behavior on social networks.

Hypothesis 2 (H2): Observational learning positively affects users' content-sharing behavior.

Hypothesis 3 (H3): Users' perceived self-efficacy in managing IoT security is positively associated with responsible social network content-sharing.

Hypothesis 4 (H4): Outcome expectations regarding privacy and security significantly affect users' content-sharing decisions on social networks.

Hypothesis 5 (H5): Reinforcement through feedback on social media (e.g., comments or shares) influences content-sharing behavior.

Hypothesis 6 (H6): Behavioral capability to implement security measures in IoT devices positively affects content-sharing behavior on social networks.

3 Methods

3.1 Data Collection and Sample Size

To determine the sample size of this study (n), we used the Cochran (1963) formula; therefore, with p = 0.5 (maximum variability), we desired a 95% confidence level and ±5% precision. The minimum sample size of this study was as follows:

$$n = \frac{Z^2 pq}{e^2} = \frac{(1.96)^2(0.5)(0.5)}{0.05^2} = 385$$

This study adopted a quantitative method involving 393 respondents based on the random sampling technique in Vietnam. The questionnaire was shared in the Facebook Messenger group and the Zalo group with a request to fill them out. The data collection period extended from October to December 2023. In the questionnaire with a cover letter, the purpose of the survey was explained, confidentiality was assured and the consent of the respondents was obtained.

3.2 Measurement of Variables

The instruments used in this study were adopted from reputable sources. Before starting the main phase of data collection, the questionnaire was rigorously evaluated by academic experts to ensure its content validity. Consequently, measurement of the variables involved self-report using multi-item scales adapted from previous research. These elements were carefully tailored to fit the unique local context and demographics of the study population. All measures were assessed using a five-point Likert scale, with a value of "1" meaning "strongly disagree" and "5" meaning "completely agree." A comprehensive presentation of these items can be found in Table 1.

3.3 Data Analysis

The data analysis procedure included three phases using IBM SPSS version 28.0 and AMOS version 28.0 software. First, an exploratory factor analysis was conducted using IBM SPSS on the scales used to measure variables. Construct reliability was assessed using Cronbach's alpha and corrected item-to-total correlations. To examine the theoretical framework of the study, covariance-based structural equation modeling (CB-SEM) was applied using AMOS software. The maximum likelihood (ML) technique was adopted to estimate the parameters as described by Hair et al. [85] suggested. The process included two different analysis phases: the evaluation of the measurement model and the examination of the structural model. The measurement model addresses the constructs' reliability, validity, and overall fitness, while the structural model focuses on hypothesis testing.

4 Results

4.1 Reliability and Validity Analysis

The analysis of unrotated factor solutions [86] disclosed that a solitary factor accounted for 28.36% of the variance in the examined variables, less than the 50% threshold. Consequently, the likelihood of Common Method Variance (CMV) influencing the sample data of this study was not critical [87]. The results revealed that the values of Cronbach's alpha coefficient (α) were distributed within a range of 0.741–0.927, and all corrected item-to-total correlations exceeded the threshold of 0.5, signifying a robust internal consistency in terms of reliability [88]. Table 2 presents the mean scores, standard deviations, and α coefficients for the various variables.

Table 3 shows that the average variance extracted (AVE) exceeded the threshold of 0.5, and the composite reliability (CR) exceeded 0.7. Furthermore, the standardized

Table 1 Variables and items

Variables	Items	Measurement	Sources
IoT security awareness (IOA)	SA1	I am aware of the security threats associated with IoT devices	[77, 78]
	SA2	I regularly update myself on new security measures for IoT devices	
	SA3	I understand the consequences of IoT security breaches	
Observational learning (OL)	OL1	I often observe my peers' behavior regarding IoT security before deciding what to do	[13, 79]
	OL2	I learn about IoT security measures by watching others' experiences on social media	
	OL3	I am influenced by how my social network contacts manage their IoT security	
Self-efficacy (SE)	SE1	I am confident in my ability to control my privacy settings on social networks	[14, 80, 81]
	SE2	I feel capable of managing the security settings of my IoT devices	
	SE3	I am confident in making decisions about what to share on social networks considering IoT security	
Outcome expectations (OE)	OE1	I believe that taking IoT security measures will protect my personal information	[81, 82]
	OE2	I think that secure IoT practices will prevent potential cyberattacks	
	OE3	I feel that being cautious with IoT security will have a positive impact on my privacy	
Reinforcement (RE)	RE1	Positive feedback on social media encourages me to share more content	[83]
	RE2	I am more likely to share content if it has been positively received by others	
	RE3	The reactions of my social network influence my decision to share content related to IoT	
Behavioral capability (BC)	BC1	I feel confident in identifying potential security threats in IoT devices	[14]
	BC2	I am knowledgeable about the steps to take to secure my IoT devices	

(continued)

Table 1 (continued)

Variables	Items	Measurement	Sources
	BC3	I can effectively implement security measures for my IoT devices	
Content-sharing behavior (SB)	SB1	I am careful about the type of personal information I share on social networks	[84]
	SB2	I think twice before sharing information online when it's related to my IoT devices	
	SB3	I avoid sharing content that might compromise my IoT device security	

Table 2 Reliability analysis

Variable	Mean	Standard deviation	Cronbach's alpha
SE	4.733	0.533	0.927
SB	4.777	0.483	0.92
OE	4.782	0.478	0.91
BC	4.813	0.44	0.871
OL	4.025	0.526	0.787
SA	4.146	0.609	0.77
RE	4.364	0.549	0.741

factor loadings consistently exceeded the value of 0.5. These results indicate satisfactory convergent validity, as Hair et al. [89] suggested. In addition, discriminant validity was checked through an assessment that determined whether the square root of the AVE for each construct was superior to its correlation coefficients with other constructs, following the methodology suggested by Fornell and Larcker [90].

4.2 Measurement Model

Measurement model fit was assessed by confirmatory factor analysis (CFA) using maximum likelihood estimation applied to all constructs using IBM AMOS version 28.0. As shown in Fig. 1, the resulting fit indices achieved excellent values: the χ^2/df ratio was 1.344, the Goodness of Fit Index (GFI) was 0.948, the Comparative Fit Index (CFI) was 0.988, and the Tucker Lewis Index (TLI) at 0.985, Root Mean Square Error of Approximation (RMSEA) at 0.030 and Standardized Root Mean Square Residual (SRMR) at 0.035, according to the criteria established by Hu and Bentler [91]. These statistical measurements demonstrated a competent approximation of the sample data.

Table 3 Validity analysis

	CR	AVE	MSV	SE	SB	OE	BC	OL	SA	RE
SE	0.928	0.81	0.191	0.9						
SB	0.923	0.8	0.335	0.376***	0.894					
OE	0.919	0.792	0.335	0.437***	0.579***	0.89				
BC	0.886	0.723	0.323	0.424***	0.475***	0.568***	0.85			
OL	0.789	0.555	0.008	0.022^{NS}	0.004^{NS}	−0.016^{NS}	0.008^{NS}	0.745		
SA	0.774	0.535	0.008	−0.027^{NS}	−0.012^{NS}	−0.037^{NS}	−0.035^{NS}	0.087^{NS}	0.731	
RE	0.751	0.506	0.017	0.091^{NS}	0.132*	0.120*	−0.03^{NS}	−0.016^{NS}	0.081^{NS}	0.711

NS: Not significantly
* $p < 0.050$
*** $p < 0.001$

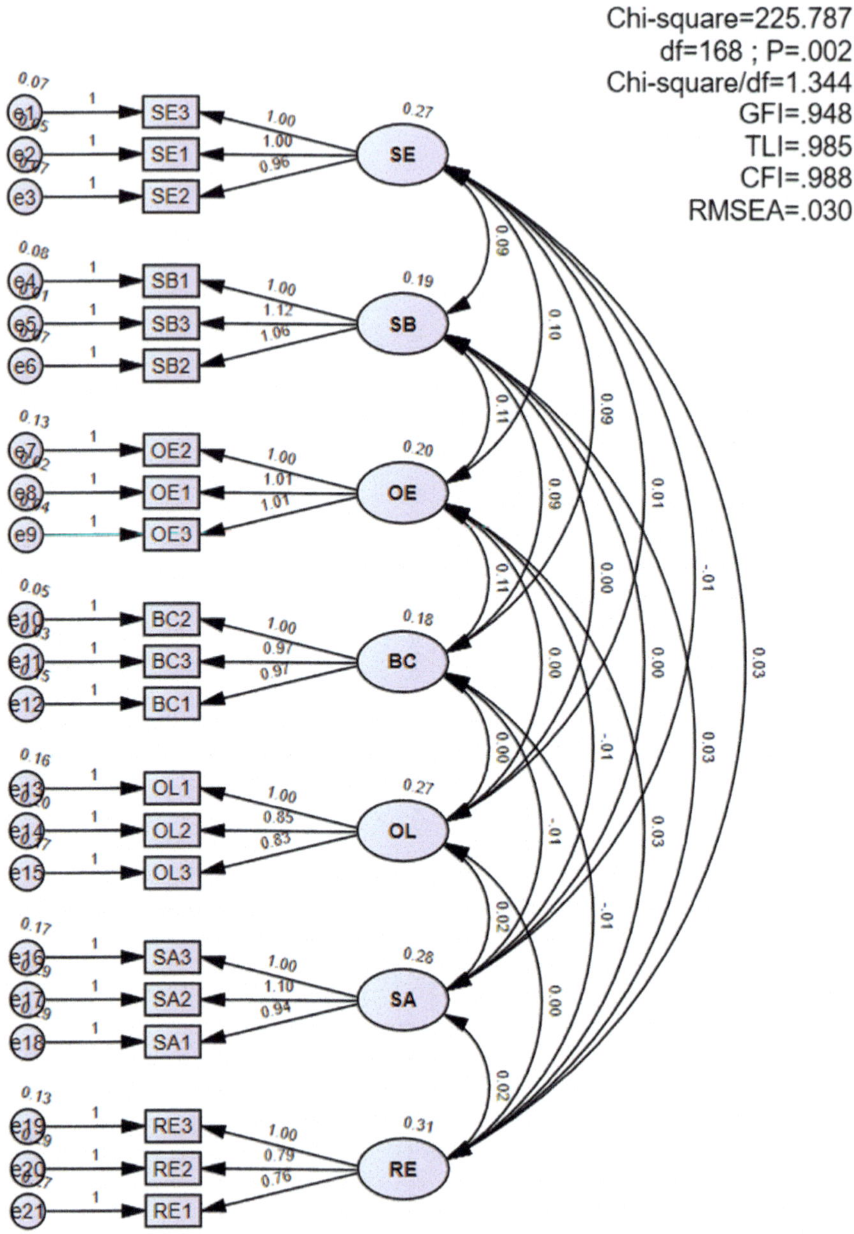

Fig. 1 Confirmatory factor analysis

4.3 Structural Models

The CFA demonstrated that the suggested model satisfied the validity and reliability requirements. Following the validation of the measurement model, the relationships between the variables were examined using structural models. A structural analysis was performed, and the results of the theoretical model (Fig. 2) showed an excellent model fit. The results of the proposed model suggest excellent explanatory power for content-sharing behavior on social networks (adjusted $R^2 = 0.383$), meaning that the independent variables affect 38.3% of the variation in SB.

Table 4 shows that SE ($\beta = 0.088$, CR = 2.071, $p < 0.05$) had a positive and significant effect on SB, supporting the third hypothesis. Similarly, OE ($\beta = 0.396$, CR = 6.816, $p < 0.001$) had a significant positive relationship with SB (Hypothesis 4). BC ($\beta = 0.209$, CR = 3.498, $p < 0.001$) also had a positive relationship with SB (hypothesis 6). The relationship between SA and SB was not significant ($\beta = 0.005$, CR = 0.116, $p > 0.05$), implying that the first hypothesis was rejected. On the other hand, Hypotheses 2 (OL - > SB) and 5 (RE -> SB) were rejected because the p-value was higher than 0.05.

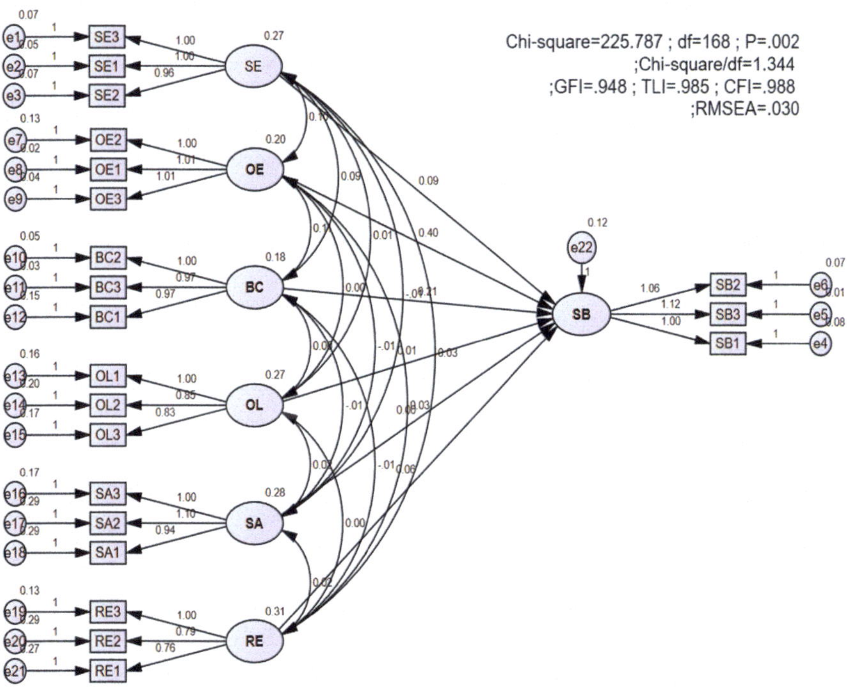

Fig. 2 Structural models

Table 4 Hypotheses testing

Hypothesis	Path	Beta	S.E	C.R	P-value	Decision
H1	SA -> SB	0.005	0.039	0.116	0.908	Rejected
H2	OL -> SB	0.006	0.039	0.166	0.868	Rejected
H3	SE -> SB	0.088	0.042	2.071	0.038	Supported
H4	OE -> SB	0.396	0.058	6.816	0.000	Supported
H5	RE -> SB	0.062	0.038	1.641	0.101	Rejected
H6	BC -> SB	0.209	0.06	3.498	0.000	Supported

5 Discussion

The results of this study illuminate the intricate relationship among IoT security awareness, social cognitive factors, and content-sharing behaviors on social networks. Rooted in SCT, our findings reveal a nuanced understanding of digital behaviors in the context of the IoT and social media interactions. IoT security awareness did not show a significant relationship with social network content-sharing behavior, suggesting that mere awareness of security threats in IoT may not directly influence users' content-sharing behaviors. This aligns with Bandura's SCT, which posits that awareness alone may not suffice to change behavior unless coupled with self-efficacy and outcome expectations [13, 14].

Observational learning and reinforcement also did not significantly influence SB. This could indicate that the modeling of security behaviors and feedback mechanisms on social networks may not be prominent factors in shaping content-sharing decisions. This finding diverges from the traditional SCT understanding, where observational learning is considered a key mechanism for behavioral acquisition [13].

Conversely, self-efficacy, outcome expectations, and behavioral capability significantly influenced SB. SE's positive effect of SE corroborates Bandura's assertion that confidence in one's abilities is a critical determinant of behavior [68]. The strong influence of OE on SB highlights the role of perceived consequences in shaping online behaviors, resonating with SCT's emphasis on outcome expectations [14]. BC's impact of BC suggests that possessing the skills and knowledge to implement security measures in IoT devices encourages responsible content-sharing, aligning with SCT's focus on behavioral capability [14].

This study contributes to the Social Cognitive Theory by highlighting the importance of self-efficacy, outcome expectations, and behavioral capability in the context of IoT security and social network content sharing. It extends SCT's applicability to digital behaviors, particularly in understanding how cognitive and behavioral factors influence online security and sharing practices [14, 16]. This extension is crucial, as it bridges traditional SCT concepts with contemporary digital contexts, thereby enhancing our theoretical understanding of user interactions in the increasingly prevalent digital landscape [15, 18].

For stakeholders in the IoT and social media industries, these findings underscore the need to focus on enhancing user self-efficacy, clarifying outcome expectations, and improving behavioral capabilities related to IoT security. This could involve targeted educational programs, user-friendly security tools, and clear communication regarding the benefits of secure online behavior [68, 77]. Programs designed to enhance self-efficacy and behavioral capability are particularly relevant in the context of rapid technological advancements where users often face new and complex security challenges [69, 78]. Furthermore, emphasizing the positive outcomes of secure behaviors could significantly influence users' decision-making processes, in line with SCT principles [81, 82]. By addressing these issues, stakeholders can create more effective strategies for promoting safer online environments, which is essential in the era of ubiquitous digital connectivity and IoT proliferation [8, 21].

The study's primary limitation is its cross-sectional nature and the specific context of Vietnam, which may limit the generalizability of the findings. Future research should consider longitudinal designs and diverse geographical settings to explore these relationships further. Additionally, the insignificant findings for SA, OL, and RE suggest a complex interplay between factors influencing content-sharing behavior. Future research could explore additional variables, such as personal attitudes, social norms, and specific types of IoT devices, to gain a more comprehensive understanding of these dynamics.

6 Conclusion

This study investigated the relationship between IoT security awareness, social cognitive factors, and content-sharing behaviors on social networks, and applied Social Cognitive Theory as major framework. The findings revealed that while IoT security awareness did not significantly influence social network content-sharing behavior, factors such as self-efficacy, outcome expectations, and behavioral capability showed a significant positive impact. These results extend SCT's applicability of SCT to the digital realm, particularly in the context of IoT security and social media, underscoring the importance of self-belief, perceived consequences, and practical skills in shaping online behaviors.

The implications of this study are two-fold. Theoretically, it broadens the scope of SCT by applying its constructs to the dynamics of digital behavior, offering insights into the complex interplay of cognitive and behavioral factors in online security and content sharing. Practically, it emphasizes the need for targeted educational and technological interventions to enhance online practices. However, the research's limitations, including its cross-sectional nature and specific geographical focus, suggest the need for further exploration in diverse contexts and longitudinal studies. Future research should also consider additional variables, such as personal attitudes and social norms, to deepen our understanding of digital behavior in the rapidly evolving landscape of the IoT and social networks.

References

1. Hassija, V., Chamola, V., Saxena, V., Jain, D., Goyal, P., Sikdar, B.: A survey on IoT security: application areas, security threats, and solution architectures. IEEE Access. **7**, 82721–82743 (2019). https://doi.org/10.1109/ACCESS.2019.2924045
2. Rani, S., Kataria, A., Sharma, V., Ghosh, S., Karar, V., Lee, K., Choi, C.: Threats and corrective measures for IoT security with observance of cybercrime: a survey. Wirel. Commun. Mob. Comput. **2021**, 5579148 (2021). https://doi.org/10.1155/2021/5579148
3. Nik Zulkipli, N.H., Alenezi, A., B. Wills, G.: IoT forensic: bridging the challenges in digital forensic and the internet of things: In: Proceedings of the 2nd International Conference on Internet of Things, Big Data and Security. pp. 315–324. SCITEPRESS-Science and Technology Publications, Porto, Portugal (2017). https://doi.org/10.5220/0006308703150324
4. Roman, R., Najera, P., Lopez, J.: Securing the internet of things. Computer **44**, 51–58 (2011). https://doi.org/10.1109/MC.2011.291
5. Zhang, X., Tang, T. (Ya), Chen, M., Wang, F.: Effects of employee identity salience on content sharing in social networks. MIP **40**, 89–104 (2022). https://doi.org/10.1108/MIP-09-2021-0304
6. Abomhara, M., Koien, G.M.: Security and privacy in the internet of things: current status and open issues. In: 2014 International Conference on Privacy and Security in Mobile Systems (PRISMS). pp. 1–8. IEEE, Aalborg, Denmark (2014). https://doi.org/10.1109/PRISMS.2014.6970594.
7. Nogueira Cortimiglia, M., Ghezzi, A., Renga, F.: Social applications: revenue models, delivery channels, and critical success factors-an exploratory study and evidence from the Spanish-speaking market. J. Theor. Appl. Electron. Commer. Res. **6**, 18–19 (2011). https://doi.org/10.4067/S0718-18762011000200010
8. Samaila, M.G., Sequeiros, J.B.F., Simoes, T., Freire, M.M., Inacio, P.R.M.: IoT-HarPSecA: a framework and roadmap for secure design and development of devices and applications in the IoT space. IEEE Access. **8**, 16462–16494 (2020). https://doi.org/10.1109/ACCESS.2020.2965925
9. Kempe, D., Kleinberg, J., Tardos, É.: Maximizing the spread of influence through a social network. In: Proceedings of the Ninth ACM SIGKDD International Conference on Knowledge Discovery and Data Mining. pp. 137–146. ACM, Washington, D.C. (2003). https://doi.org/10.1145/956750.956769
10. Berjozkina, G., Garanti, Z.: Emerging influencers promoting travel: the case of local tourism in latvia. JRESD. **1**, 51 (2020). https://doi.org/10.17770/jresd2020vol1.12.5402
11. Mishra, P.: Secured novel lightweight IoT end device architecture using confidentiality, integrity, authenticity & availability based tight security approach. TURCOMAT. **12** (2021). https://doi.org/10.17762/turcomat.v12i10.5543
12. Haq, E.-U., Braud, T., Yau, Y.-P., Lee, L.-H., Keller, F.B., Hui, P.: Screenshots, symbols, and personal thoughts: the role of instagram for social activism. In: Proceedings of the ACM Web Conference 2022, pp. 3728–3739. ACM, Virtual Event, Lyon France (2022). https://doi.org/10.1145/3485447.3512268
13. Bandura, A.: Social cognitive theory: an agentic perspective. Asian J. Soc. Psycho. **2**, 21–41 (1999). https://doi.org/10.1111/1467-839X.00024
14. Bandura, A.: Health promotion by social cognitive means. Health Educ. Behav. **31**, 143–164 (2004). https://doi.org/10.1177/1090198104263660
15. Chen, J., Zhu, Q.: Interdependent strategic security risk management with bounded rationality in the internet of things. IEEE Trans. Inform. Forensic Secur. **14**, 2958–2971 (2019). https://doi.org/10.1109/TIFS.2019.2911112
16. Van Deursen, A.J.A.M., Van Der Zeeuw, A., De Boer, P., Jansen, G., Van Rompay, T.: Digital inequalities in the internet of things: differences in attitudes, material access, skills, and usage. Inf. Commun. Soc. **24**, 258–276 (2021). https://doi.org/10.1080/1369118X.2019.1646777
17. May Amy, Y.C., Tan, G.G., Carter, S.: The conundrum of internet of things adoption in higher educational institutions. RBAOS. **2**, 67–94 (2020). https://doi.org/10.32770/rbaos.vol267-94

18. Al-Ameen, M.N., Chauhan, A., Ahsan, M.A.M., Kocabas, H.: A look into user's privacy perceptions and data practices of IoT devices. ICS. **29**, 573–588 (2021). https://doi.org/10.1108/ICS-08-2020-0134
19. Karie, N.M., Sahri, N.M., Yang, W., Valli, C., Kebande, V.R.: A review of security standards and frameworks for IoT-based smart environments. IEEE Access. **9**, 121975–121995 (2021). https://doi.org/10.1109/ACCESS.2021.3109886
20. Ding, A.Y., De Jesus, G.L., Janssen, M.: Ethical hacking for boosting IoT vulnerability management: a first look into bug bounty programs and responsible disclosure. In: Proceedings of the Eighth International Conference on Telecommunications and Remote Sensing. pp. 49–55. ACM, Rhodes Greece (2019). https://doi.org/10.1145/3357767.3357774
21. Zhu, G., Li, X., Zheng, C., Wang, L.: Multimedia fusion privacy protection algorithm based on IoT data security under network regulations. Comput. Intell. Neurosci. **2022**, 1–11 (2022). https://doi.org/10.1155/2022/3574812
22. Hameed, S., Khan, F.I., Hameed, B.: Understanding security requirements and challenges in internet of things (IoT): a review. J. Comput. Netw. Commun. **2019**, 1–14 (2019). https://doi.org/10.1155/2019/9629381
23. Chinnasamy, P.: Secure smart green house farming using blockchain technology. TURCOMAT. **12**, 2858–2865 (2021). https://doi.org/10.17762/turcomat.v12i6.5793.
24. Alraja, M.N., Farooque, M.M.J., Khashab, B.: The effect of security, privacy, familiarity, and trust on users' attitudes toward the use of the IoT-based healthcare: the mediation role of risk perception. IEEE Access. **7**, 111341–111354 (2019). https://doi.org/10.1109/ACCESS.2019.2904006
25. Romero-Rodríguez, J.-M., Alonso-García, S., Marín-Marín, J.-A., Gómez-García, G.: Considerations on the implications of the internet of things in Spanish universities: the usefulness perceived by professors. Future Internet. **12**, 123 (2020). https://doi.org/10.3390/fi12080123
26. Nemec Zlatolas, L., Feher, N., Hölbl, M.: Security perception of IoT devices in smart homes. JCP. **2**, 65–73 (2022). https://doi.org/10.3390/jcp2010005
27. Mosenia, A., Jha, N.K.: A comprehensive study of security of internet-of-things. IEEE Trans. Emerg. Topics Comput. **5**, 586–602 (2017). https://doi.org/10.1109/TETC.2016.2606384
28. Zheng, S., Apthorpe, N., Chetty, M., Feamster, N.: User perceptions of smart home IoT privacy. Proc. ACM Hum.-Comput. Interact. **2**, 1–20 (2018). https://doi.org/10.1145/3274469
29. McDermott, C.D., Isaacs, J.P., Petrovski, A.V.: Evaluating awareness and perception of botnet activity within consumer internet-of-things (IoT) networks. Informatics. **6**, 8 (2019). https://doi.org/10.3390/informatics6010008
30. Koohang, A., Sargent, C.S., Nord, J.H., Paliszkiewicz, J.: Internet of things (IoT): from awareness to continued use. Int. J. Inf. Manage. **62**, 102442 (2022). https://doi.org/10.1016/j.ijinfomgt.2021.102442
31. Kim, A., Moravec, P.L., Dennis, A.R.: Combating fake news on social media with source ratings: the effects of user and expert reputation ratings. J. Manag. Inf. Syst. **36**, 931–968 (2019). https://doi.org/10.1080/07421222.2019.1628921
32. Bhagat, S., Kim, D.J.: Examining users' news sharing behaviour on social media: role of perception of online civic engagement and dual social influences. Behav. Inf. Technol. **42**, 1194–1215 (2023). https://doi.org/10.1080/0144929X.2022.2066019
33. Kümpel, A.S., Karnowski, V., Keyling, T.: News sharing in social media: a review of current research on news sharing users, content, and networks. Soc. Media Soc. **1**, 2056305115610141 (2015). https://doi.org/10.1177/2056305115610141
34. Thompson, N., Wang, X., Daya, P.: Determinants of news sharing behavior on social media. J. Comput. Inf. Syst. **60**, 593–601 (2020). https://doi.org/10.1080/08874417.2019.1566803
35. Moghavvemi, S., Sharabati, M., Paramanathan, T., Rahin, N.M.: The impact of perceived enjoyment, perceived reciprocal benefits and knowledge power on students' knowledge sharing through Facebook. Int. J. Manag. Educ. **15**, 1–12 (2017). https://doi.org/10.1016/j.ijme.2016.11.002
36. Karnowski, V., Leonhard, L., Kümpel, A.S.: Why users share the news: a theory of reasoned action-based study on the antecedents of news-sharing behavior. Commun. Res. Rep. **35**, 91–100 (2018). https://doi.org/10.1080/08824096.2017.1379984

37. Apuke, O.D., Omar, B.: Fake news and COVID-19: modelling the predictors of fake news sharing among social media users. Telemat. Inform. **56**, 101475 (2021). https://doi.org/10.1016/j.tele.2020.101475
38. Smith, B.G., Gallicano, T.D.: Terms of engagement: analyzing public engagement with organizations through social media. Comput. Hum. Behav. **53**, 82–90 (2015). https://doi.org/10.1016/j.chb.2015.05.060
39. Tang, T. (Ya), Fang, E. (Er), Wang, F.: Is neutral really neutral? The effects of neutral user-generated content on product sales. J. Market. **78**, 41–58 (2014). https://doi.org/10.1509/jm.13.0301
40. Aoun, S.B., Jerbi, H.: Use of social media networks in learning and its impact on students' academic skills: case study of the computer science college at UOH. Int. J. Adv. Appl. Sci. **4**, 154–159 (2017). https://doi.org/10.21833/ijaas.2017.08.022
41. Gross, R., Acquisti, A.: Information revelation and privacy in online social networks. In: Proceedings of the 2005 ACM Workshop on Privacy in the Electronic Society. pp. 71–80. ACM, Alexandria VA USA (2005). https://doi.org/10.1145/1102199.1102214
42. Acquisti, A., Gross, R.: Imagined communities: awareness, information sharing, and privacy on the facebook. In: Danezis, G., Golle, P. (eds.) Privacy Enhancing Technologies. pp. 36–58. Springer Berlin Heidelberg, Berlin, Heidelberg (2006). https://doi.org/10.1007/11957454_3
43. Abdelhamid, M., Gaia, J., Sanders, G.L.: Putting the focus back on the patient: how privacy concerns affect personal health information sharing intentions. J. Med. Internet Res. **19**, e169 (2017). https://doi.org/10.2196/jmir.6877
44. Chai, S.: Does cultural difference matter on social media? An examination of the ethical culture and information privacy concerns. Sustainability. **12**, 8286 (2020). https://doi.org/10.3390/su12198286
45. Kim, J.: Protecting metadata of access indicator and region of interests for image files. Secur. Commun. Netw. **2020**, 1–10 (2020). https://doi.org/10.1155/2020/4836109
46. Bright, L.F., Logan, K.: Is my fear of missing out (FOMO) causing fatigue? Advertising, social media fatigue, and the implications for consumers and brands. INTR. **28**, 1213–1227 (2018). https://doi.org/10.1108/IntR-03-2017-0112
47. Baker-Eveleth, L., Stone, R., Eveleth, D.: Understanding social media users' privacy-protection behaviors. ICS. **30**, 324–345 (2022). https://doi.org/10.1108/ICS-07-2021-0099
48. Bright, L.F., Lim, H.S., Logan, K.: "Should I post or ghost?": examining how privacy concerns impact social media engagement in US consumers. Psychol. Mark. **38**, 1712–1722 (2021). https://doi.org/10.1002/mar.21499
49. Mitchell, D., El-Gayar, O.: The effect of privacy policies on information sharing behavior on social networks: a systematic literature review. Presented at the Hawaii International Conference on System Sciences (2020). https://doi.org/10.24251/HICSS.2020.517
50. Cain, J.A., Imre, I.: Everybody wants some: collection and control of personal information, privacy concerns, and social media use. New Media Soc. **24**, 2705–2724 (2022). https://doi.org/10.1177/14614448211000327
51. Culnan, M.J., Armstrong, P.K.: Information privacy concerns, procedural fairness, and impersonal trust: an empirical investigation. Organ. Sci. **10**, 104–115 (1999). https://doi.org/10.1287/orsc.10.1.104
52. Shojaifar, A., Fricker, S.A.: SMEs' confidentiality concerns for security information sharing. In: Clarke, N. and Furnell, S. (eds.) Human Aspects of Information Security and Assurance. pp. 289–299. Springer International Publishing, Cham (2020). https://doi.org/10.1007/978-3-030-57404-8_22
53. Obar, J.A., Oeldorf-Hirsch, A.: The biggest lie on the internet: ignoring the privacy policies and terms of service policies of social networking services. Inf. Commun. Soc. **23**, 128–147 (2020). https://doi.org/10.1080/1369118X.2018.1486870
54. Yuniar, A.D.: Thin privacy boundaries: proximity and accessibility of E-commerce privacy policy in young consumers of Indonesia. IJSE. **51**, 1182–1194 (2024). https://doi.org/10.1108/IJSE-11-2022-0740

55. Boerman, S.C., Kruikemeier, S., Zuiderveen Borgesius, F.J.: Exploring motivations for online privacy protection behavior: insights from panel data. Commun. Res. **48**, 953–977 (2021). https://doi.org/10.1177/0093650218800915
56. Malik, A., Hiekkanen, K., Dhir, A., Nieminen, M.: Impact of privacy, trust and user activity on intentions to share facebook photos. JICES. **14**, 364–382 (2016). https://doi.org/10.1108/JICES-06-2015-0022
57. Chai, S., Bagchi-Sen, S., Morrell, C., Rao, H.R., Upadhyaya, S.J.: Internet and online information privacy: an exploratory study of preteens and early teens. IEEE Trans. Profess. Commun. **52**, 167–182 (2009). https://doi.org/10.1109/TPC.2009.2017985
58. Putri, N., Prasetya, Y., Handayani, P.W., Fitriani, H.: TikTok shop: how trust and privacy influence generation Z's purchasing behaviors. Cogent Soc. Sci. **10**, 2292759 (2024). https://doi.org/10.1080/23311886.2023.2292759
59. Lin, X., Wang, X.: Examining gender differences in people's information-sharing decisions on social networking sites. Int. J. Inf. Manage. **50**, 45–56 (2020). https://doi.org/10.1016/j.ijinfomgt.2019.05.004
60. Dhami, A., Agarwal, N., Chakraborty, T.K., Singh, B.P., Minj, J.: Impact of trust, security and privacy concerns in social networking: an exploratory study to understand the pattern of information revelation in Facebook. In: 2013 3rd IEEE International Advance Computing Conference (IACC). pp. 465–469. IEEE, Ghaziabad (2013). https://doi.org/10.1109/IAdCC.2013.6514270
61. Oliveira, T., Araujo, B., Tam, C.: Why do people share their travel experiences on social media? Tour. Manage. **78**, 104041 (2020). https://doi.org/10.1016/j.tourman.2019.104041
62. Reich, S., Bachl, M.: Do sexist comments hinder participation in online political discussions?: A preregistered experiment. J. Media Psychol. **35**, 355–361 (2023). https://doi.org/10.1027/1864-1105/a000373
63. Bao, Z., Han, Z.: What drives users' participation in online social Q&A communities? An empirical study based on social cognitive theory. AJIM. **71**, 637–656 (2019). https://doi.org/10.1108/AJIM-01-2019-0002
64. Jin, J., Li, Y., Zhong, X., Zhai, L.: Why users contribute knowledge to online communities: an empirical study of an online social Q&A community. Inf. Manag. **52**, 840–849 (2015). https://doi.org/10.1016/j.im.2015.07.005
65. Negara, H.R.P., Wahyudin, Nurlaelah, E., Herman, T.: Improving students' mathematical reasoning abilities through social cognitive learning using GeoGebra. Int. J. Emerg. Technol. Learn. **17**, 118–135 (2022). https://doi.org/10.3991/ijet.v17i18.32151
66. Miao, B., Liang, Y., Suo, Y.: The influence of organizational social network on enterprise collaborative innovation—mediating role of knowledge sharing and moderating effect of digital construction. IEEE Access. **11**, 5110–5122 (2023). https://doi.org/10.1109/ACCESS.2023.3236398
67. Anderson-Bill, E.S., Winett, R.A., Wojcik, J.R.: Social cognitive determinants of nutrition and physical activity among web-health users enrolling in an online intervention: the influence of social support, self-efficacy, outcome expectations, and self-regulation. J. Med. Internet Res. **13**, e28 (2011). https://doi.org/10.2196/jmir.1551
68. Bandura, A.: Social cognitive theory of self-regulation. Organ. Behav. Hum. Decis. Process. **50**, 248–287 (1991). https://doi.org/10.1016/0749-5978(91)90022-L
69. Herath, T., Yim, M.-S., D'Arcy, J., Nam, K., Rao, H.R.: Examining employee security violations: moral disengagement and its environmental influences. ITP. **31**, 1135–1162 (2018). https://doi.org/10.1108/ITP-10-2017-0322
70. Wang, Y., Lin, Y., Zhong, R.Y., Xu, X.: IoT-enabled cloud-based additive manufacturing platform to support rapid product development. Int. J. Prod. Res. **57**, 3975–3991 (2019). https://doi.org/10.1080/00207543.2018.1516905
71. Chow, W.S., Chan, L.S.: Social network, social trust and shared goals in organizational knowledge sharing. Inf. Manag. **45**, 458–465 (2008). https://doi.org/10.1016/j.im.2008.06.007
72. Sylla, T., Chalouf, M.A., Krief, F., Samaké, K.: Context-aware security in the internet of things: a survey. IJAACS. **14**, 231 (2021). https://doi.org/10.1504/IJAACS.2021.117808

73. Yang, W., Zhang, J., Wang, C., Mo, X.: Situation prediction of large-scale internet of things network security. EURASIP J. Info. Secur. **2019**, 13 (2019). https://doi.org/10.1186/s13635-019-0097-z
74. Nowak, M., Nowak, S., Domanska, J.: Cognitive routing for improvement of IoT security. In: 2019 IEEE International Conference on Fog Computing (ICFC). pp. 41–46. IEEE, Prague, Czech Republic (2019). https://doi.org/10.1109/ICFC.2019.00014
75. Lei, W., Wen, H., Hou, W., Xu, X.: New security state awareness model for IoT devices With edge intelligence. IEEE Access. **9**, 69756–69765 (2021). https://doi.org/10.1109/ACCESS.2021.3075220
76. Alarefi, M.: Adoption of IoT by telecommunication companies in GCC: The role of blockchain. https://doi.org/10.5267/j.dsl. **12**, 55–68 (2023). https://doi.org/10.5267/j.dsl.2022.10.006.
77. Egelman, S., Peer, E.: Scaling the security wall: developing a security behavior intentions scale (SeBIS). In: Proceedings of the 33rd Annual ACM Conference on Human Factors in Computing Systems. pp. 2873–2882. ACM, Seoul Republic of Korea (2015). https://doi.org/10.1145/2702123.2702249
78. McCormac, A., Calic, D., Butavicius, M., Parsons, K., Zwaans, T., Pattinson, M.: A reliable measure of information security awareness and the identification of bias in responses. AJIS. **21** (2017). https://doi.org/10.3127/ajis.v21i0.1697
79. Campbell, S.L., Ronfeldt, M.: Observational evaluation of teachers: measuring more than we bargained for? Am. Educ. Res. J. **55**, 1233–1267 (2018). https://doi.org/10.3102/0002831218776216
80. Iroegbu, M.N.: Self efficacy and work performance: a theoretical framework of albert bandura's model, review of findings, implications and directions for future research. Psychol. Behav. Sci. **4**, 170 (2015). https://doi.org/10.11648/j.pbs.20150404.15
81. Chen, C.-J., Yeh, M.-C., Tang, F.-I., Yu, S.: The smoking outcome expectation scale and anti-smoking self-efficacy scale for early adolescents: instrument development and validation. J. Sch. Nurs. **31**, 363–373 (2015). https://doi.org/10.1177/1059840514560352
82. Resnick, B.: Reliability and validity of the outcome expectations for exercise scale-2. J. Aging Phys. Act. **13**, 382–394 (2005). https://doi.org/10.1123/japa.13.4.382
83. Compeau, D.R., Higgins, C.A.: Computer self-efficacy: development of a measure and initial test. MIS Q. **19**, 189 (1995). https://doi.org/10.2307/249688
84. Torun, E.D.: Educational use of social media in higher education: gender and social networking sites as the predictors of consuming, creating, and sharing content. Acta Educ Gen. **10**, 112–132 (2020). https://doi.org/10.2478/atd-2020-0013
85. Hair, J.F., Babin, B.J., Krey, N.: Covariance-based structural equation modeling in the journal of advertising: review and recommendations. J. Advert. **46**, 163–177 (2017). https://doi.org/10.1080/00913367.2017.1281777
86. Podsakoff, P.M., MacKenzie, S.B., Lee, J.-Y., Podsakoff, N.P.: Common method biases in behavioral research: a critical review of the literature and recommended remedies. J. Appl. Psychol. **88**, 879–903 (2003). https://doi.org/10.1037/0021-9010.88.5.879
87. Malhotra, N.K., Kim, S.S., Patil, A.: Common method variance in IS research: a comparison of alternative approaches and a reanalysis of past research. Manage. Sci. **52**, 1865–1883 (2006). https://doi.org/10.1287/mnsc.1060.0597
88. George, D., Mallery, P.: IBM SPSS statistics 26 step by step: a simple guide and reference. Routledge (2019). https://doi.org/10.4324/9780429056765
89. Hair, J.F. ed: Multivariate data analysis. Pearson Prentice Hall, Upper Saddle River, NJ (2010)
90. Fornell, C., Larcker, D.F.: Evaluating structural equation models with unobservable variables and measurement error. J. Mark. Res. **18**, 39–50 (1981). https://doi.org/10.1177/002224378101800104
91. Hu, L.T., Bentler, P.M.: Cutoff criteria for fit indexes in covariance structure analysis: conventional criteria versus new alternatives. Struct. Equ. Model. **6**, 1–55 (1999). https://doi.org/10.1080/10705519909540118

IoT-Enabled Multimedia Applications in Healthcare, Telemedicine, and Imaging

Tarun Kumar Vashishth, Vikas Sharma, Kewal Krishan Sharma, Bhupendra Kumar, Sachin Chaudhary, and Sanyam Agarwal

Abstract The fusion of Internet of Things (IoT) technology with multimedia applications has significantly impacted the healthcare sector, revolutionizing patient care, telemedicine, and medical imaging. This research explores the diverse applications and transformative potential of IoT-enabled multimedia in healthcare settings. From remote patient monitoring to telemedicine consultations, the integration of multimedia data such as images, videos, and audio has facilitated real-time communication and decision-making. The study delves into the challenges, opportunities, and advancements in leveraging IoT-enabled multimedia for improved diagnostics, treatment, and overall healthcare delivery. By analyzing key use cases and technological considerations, this research aims to provide insights into the evolving landscape of healthcare applications within the IoT and multimedia convergence.

Keywords IoT · Multimedia applications · Healthcare · Telemedicine · Medical imaging · Remote patient monitoring · Real-time communication · Diagnostics · Treatment · Internet of Things · Healthcare technology

1 Introduction

The integration of the Internet of Things (IoT) in healthcare is reshaping the way medical services are delivered, making them more accessible, efficient, and data-driven. IoT refers to the interconnected network of devices, sensors, and systems that collect, process, and share data seamlessly, enabling real-time monitoring, analysis, and decision-making. Within the healthcare sector, IoT has evolved to support applications ranging from remote patient monitoring (RPM) and telemedicine to smart hospital management and personalized medicine. By connecting patients, healthcare

T. K. Vashishth (✉) · V. Sharma · K. K. Sharma · B. Kumar · S. Chaudhary
School of Computer Science and Applications, IIMT University, Meerut, U.P, India
e-mail: tarunvashishth@gmail.com

S. Agarwal
ACE College of Engineering and Management, Agra, U.P, India

R. Sharma et al. (eds.), *Multimedia Technologies in the Internet of Things Environment, Volume 4*, Studies in Big Data 173, https://doi.org/10.1007/978-981-96-4356-1_7

providers, and medical devices, IoT enhances the quality of care, reduces costs, and facilitates preventive healthcare practices [1].

Multimedia technologies, including audio, video, and imaging, play a pivotal role in this transformation. The integration of multimedia with IoT allows for richer and more interactive healthcare experiences. For instance, video-enabled telemedicine platforms enable remote consultations, while IoT-connected imaging devices enhance diagnostic accuracy and speed by transmitting high-quality medical images in real-time [2]. The seamless fusion of IoT and multimedia technologies ensures that critical health information is captured, transmitted, and analyzed efficiently, empowering clinicians to make informed decisions with greater confidence.

1.1 Overview of IoT in Healthcare

Definition and Scope of IoT in Healthcare

The Internet of Things (IoT) in healthcare refers to the interconnected network of physical devices, sensors, wearables, and systems that collect, exchange, and analyze data to enhance patient care, improve clinical workflows, and optimize operational efficiencies. These IoT-enabled devices are capable of monitoring a patient's health in real time, transmitting vital data (such as heart rate, blood pressure, oxygen levels, and glucose levels) to healthcare providers, and alerting them in case of abnormal readings or emergencies. The scope of IoT in healthcare is vast, encompassing areas such as remote patient monitoring (RPM), telemedicine, medical device integration, chronic disease management, smart hospitals, and personalized medicine [3]. IoT applications also extend to hospital asset management, staff coordination, and improving hospital efficiency by tracking inventory and equipment usage. By connecting patients, healthcare providers, and healthcare systems, IoT facilitates seamless communication and enables continuous patient care, making healthcare more proactive, patient-centric, and efficient.

Evolution of IoT Technologies in the Medical Field

The evolution of IoT in healthcare has been significantly shaped by advancements in sensor technologies, communication systems, and data analytics. In the early stages, IoT in healthcare was limited to basic monitoring devices that could track physiological parameters like heart rate, body temperature, and blood pressure. Over time, the scope expanded with the introduction of more sophisticated sensors capable of measuring a wide array of biomarkers and transmitting these data wirelessly to healthcare systems for analysis [4]. As wireless communication technologies, such as Bluetooth and Wi-Fi, matured, IoT devices became more capable of transmitting data over longer distances, enabling remote patient monitoring. The evolution was further accelerated by cloud computing, which allowed vast amounts of healthcare data to be stored, analyzed, and shared in real time. The introduction of advanced technologies such as artificial intelligence (AI) and machine learning (ML) enabled

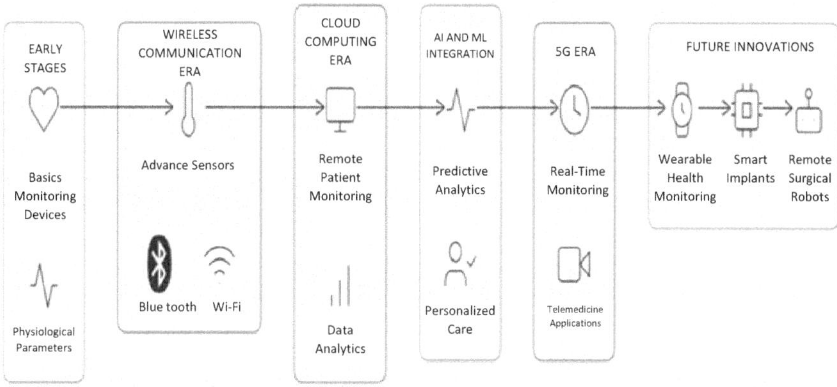

Fig. 1 IoT evolution in healthcare

predictive analytics, making it possible to detect early signs of diseases and provide personalized care [5]. The arrival of 5G technology further enhanced the capabilities of IoT, improving the speed, reliability, and capacity of data transmission, which is crucial for real-time monitoring and telemedicine applications. Today, IoT technologies continue to advance, with new innovations such as wearable health monitors, smart implants, and remote surgical robots, shaping the future of healthcare delivery (Fig. 1).

1.2 Multimedia Integration in Healthcare

The Role of Multimedia (Audio, Video, and Imaging) in Enhancing Healthcare Delivery

Multimedia technologies—comprising audio, video, and imaging—have become essential in transforming the way healthcare is delivered. These technologies facilitate enhanced communication, diagnosis, and treatment, particularly in the context of telemedicine, remote consultations, and medical imaging [6].

Audio: In healthcare, audio plays a crucial role in enabling real-time communication between healthcare providers and patients, especially in telemedicine consultations. Audio-based technologies facilitate the exchange of medical information, such as symptoms or instructions, between doctors and patients, enhancing the virtual care experience. Furthermore, audio is often combined with voice recognition systems to document patient interactions or monitor patients' verbal responses in real time.

Video: Video technologies are central to telemedicine, allowing healthcare professionals to conduct virtual consultations with patients, irrespective of geographic barriers. Video conferencing enables doctors to visually assess symptoms, interact with patients in real time, and provide immediate feedback or prescriptions. Additionally, video is used in surgical procedures for remote guidance, as well as in patient monitoring through surveillance systems in critical care units [7].

Imaging: Medical imaging technologies, including X-rays, CT scans, MRI, ultrasound, and endoscopic imaging, are essential in diagnosing various conditions. The integration of IoT with imaging technologies allows healthcare providers to access high-resolution images remotely, enhancing diagnostic accuracy and facilitating timely interventions. IoT-enabled imaging devices can capture, store, and share images in real time, allowing radiologists and specialists to make decisions quickly and efficiently.

1.3 Importance of IoT in Enabling Multimedia Applications

IoT plays a pivotal role in enabling and enhancing multimedia applications in healthcare by ensuring seamless integration of diverse data sources and real-time communication between patients, healthcare providers, and medical systems. By embedding sensors and connectivity into multimedia devices, IoT enables real-time data collection, transmission, and analysis, facilitating a more holistic approach to patient care. For example, IoT-enabled imaging devices can automatically transmit diagnostic images to radiologists or specialists, allowing for faster interpretations and reducing the time it takes to make treatment decisions. Similarly, IoT-based telemedicine platforms integrate video and audio technologies with patient data collected through wearable devices, enabling comprehensive virtual consultations where clinicians can evaluate patients based on both physical symptoms and real-time health metrics (such as heart rate, oxygen saturation, or glucose levels). IoT also enables the synchronization of multimedia data across different healthcare systems and devices. For instance, a patient's medical history, images, videos, and vital signs can be integrated into a single electronic health record (EHR), which can be accessed and reviewed by multiple healthcare providers. This interconnectedness enhances diagnostic accuracy, streamlines decision-making, and ensures that patients receive coordinated care. Ultimately, the combination of IoT and multimedia technologies is critical to improving healthcare delivery, expanding access to care, and providing patients with personalized, timely, and effective treatment. By integrating these technologies, healthcare systems can move towards more proactive and preventative care models, ultimately leading to improved health outcomes and greater efficiency in the healthcare industry [8].

2 IoT-Enabled Multimedia Applications in Healthcare

2.1 Remote Patient Monitoring (RPM)

Remote Patient Monitoring (RPM) is one of the most transformative applications of IoT in healthcare. IoT devices, such as wearable health trackers, sensors, and smart devices, enable continuous, real-time monitoring of patients' vital signs, such as heart rate, blood pressure, oxygen saturation, glucose levels, and body temperature. These devices collect data and transmit it wirelessly to healthcare providers, allowing clinicians to monitor the patient's health status remotely without requiring physical visits. For instance, wearable ECG monitors can detect arrhythmias, glucose sensors can monitor blood sugar levels in diabetic patients, and smart thermometers can alert providers to fever or infection. RPM enables early detection of potential health issues, facilitates proactive care, and allows patients to manage chronic conditions without needing frequent in-person visits, reducing hospital readmissions and improving patient outcomes. Multimedia technologies, including video, audio, and imaging, complement IoT-enabled RPM by enhancing how data is transmitted and analyzed. For example, video conferencing systems can be integrated into RPM platforms to provide patients with virtual consultations, where they can discuss their symptoms and share their health status through video. Additionally, multimedia data, such as audio recordings of patient feedback or videos of their physical symptoms, can be used by healthcare providers to supplement real-time health metrics gathered through IoT devices. Furthermore, IoT devices can send images, such as digital photographs of wounds or skin conditions, to healthcare providers for evaluation. AI-powered platforms can analyze this multimedia data alongside real-time vital signs, providing healthcare providers with actionable insights for more accurate diagnoses and treatment plans. By integrating multimedia with IoT in RPM, healthcare professionals can gain a holistic view of a patient's condition, improving decision-making and patient care.

2.2 Telemedicine and Virtual Consultations

Telemedicine refers to the practice of delivering healthcare services remotely using technology, and it has seen rapid growth, especially with advancements in IoT and multimedia [9]. It enables healthcare providers to diagnose, treat, and monitor patients without the need for physical presence, reducing barriers such as geographic distance, transportation, and time constraints.

Integration of Video and Audio Technologies for Remote Consultations

One of the key components of telemedicine is the integration of video and audio technologies, which are used for virtual consultations between patients and healthcare providers. Video technology allows physicians to visually assess patients, examine symptoms, and make decisions based on real-time visual feedback. For instance, video can be used to observe a patient's skin condition, wound healing, or signs of distress, which aids in accurate diagnoses. It also enables face-to-face consultations where doctors can engage in detailed conversations with patients, asking questions, guiding them through their symptoms, and recommending treatment options. Along with video, audio communication allows for interactive dialogue between the patient and healthcare professional. Through clear audio systems, doctors can listen to patients' concerns, ask follow-up questions, and assess their condition in real time. This two-way communication (video and audio) allows for a dynamic interaction that closely mimics an in-person visit, ensuring the exchange of crucial health information and reducing the chances of miscommunication. Moreover, in telemedicine platforms, video and audio systems are sometimes paired with IoT-enabled health devices that provide real-time medical data [10]. For example, a patient might use a connected blood pressure cuff during a telemedicine session, and the doctor can view the results instantly through the video consultation, helping the physician to make informed decisions on the spot. This seamless integration of multimedia data with remote communication enhances the quality of care, improves patient engagement, and allows for quicker diagnoses, even from a distance.

IoT-Enabled Platforms for Seamless Doctor-Patient Interactions

The integration of IoT in telemedicine is transforming how virtual consultations take place by enabling real-time monitoring and automated data sharing between patients and healthcare providers [11]. IoT devices, such as wearables and home-based medical equipment, collect continuous health data, including vital signs, and send this information to telemedicine platforms. This allows physicians to monitor patients remotely and make data-driven decisions during virtual consultations. For example, an IoT-enabled smartwatch might measure a patient's heart rate, oxygen saturation, and physical activity levels. These data can be sent to the telemedicine system, where the doctor can view the information in real time during a virtual consultation. Similarly, glucose monitors can provide continuous glucose readings for diabetic patients, which are immediately accessible to doctors in a telemedicine session. Furthermore, IoT platforms enable the seamless exchange of data and multimedia during consultations. A patient might share images or videos of their symptoms, such as rashes or wounds, through the telemedicine system, and the doctor can assess these images while interacting with the patient through video or audio. Some advanced telemedicine systems also incorporate AI to analyze the patient's health data, providing preliminary assessments or recommendations to assist healthcare providers in their decision-making process. By streamlining data collection, analysis, and communication, IoT-enabled platforms facilitate efficient, effective, and personalized doctor-patient interactions. These platforms offer real-time connectivity, reduced wait times, and increased access to care for patients, especially in remote or underserved areas, improving the overall healthcare experience for both parties (Fig. 2).

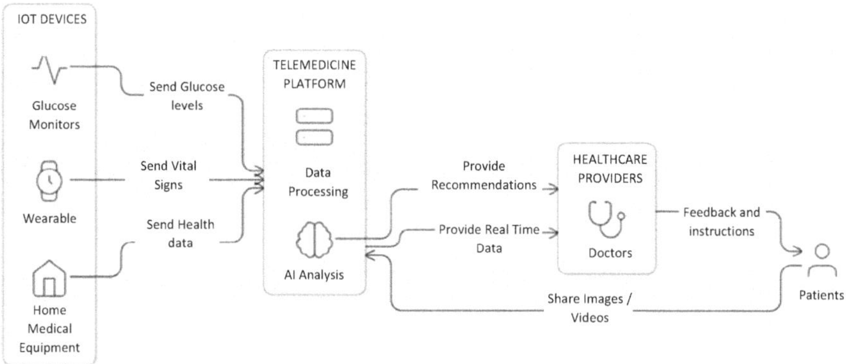

Fig. 2 IoT platforms for seamless doctor-patient interactions

2.3 *Smart Hospitals and Connected Devices*

Smart hospitals, powered by IoT technologies and connected devices, are transforming healthcare by optimizing workflows, enhancing communication, and improving patient care. IoT enables automation in various hospital operations, such as tracking patients, medical equipment, and supplies, ensuring efficient resource utilization. Real-time monitoring of vital signs through IoT devices allows healthcare providers to respond promptly to changes in a patient's condition, improving patient safety. Additionally, multimedia systems such as video conferencing and instant messaging facilitate seamless communication among medical staff, enabling faster decision-making. IoT-enabled platforms also support the real-time transmission of medical imaging, allowing specialists to collaborate and make informed diagnoses without delays. Moreover, these technologies enhance patient engagement by providing access to health information, entertainment, and educational content, improving the overall patient experience. The integration of IoT and multimedia technologies in smart hospitals leads to more efficient operations, better communication, and improved patient outcomes [12].

3 IoT in Medical Imaging and Diagnostics

3.1 *Real-Time Imaging and Data Sharing*

IoT in medical imaging is transforming diagnostic processes by enabling real-time imaging and data sharing, significantly improving both the speed and accuracy of patient care. IoT-enabled imaging devices, such as digital X-rays, MRI machines, and CT scanners, are equipped with sensors that can capture high-resolution images of internal body structures. These devices are connected to healthcare networks through IoT technologies, allowing for the immediate transmission of the acquired images to doctors, specialists, or radiologists, regardless of their location. The ability to share medical images in real time has dramatically reduced delays associated with manual image development and physical transportation of films or disks, which were common in traditional imaging systems. With IoT, healthcare professionals can access and review images within moments of their capture, speeding up the decision-making process, particularly in emergency situations. Furthermore, the integration of IoT enables these medical images to be stored in centralized cloud-based systems, making them easily accessible to authorized medical personnel at any time [13]. This cloud storage solution not only improves accessibility but also facilitates better collaboration across medical teams, as they can share and discuss images instantly. For example, a patient undergoing an MRI scan in one hospital can have their results instantly shared with a specialist at another healthcare facility, allowing for quicker consultations, second opinions, and referrals without the need for the patient to travel. This real-time sharing of images enhances diagnostic accuracy by enabling medical professionals to collaborate more effectively, ensuring that no important detail goes unnoticed. Moreover, the integration of multimedia formats, such as digital images, videos, and even 3D imaging, into medical diagnostics enhances the richness of the data being shared. For instance, radiologists can use 3D imaging techniques to provide more detailed views of complex anatomical structures, which may be essential for diagnosing intricate conditions such as tumors or vascular blockages. The use of these advanced formats in tandem with IoT devices allows for a deeper, more comprehensive understanding of the patient's condition. As a result, the healthcare system becomes more agile, responsive, and patient-centered, providing faster diagnoses and enabling immediate treatment plans to be implemented.

3.2 Advanced Analytics with Multimedia Data

The integration of Artificial Intelligence (AI) with Internet of Things (IoT)-enabled medical imaging systems is revolutionizing the field of diagnostics by significantly enhancing the analysis of image and video data. By combining the real-time capabilities of IoT with the advanced processing power of AI, healthcare providers can analyze complex medical images with unparalleled precision and speed. AI algorithms, particularly those based on machine learning and deep learning, are able to automatically detect, classify, and interpret various patterns in medical images that may be difficult for human eyes to spot. For example, AI can detect early-stage cancers, brain anomalies, or heart conditions by analyzing thousands of images in real time, reducing the chances of diagnostic errors and providing more accurate results. These AI-driven systems can even learn from vast datasets, improving their accuracy over time and helping healthcare providers make more informed clinical decisions. AI and IoT integration extend to video data as well, where video feeds from endoscopes, laparoscopes, or ultrasound devices can be analyzed to identify abnormalities or track the progression of diseases in real-time. For instance, during minimally invasive surgery, IoT-enabled cameras can capture high-resolution video footage, which is then analyzed by AI algorithms to provide real-time feedback to the surgeon. This dynamic analysis not only helps in accurate diagnosis but also assists in guiding surgeries, reducing complications, and enhancing the overall quality of care. In addition to enhancing the accuracy and efficiency of image and video analysis, predictive analytics powered by IoT-enabled imaging tools is becoming a key part of healthcare diagnostics. Predictive analytics uses historical and real-time data to identify patterns that can foresee future health conditions. For example, AI algorithms can analyze medical images alongside patient data such as medical history, lifestyle factors, and genetic information to predict the likelihood of a disease developing in the future. IoT-enabled imaging devices, integrated with sensors and monitoring equipment, continuously collect and transmit patient data, providing healthcare professionals with up-to-date information. When combined with predictive analytics, these data allow for early diagnosis of conditions before symptoms become severe, enabling proactive treatment plans. For instance, AI algorithms integrated into IoT-enabled mammography devices can assess changes in breast tissue over time, predicting the risk of breast cancer before it becomes clinically apparent. Similarly, predictive analytics tools using medical imaging data can help detect conditions like stroke or heart disease by identifying early biomarkers or signs of the condition. Early detection is critical in improving patient outcomes, as it allows for the implementation of preventive measures, reduces the need for invasive treatments, and ultimately saves lives.

4 Technological Framework and Architecture

4.1 IoT Infrastructure for Multimedia Applications

The IoT infrastructure required for multimedia applications in healthcare involves a comprehensive system composed of various hardware, software, communication protocols, and data standards to ensure seamless operation and real-time functionality [14]. The architecture must be robust, scalable, and secure to handle the vast amounts of multimedia data, such as images, videos, and audio, generated from IoT-enabled medical devices and sensors.

Hardware and Software Requirements: The hardware required for IoT-enabled multimedia applications includes a range of devices such as sensors, medical imaging devices, cameras, wearables, and mobile devices. These devices are equipped with high-resolution imaging capabilities, real-time data capture features, and sensors capable of collecting patient health metrics like heart rate, oxygen levels, and temperature. For instance, IoT-enabled medical imaging devices like digital X-ray machines, MRI scanners, or ultrasound equipment are critical components in capturing detailed medical images. Additionally, patient monitoring systems (e.g., wearable ECG or blood pressure monitors) collect physiological data that can be integrated with multimedia applications [15]. To support these hardware devices, a powerful and efficient software infrastructure is necessary to process, analyze, store, and transmit the data. Software components include data management systems, cloud storage platforms, image and video processing software, AI-based analytics tools, and security software to ensure the confidentiality and integrity of patient data [16]. The cloud-based storage system ensures that large volumes of medical data, such as high-resolution images and videos, can be stored, retrieved, and shared securely across healthcare institutions. Software platforms need to have capabilities for real-time streaming, advanced image processing (e.g., for enhancing and analyzing medical images), and integration with electronic health records (EHR) systems to provide a holistic view of patient data [17].

Communication Protocols and Data Standards: The success of IoT-enabled multimedia applications heavily depends on the communication protocols and data standards that enable efficient and secure exchange of information between devices, cloud platforms, and end-users (e.g., healthcare professionals, patients) [18]. Several communication protocols are commonly used in IoT systems to ensure that multimedia data can be transmitted reliably and securely:

MQTT (Message Queuing Telemetry Transport): MQTT is a lightweight and efficient messaging protocol widely used in IoT applications. It is particularly useful in scenarios where bandwidth, high latency, low or intermittent connectivity are present, such as remote patient monitoring. It works on a publish/subscribe model, which ensures that messages (e.g., images, video, sensor data) are sent and received with minimal overhead. MQTT is especially effective for transmitting small, frequent updates such as real-time patient data, ensuring that healthcare professionals have access to the most up-to-date information (Fig. 3).

CoAP (Constrained Application Protocol): CoAP is another protocol designed for constrained environments, which is ideal for IoT systems where low power consumption and low data rates are essential. CoAP operates over UDP (User Datagram Protocol) rather than TCP, allowing for faster, more efficient data transfers. It is particularly well-suited for applications that require real-time interaction, such as the transmission of medical sensor data and multimedia information in remote patient monitoring systems.

HTTPS and TLS: For secure communication, HTTPS (Hypertext Transfer Protocol Secure) is often used in conjunction with Transport Layer Security (TLS) protocols to ensure encrypted and secure transmission of multimedia data between devices, healthcare systems, and cloud platforms. This is particularly important in healthcare, where patient privacy and data confidentiality are paramount.

Data Standards: To ensure that the data shared across systems are compatible and can be easily interpreted by various devices and software, certain data standards are employed. Common data formats for medical images include DICOM (Digital Imaging and Communications in Medicine), which is widely used in the healthcare industry for handling, storing, and transmitting medical imaging data like CT scans, MRIs, and X-rays. DICOM ensures that imaging data is exchanged in a standardized

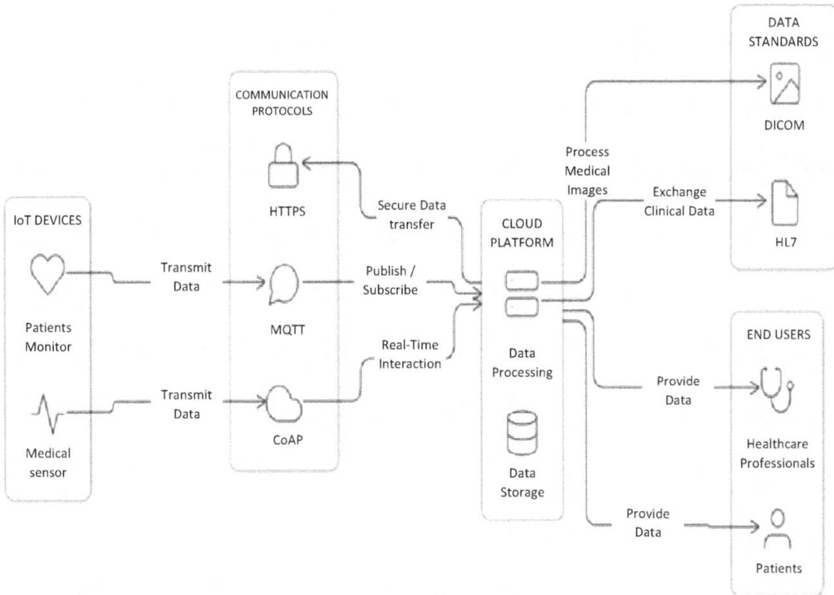

Fig. 3 IoT communication protocols and data standards

manner, allowing different devices and software systems to interpret and process it uniformly. Other standards like HL7 (Health Level 7) are used for exchanging clinical data, ensuring interoperability between different healthcare information systems such as EHRs.

The use of these communication protocols and data standards ensures seamless integration of diverse IoT devices and multimedia applications in healthcare settings. It allows real-time data from medical devices and sensors to be transmitted efficiently and securely, enabling better decision-making, remote monitoring, and enhanced patient care. Proper integration of these components into a cohesive IoT infrastructure forms the backbone of advanced healthcare systems that leverage IoT and multimedia technologies to improve diagnostic processes, treatment plans, and overall patient outcomes.

4.2 Data Integration and Interoperability

In healthcare systems, the integration and interoperability of data are essential to the success of IoT-enabled multimedia applications. With the proliferation of IoT devices in medical environments—such as wearables, imaging equipment, and remote patient monitoring systems—data are generated from multiple sources. These diverse data sets, including real-time vital signs, medical images, and video consultations, must be integrated and shared seamlessly across various systems and platforms to ensure a comprehensive and accurate representation of patient health. Data integration refers to the process of combining these disparate data from different sources, whereas interoperability refers to the ability of different systems, applications, and devices to communicate and exchange information effectively. The integration process involves several steps, such as data aggregation, normalization, and synchronization. IoT-enabled devices often collect data in various formats, including sensor data (e.g., heart rate, blood pressure), medical images (e.g., X-rays, MRIs), and video (e.g., telemedicine consultations). These data must be processed to ensure consistency across different platforms, as healthcare institutions may use varied software solutions. A key challenge in integrating such data is the need to standardize formats across the devices. For example, the integration of imaging data requires systems that support DICOM (Digital Imaging and Communications in Medicine) standards, while sensor data may require HL7 (Health Level 7) or FHIR (Fast Healthcare Interoperability Resources) standards for electronic health records (EHRs). Data integration platforms or middleware are typically used to combine these different data types and store them in centralized or cloud-based systems, where they can be accessed by authorized healthcare providers. Interoperability ensures that different systems can work together and communicate effectively. This is crucial in healthcare, where a variety of devices, databases, and software applications are used across various institutions, such as hospitals, clinics, laboratories, and insurance providers. For IoT-enabled healthcare systems, interoperability often involves using open data standards and application programming interfaces (APIs) to ensure smooth communication between devices and systems. APIs allow different applications to request and share data without needing to understand the underlying technology of the other system. Additionally, cloud platforms with built-in interoperability features are increasingly used to facilitate communication between IoT devices and the various systems they connect to. Achieving true interoperability is critical for enabling coordinated care. For instance, a patient's data from remote monitoring devices can be integrated with their medical history from an EHR, allowing healthcare providers to have a holistic view of the patient's health. This integration leads to better-informed decision-making, reduces the chances of errors, and enables more accurate diagnoses, treatments, and monitoring. Furthermore, interoperability supports continuity of care, as patient data can be shared between various healthcare providers, regardless of the specific devices or systems in use, facilitating collaborative efforts across healthcare teams.

4.3 Cybersecurity and Privacy Concerns

As IoT-enabled multimedia applications in healthcare continue to evolve, ensuring robust cybersecurity and addressing privacy concerns become critical priorities. IoT devices, which collect and transmit sensitive medical data, are vulnerable to a wide range of cybersecurity threats, including unauthorized access, data breaches, and cyberattacks. Given the sensitive nature of healthcare data, including personal health information (PHI), financial records, and medical images, the protection of these data is paramount to maintaining patient trust and complying with legal and regulatory standards, such as the Health Insurance Portability and Accountability Act (HIPAA) in the United States or the General Data Protection Regulation (GDPR) in the European Union.

Cybersecurity Risks: One of the key challenges is the security of IoT devices themselves. Many IoT devices in healthcare, including wearables, smart thermometers, and imaging equipment, are often deployed with limited security features due to their resource constraints (e.g., processing power, battery life, and memory). These vulnerabilities can be exploited by cybercriminals to gain unauthorized access to devices, manipulate data, or even disrupt medical procedures. For example, a hacker could potentially manipulate the data from a patient's heart rate monitor, leading to inaccurate readings and potentially life-threatening consequences. Furthermore, connected medical devices that communicate through wireless networks are susceptible to data interception or man-in-the-middle attacks, where attackers can access or alter data being transmitted between devices and systems.

Privacy Concerns: Another major concern is the privacy of patient data. IoT-enabled healthcare systems collect large volumes of personal health data, which include sensitive information such as diagnostic images, video consultations, and real-time physiological measurements. These data, when not adequately protected, can be exposed to unauthorized third parties. Given that healthcare data is often targeted by hackers due to its high value on the black market, ensuring privacy is a top priority. Unauthorized sharing or access to these data could lead to identity theft, financial fraud, or even discrimination based on health conditions.

Security Measures: To address these risks, healthcare organizations must implement comprehensive security measures that safeguard patient data at every stage of its lifecycle—from collection and transmission to storage and access. One of the fundamental practices is encryption, which ensures that data transmitted between IoT devices, servers, and healthcare applications are unreadable to unauthorized individuals. Advanced encryption protocols, such as Secure Socket Layer (SSL) and Transport Layer Security (TLS), are commonly used to encrypt data during transmission, while stronger encryption algorithms are used to protect data stored in databases. In addition to encryption, strong authentication mechanisms, such as multi-factor authentication (MFA) and biometrics, are essential for verifying the identity of users and ensuring that only authorized individuals can access patient data. Regular software updates and patches must be applied to IoT devices and healthcare systems to address known vulnerabilities and prevent attacks.

Data Anonymization and De-identification: An effective method for enhancing patient privacy in IoT-enabled healthcare systems is th e use of data anonymization and de-identification techniques. These methods remove personally identifiable information (PII) from the datasets, ensuring that even if data are intercepted or accessed without authorization, it cannot be traced back to an individual. Anonymization and de-identification are particularly important when dealing with large datasets used for AI and machine learning models, as they allow organizations to use data for research and analytics without compromising patient privacy [19].

5 Challenges in IoT-Enabled Multimedia Healthcare

5.1 Technological Challenges

The integration of IoT-enabled multimedia applications in healthcare presents numerous technological challenges that need to be addressed to ensure seamless deployment and effective utilization [20]. These challenges span across various domains, including device interoperability, data management, network infrastructure, and integration of emerging technologies.

Device Interoperability and Standardization: One of the primary technological challenges is ensuring interoperability between various IoT devices and healthcare systems. As healthcare institutions deploy a wide range of IoT devices such as wearable sensors, smart beds, diagnostic imaging equipment, and telemedicine platforms, these devices often operate on different hardware, software, and communication protocols [21]. The lack of standardized protocols and data formats can lead to compatibility issues, where devices cannot communicate or exchange data efficiently. For example, medical devices may generate data in proprietary formats, making it difficult to integrate with other systems that rely on different standards like HL7, DICOM, or FHIR. The absence of universally accepted standards hinders the efficient sharing and consolidation of patient data across different systems, potentially delaying diagnosis and treatment. Overcoming this challenge requires developing open standards, implementing data conversion protocols, and investing in middleware solutions that can integrate devices across diverse systems.

Data Overload and Storage: With the continuous stream of data generated by IoT-enabled devices, healthcare systems face the challenge of managing and storing large volumes of information. For instance, real-time monitoring systems that track a patient's vital signs or smart imaging devices producing high-resolution medical images can generate immense data that needs to be stored and processed. The storage and processing of these data can place a strain on healthcare infrastructure, especially in smaller healthcare organizations with limited IT resources. Cloud storage and edge computing can help alleviate some of these issues by offering scalable solutions for data storage and processing closer to the device, reducing latency. However, maintaining the security, privacy, and compliance of these data in the cloud or edge environments introduces further challenges, especially when managing sensitive patient information.

Network Connectivity and Bandwidth: IoT-enabled healthcare applications rely heavily on stable and high-speed network connectivity for the smooth transmission of large data sets, such as medical images or video consultations. The reliance on wireless networks (Wi-Fi, 5G, etc.) can introduce challenges in terms of bandwidth limitations, latency, and reliability. Healthcare facilities, especially those in rural or underserved areas, may face network coverage issues that hinder the effectiveness of IoT solutions. Additionally, real-time medical data, such as video consultations or remote surgeries, require low-latency networks to avoid delays that could jeopardize patient safety. The high demand for bandwidth to transmit multimedia data, along with the need for reliable network infrastructure, presents significant technological hurdles.

Integration of Emerging Technologies (AI, ML, and Blockchain): IoT in healthcare often involves the integration of cutting-edge technologies such as artificial intelligence (AI), machine learning (ML), and blockchain for enhanced functionality. For instance, AI can analyze medical images captured by IoT devices, and ML algorithms can provide predictive analytics for patient care. However, integrating these advanced technologies with existing healthcare infrastructure requires specialized expertise and resources. Developing AI and ML models that work seamlessly with IoT data, ensuring that they are accurate, reliable, and explainable, is a significant challenge. Furthermore, implementing blockchain technology for secure data sharing and transactions adds complexity to the existing systems, as it requires modifications to the underlying architecture and new governance models. Therefore, ensuring smooth integration of emerging technologies while maintaining operational continuity is a major technological challenge.

5.2 *Human and Operational Challenges*

While technological factors are critical to the success of IoT-enabled multimedia applications in healthcare, human and operational challenges are equally important in ensuring these systems are effectively utilized and provide real value. These challenges include the adoption and training of healthcare staff, maintaining patient trust, and addressing operational workflow inefficiencies.

Training and Skill Gaps: One of the main human challenges in adopting IoT-enabled multimedia applications is ensuring that healthcare professionals are adequately trained to use the new technology effectively. Healthcare providers, including doctors, nurses, and technicians, must understand how to operate IoT devices, interpret the data they generate, and respond appropriately. For example, interpreting data from remote patient monitoring devices requires specific knowledge of the device's outputs and how they correlate with patient conditions. Without proper training, healthcare workers may struggle to leverage these technologies, potentially leading to errors in treatment or diagnosis. Similarly, professionals working with medical imaging systems need to be familiar with interpreting IoT-enabled images, which may differ from traditional methods. Addressing these skill gaps through comprehensive training programs and continuing education initiatives is essential for successful implementation.

Resistance to Change and Adoption Challenges: Healthcare organizations, especially large ones, often face resistance to adopting new technologies due to existing workflows, processes, and organizational culture. Staff may be hesitant to adopt IoT-enabled solutions, especially if they perceive them as complex, time-consuming, or unnecessary. Additionally, there may be concerns about the potential for technology to replace human jobs or alter traditional roles. Overcoming this resistance requires effective change management strategies, clear communication about the benefits of IoT integration, and ensuring that the technology complements rather than replaces the human element in healthcare [22]. Involving healthcare professionals in the decision-making process during the implementation phase can also improve buy-in and reduce resistance.

Maintaining Patient Trust and Privacy Concerns: IoT-enabled healthcare applications often involve the continuous collection of sensitive patient data, such as biometric readings, health histories, and medical images. Patients may be hesitant to adopt remote monitoring or telemedicine services due to concerns about the security and privacy of their data. Ensuring robust cybersecurity measures and transparent privacy policies is critical to maintaining patient trust. Additionally, patients may worry about the accuracy of data provided by IoT devices, especially in the case of wearable health trackers or home-based monitoring equipment. Addressing these concerns through clear communication, providing data accuracy guarantees, and allowing patients to control how their data is shared can help mitigate mistrust.

Operational Workflow Integration: Incorporating IoT-enabled multimedia applications into existing healthcare workflows can be complex, as it requires adjustments to operational practices and coordination between multiple departments. For example, real-time monitoring of patient vitals may require the integration of data from various sensors into patient electronic health records (EHRs), which must then be analyzed by clinicians for decision-making. The smooth integration of IoT devices into these workflows requires collaboration across departments, clear protocols for handling data, and streamlined communication channels. Operational inefficiencies may arise if the technology is not well-integrated or if staff are overwhelmed with additional tasks, such as managing multiple devices or interpreting complex data. Optimizing workflows to incorporate these IoT systems requires careful planning, resource allocation, and ongoing support to ensure that the technology enhances productivity without adding unnecessary complexity (Fig. 4).

Regulatory and Compliance Issues: Another operational challenge is ensuring that IoT-enabled multimedia healthcare applications comply with regulatory standards such as HIPAA, GDPR, and other local data protection laws. These regulations govern how patient data is stored, shared, and transmitted, and non-compliance can result in severe legal consequences. Healthcare providers must ensure that IoT devices and platforms adhere to these privacy and security regulations. Additionally, IoT devices themselves must be validated and certified for clinical use, which requires rigorous testing and regulatory approvals. Ensuring compliance across multiple devices, platforms, and applications adds an operational layer of complexity, especially as IoT technology evolves faster than regulatory frameworks.

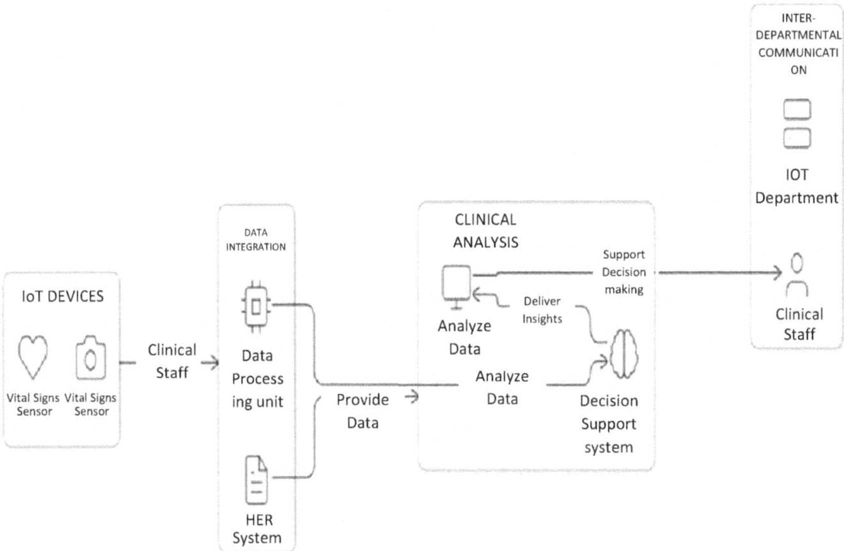

Fig.4 IoT-enabled multimedia operational workflow

6 Future Trends and Opportunities

The integration of Internet of Things (IoT)-enabled multimedia technologies in healthcare has already proven transformative, but the future holds even greater potential. As technology continues to evolve, there will be significant advancements in IoT devices, connectivity, data processing, and artificial intelligence (AI), all of which will continue to reshape the healthcare landscape. The next few years will likely bring about new trends and opportunities that will further improve patient care, operational efficiency, and healthcare outcomes. Below are the key future trends and opportunities in IoT-enabled multimedia applications in healthcare.

6.1 AI and Machine Learning Integration

The fusion of IoT and AI will be one of the defining trends in healthcare's future. As IoT devices continue to collect vast amounts of real-time data, AI and machine learning algorithms will increasingly play a crucial role in processing and analyzing this data. AI can significantly enhance diagnostic accuracy by identifying patterns in medical images and patient vitals, offering predictive insights for early disease detection. Machine learning models trained on large datasets of medical images and real-time patient data can continuously improve their diagnostic capabilities, enabling better clinical decision-making.

For example, AI-enabled image processing tools could help radiologists detect early signs of conditions like cancer, heart disease, or neurological disorders from medical imaging data captured by IoT devices, such as X-rays, MRIs, or CT scans. Furthermore, AI-powered systems integrated with wearable IoT devices can analyze vital signs and other health metrics in real time, alerting healthcare providers to potential health risks, enabling quicker interventions.

6.2 5G Connectivity and Low-Latency Communication

The advent of 5G networks will revolutionize healthcare by providing the high-speed, low-latency connectivity necessary for real-time, remote healthcare services. 5G will enable IoT devices to transmit large data packets—such as high-definition medical images or real-time video consultations—without delays. This enhanced connectivity will be especially critical in telemedicine, where doctors will be able to conduct virtual consultations with minimal lag, ensuring that video quality and diagnostic accuracy are not compromised.

With 5G's capabilities, smart healthcare devices such as wearable health monitors, remote diagnostics, and robotic surgery systems can operate more effectively, allowing healthcare professionals to make timely and well-informed decisions from remote locations [23]. The scalability of 5G networks will also pave the way for broader deployment of IoT devices in healthcare settings, particularly in rural or underserved areas, ensuring that quality care reaches a wider population.

6.3 Edge Computing for Real-Time Data Processing

Edge computing, which involves processing data closer to the source of its generation (i.e., IoT devices), is expected to play a significant role in the future of healthcare. With edge computing, healthcare providers can process real-time data from wearable devices, sensors, and medical imaging equipment at the point of collection, rather than relying on distant cloud servers [24]. This reduces latency and allows for faster decision-making in critical situations, such as monitoring patients in emergency rooms or operating rooms.

Edge computing will also help reduce bandwidth demands on network infrastructure, as only essential data will be sent to centralized servers or cloud platforms. This can significantly improve the efficiency of healthcare systems, reduce costs, and improve the overall user experience for both healthcare providers and patients. Edge computing will be vital for IoT applications in real-time remote monitoring, robotic surgeries, and real-time imaging analysis, where immediate feedback and actions are required.

6.4 Personalization and Precision Medicine

The future of IoT-enabled multimedia healthcare lies in the evolution toward personalized and precision medicine, where treatment plans and healthcare services are tailored to the individual needs of patients based on their unique health data. IoT devices that monitor health parameters in real time will enable more accurate assessments of a patient's current condition, making it possible to customize treatment plans and interventions.

For example, wearable devices that track a patient's heart rate, blood pressure, glucose levels, and other vital signs, combined with data from electronic health records (EHR), will allow healthcare professionals to offer more precise treatments. Precision medicine, supported by IoT-enabled platforms, will provide better-targeted therapies based on genetic, lifestyle, and environmental factors, thereby improving patient outcomes and reducing the chances of adverse drug reactions or ineffective treatments.

6.5 Integration of Blockchain for Data Security and Transparency

As healthcare data become increasingly digitized and shared across IoT devices, ensuring privacy and security becomes paramount. Blockchain technology, known for its ability to provide secure, immutable, and transparent records, offers significant potential in the healthcare sector. Blockchain can be integrated with IoT systems to secure patient data, prevent unauthorized access, and ensure that sensitive information is transmitted only between authorized parties.

In a healthcare setting, blockchain could be used to create a secure, decentralized repository of patient data, making it easy for healthcare providers to access the right information at the right time, without compromising security. Additionally, blockchain's transparency ensures that all transactions related to patient data—such as accessing medical records, processing insurance claims, or conducting telemedicine consultations—are recorded in a tamper-proof manner, which fosters trust and accountability between patients, healthcare providers, and insurance companies.

6.6 Enhanced Virtual and Augmented Reality Applications

Virtual reality (VR) and augmented reality (AR) technologies are set to play an increasingly important role in healthcare, especially in training, diagnostics, and therapy. IoT-enabled VR and AR systems can offer immersive experiences that improve surgical training and education by simulating realistic medical scenarios

[25]. Surgeons can practice complex procedures in a virtual environment before performing them on real patients, reducing the risk of errors and improving patient safety.

In addition, AR technologies can be used in real time to overlay critical patient information, such as vital signs or medical imaging, onto the surgeon's field of view during an operation, improving the accuracy and efficiency of procedures. These immersive technologies, powered by IoT devices and real-time data streaming, will likely revolutionize the way healthcare professionals learn, collaborate, and perform medical procedures in the future [26].

6.7 Expanded Use of Wearables and Consumer Health Devices

The use of wearables and consumer health devices is expanding rapidly, and this trend is expected to continue as more advanced devices become available. These devices, which include fitness trackers, smart watches, and patches, enable individuals to monitor their own health and make data-driven decisions about their wellness. In the future, these devices will become more sophisticated, offering continuous monitoring of a wide range of health parameters such as blood glucose levels, respiratory rates, and even mental health indicators like stress levels.

For healthcare providers, the data collected from these devices can be integrated with EHR systems to offer a more comprehensive view of a patient's health. IoT-enabled wearables will also allow healthcare professionals to offer personalized, proactive care, addressing potential health issues before they become serious problems. The rise of consumer health technology will empower patients to take a more active role in their healthcare and improve overall public health outcomes.

6.8 Global Healthcare Access and Equity

IoT-enabled multimedia applications in healthcare present an exciting opportunity to bridge the gap in healthcare access, especially in underserved and rural regions. By leveraging telemedicine, remote patient monitoring, and mobile health technologies, healthcare providers can offer consultations and diagnostic services to patients regardless of their location. This opens up new possibilities for expanding healthcare access to populations that have traditionally been underserved due to geographic, economic, or social barriers.

IoT-powered solutions can reduce the need for patients to travel long distances for medical consultations, saving time and money while improving the efficiency of healthcare delivery. Additionally, as IoT devices become more affordable and accessible, they can empower low-income or rural populations to monitor and manage their health conditions at home, improving the overall quality of life and reducing the strain on healthcare systems.

7 Case Studies and Real-World Applications

The integration of IoT-enabled multimedia technologies has already yielded tangible improvements in healthcare delivery across the globe. By showcasing case studies and real-world applications, it is possible to demonstrate how IoT devices, multimedia data, and advanced analytics are transforming patient care, diagnosis, and operational efficiency. These case studies not only highlight successful implementations but also provide valuable insights into the challenges and best practices involved in integrating IoT-enabled multimedia solutions in healthcare settings. Below are several notable examples of IoT-enabled multimedia applications in real-world healthcare scenarios.

7.1 Case Study: Remote Patient Monitoring (RPM) in Chronic Disease Management

Overview: Remote Patient Monitoring (RPM) has proven to be one of the most significant use cases for IoT-enabled healthcare solutions, especially for patients with chronic diseases like diabetes, hypertension, and cardiovascular conditions. In this case study, a leading healthcare provider implemented IoT devices to remotely monitor patients' vital signs, such as heart rate, blood pressure, glucose levels, and oxygen saturation. These IoT devices are connected to a central platform, allowing healthcare professionals to track patient data in real time.

Implementation: Patients were given wearable devices that automatically collected and transmitted vital signs to a central server. The platform used advanced analytics to analyze the incoming data, flagging any abnormal readings and notifying healthcare providers if intervention was needed. Additionally, multimedia technologies, such as video conferencing, were integrated for regular check-ins between patients and healthcare providers.

Results: The RPM system led to improved disease management for patients with chronic conditions, as it allowed for continuous monitoring and timely interventions. Healthcare providers were able to detect potential health issues early, preventing hospital readmissions and improving patient outcomes. Patients also reported a better quality of life due to the ability to manage their conditions more effectively from home. The system also reduced healthcare costs by eliminating the need for frequent in-person visits.

7.2 Case Study: Telemedicine in Rural Healthcare Delivery

Overview: Telemedicine has been particularly beneficial in rural areas, where access to healthcare is limited. A telemedicine platform, using IoT-enabled devices and multimedia tools such as video conferencing and medical imaging, was introduced to provide remote consultations and diagnostics for underserved communities.

Implementation: IoT devices, including thermometers, stethoscopes, and blood pressure monitors, were deployed to collect real-time health data from patients. These data were transmitted to healthcare providers in urban areas via secure communication channels. Video consultations enabled patients to discuss symptoms with doctors remotely, and multimedia tools allowed the transmission of medical imaging (X-rays, MRIs) for further analysis.

Results: This IoT-enabled telemedicine system allowed healthcare providers to deliver services to remote patients who would otherwise have to travel long distances for care. As a result, the initiative helped to bridge the gap in healthcare access in rural regions, improved early diagnosis, and enhanced the overall quality of healthcare. Patients benefited from convenience and timely interventions, while healthcare facilities were able to reach more patients and provide more effective care.

7.3 Case Study: Smart Hospital Solutions for Improved Workflow and Patient Care

Overview: A smart hospital initiative in a metropolitan area aimed to enhance patient care and optimize hospital operations by integrating IoT devices, multimedia applications, and automation technologies. The hospital implemented a variety of IoT solutions for patient monitoring, staff communication, and medical equipment management.

Implementation: IoT sensors were installed in patient rooms, enabling continuous monitoring of patient vital signs, such as heart rate, oxygen levels, and respiratory rate. Smart beds equipped with IoT sensors could detect patient movements and alert staff if a patient was at risk of falling. Multimedia communication systems were introduced to allow seamless interactions between healthcare providers, facilitating collaboration and quick decision-making. In addition, IoT was integrated into hospital asset management systems, ensuring that medical equipment was properly maintained and available when needed.

Results: The smart hospital system improved operational efficiency by automating routine tasks, such as patient monitoring and asset tracking, reducing human error and freeing up staff time for more critical tasks. The real-time monitoring enabled early detection of potential health issues, allowing healthcare providers to respond faster. The integration of multimedia tools in staff communication also enhanced collaboration, improving the overall workflow within the hospital. Patient satisfaction levels increased due to the improved quality of care and shorter waiting times.

7.4 Case Study: IoT-Enabled Imaging for Early Diagnosis of Cancer

Overview: One of the most promising applications of IoT-enabled multimedia in healthcare is its role in medical imaging, particularly for early detection of diseases like cancer. This case study highlights a medical center's use of IoT-connected imaging devices to enhance the speed and accuracy of cancer diagnosis.

Implementation: IoT-enabled medical imaging devices such as MRI machines, CT scanners, and X-ray machines were equipped with sensors to collect and transmit images to a centralized cloud-based platform. AI algorithms were used to analyze the images in real time, identifying potential signs of cancer or other abnormalities. The integration of high-resolution video streaming allowed radiologists to review and discuss images with oncologists remotely, facilitating quicker decision-making.

Results: The IoT-enabled imaging system significantly reduced the time needed to analyze medical images, providing faster results for patients and improving diagnostic accuracy. The use of AI algorithms also helped radiologists identify early-stage cancer that might have been overlooked in traditional manual analysis. As a result, patients were able to start treatment earlier, which improved survival rates. Additionally, the integration of video consultations and multimedia communications enhanced collaboration between specialists, leading to better treatment planning and outcomes.

7.5 Real-World Application: AI and IoT Integration in Heart Disease Monitoring

Overview: Heart disease remains one of the leading causes of death worldwide, and effective monitoring is crucial for patients with cardiovascular conditions. A health technology company has integrated IoT-enabled wearables and AI algorithms to monitor and manage heart disease more effectively.

Implementation: Patients were given IoT-enabled smartwatches equipped with sensors to monitor heart rate, ECG, blood pressure, and other critical cardiac health parameters. The data were continuously transmitted to a cloud-based platform, where AI algorithms analyzed the information in real time to detect irregularities, such as arrhythmias or potential signs of a heart attack. In case of any abnormalities, healthcare providers were alerted, and patients were immediately contacted for further evaluation.

Results: The integration of IoT devices with AI algorithms led to improved monitoring and timely intervention for heart disease patients. The system allowed healthcare providers to detect issues before they became life-threatening, reducing the risk of sudden cardiac events and improving patient survival rates. Patients also felt more secure knowing that their condition was being continuously monitored and that immediate action would be taken if necessary. This IoT solution also reduced the need for frequent in-person visits, making healthcare more convenient and efficient.

8 Conclusion

IoT-enabled multimedia applications are transforming the healthcare landscape by enhancing patient care, improving diagnostics, and streamlining healthcare operations. The integration of IoT devices with multimedia technologies such as video, audio, and imaging has enabled real-time monitoring, remote consultations, and efficient data sharing, ultimately fostering better healthcare delivery. From remote patient monitoring to advanced diagnostic tools, these innovations are making healthcare more accessible, efficient, and personalized. However, challenges such as data security, interoperability, and the need for scalable infrastructure must be addressed to fully realize the potential of IoT in healthcare. As technology continues to evolve, the future of IoT-enabled multimedia in healthcare holds immense promise, offering new opportunities to improve patient outcomes, reduce costs, and enhance the overall quality of care.

References

1. Vashishth, T.K., Sharma, V., Kumar, B., Panwar, R., Sharma, K.K., Chaudhary, S.: AI-integrated IoT in healthcare ecosystem: opportunities, challenges, and future directions. AI IoT Technol. Appl. Smart Healthc. Syst. 37–54 (2024)
2. Arun Sampaul Thomas, G., Harold Robinson, Y.: Real-Time Health System (RTHS) Centered Internet of Things (IoT) in healthcare industry: benefits, use cases and advancements in 2020. Multimed. Technol. Internet of Things Environ. **83** (2021)
3. Albahri, A.S., Alwan, J.K., Taha, Z.K., Ismail, S.F., Hamid, R.A., Zaidan, A.A., Albahri, O.S., Zaidan, B.B., Alamoodi, AH., Alsalem, M.A.: IoT-based telemedicine for disease prevention and health promotion: State-of-the-Art. J. Netw. Comput. Appl. **173**, 102873 (2021)
4. Kumar, Y., Verma, S.K., Singh, A., Kumar, K., Gupta, M. (eds.): IoT-Enabled Healthcare Systems: Applications, Benefits, and Challenges. CRC Press (2024)
5. Yigit, Y., Duran, K., Moradpoor, N., Maglaras, L., Van Huynh, N., Canberk, B.: Machine learning for smart healthcare management using IoT. In: IoT and ML for Information Management: A Smart Healthcare Perspective, pp. 135–166. Springer Nature Singapore, Singapore (2024)
6. Chauhan, S., Pahwa, K., Ahmed, S.: Telemedical and remote healthcare monitoring using IoT and machine learning. In: Computational Intelligence in Healthcare, pp. 47–66. CRC Press (2023)

7. Khan, J., Li, J.P., Ahamad, B., Parveen, S., Ul Haq, A., Khan, G.A., Sangaiah, A.K.: SMSH: secure surveillance mechanism on smart healthcare IoT system with probabilistic image encryption. IEEE Access **8**, 15747–15767 (2020)
8. Rao, T.V.N., Tabassum Sultana, S.: Applications of IoT-enabled systems in healthcare industry. In: Internet of Things-Based Machine Learning in Healthcare, pp. 147–165. Chapman and Hall/CRC (2024)
9. Goyal, S., Sharma, N., Bhushan, B., Shankar, A., Sagayam, M.: IoT enabled technology in secured healthcare: applications, challenges and future directions. Cogn. Internet Med. Things Smart Healthc. Serv. Appl. 25–48 (2021)
10. Mahajan, H.B., Junnarkar, A.A.: Smart healthcare system using integrated and lightweight ECC with private blockchain for multimedia medical data processing. Multimed. Tools Appl. **82**(28), 44335–44358 (2023)
11. Thangam, D., Malali, A.B., Subramanian, G., Park, J.Y.: Transforming healthcare through Internet of Things. In: Cloud and Fog Computing Platforms for Internet of Things, pp. 15–24. Chapman and Hall/CRC (2022)
12. Phani Praveen, S., Ali, M.H., Jaber, M.M., Buddhi, D., Prakash, C., Rani, D.R., Thirugnanam, T.: IOT-enabled healthcare data analysis in virtual hospital systems using Industry 4.0 smart manufacturing. Int. J. Pattern Recogn. Artif. Intell. **37**(02), 2356002 (2023)
13. Vashishth, T.K., Sharma, V., Sharma. K.K., Kumar, B., Chaudhary, S., Panwar, R.: Intrusion detection and prevention techniques in FL cloud-based healthcare 5.0: a comprehensive review. Pioneering Smart Healthc. 5.0 IoT, Federated Learn. Cloud Secur. 152–174 (2024)
14. Islam, M.M., Nooruddin, S., Karray, F., Muhammad, G.: Internet of Things: device capabilities, architectures, protocols, and smart applications in healthcare domain. IEEE Internet of Things J. **10**(4), 3611–3641 (2022)
15. Al Khatib, I., Shamayleh, A., Ndiaye, M.: Healthcare and the internet of medical things: applications, trends, key challenges, and proposed resolutions. In: Informatics, vol. 11, no. 3, p. 47. MDPI (2024)
16. Nissar, G., Khan, R.A., Mushtaq, S., Lone, S.A., Moon, A.H.: IoT in healthcare: a review of services, applications, key technologies, security concerns, and emerging trends. Multimed. Tools Appl. 1–62 (2024)
17. Rejeb, A., Rejeb, K., Treiblmaier, H., Appolloni, A., Alghamdi, S., Alhasawi, Y., Iranmanesh, M.: The Internet of Things (IoT) in healthcare: taking stock and moving forward. Internet of Things **22**, 100721 (2023)
18. Sekar, C., Falmari, V.R., Brindha, M.: Secure IoT-enabled sharing of digital medical records: an integrated approach with reversible data hiding, symmetric cryptosystem, and IPFS. Internet of Things **24**, 100958 (2023)
19. Vashishth, T.K., Sharma, V., Sharma, K.K., Kumar, B., Chaudhary, S., Panwar, R.: Enhancing biomedical signal processing with machine learning: A comprehensive review. Digit. Transf. Healthc. 5.0 **1**; IoT, AI Digit. Twin **277** (2024)
20. Siripurapu, S., Darimireddy, N.K., Chehri, A., Paramkusam A.V.: Technological advancements and elucidation gadgets for healthcare applications: an exhaustive methodological review-part-II (robotics, drones, 3D-printing, internet of things, virtual/augmented and mixed reality). Electronics **12**(3), 548 (2023)
21. Vashishth, T.K., Chaudhary, S., Sharma, V.: Optimum utilization of bed resources in hospitals: a stochastic approach. In: Artificial Intelligence-based Healthcare Systems, pp. 101–110. Springer Nature Switzerland, Cham (2023)
22. Talpur, M.S.H., Abro, A.A., Ebrahim, M., Kandhro, I.A., Manickam, S., Ul Arfeen, S., Dandoush, A., Uddin, M.: Illuminating healthcare management: a comprehensive review of IoT-enabled chronic disease monitoring. IEEE Access (2024)
23. Apetorgbor, M., Dulewale, N., Dhawale, C.: Smart healthcare solutions: leveraging Internet of Things technologies for advanced computing, communications and applied informatics. In: 2024 Parul International Conference on Engineering and Technology (PICET), pp. 1–7. IEEE (2024)

24. Kumar, M.S., Ganesh, D.: Improving telemedicine through IoT and cloud computing: opportunities and challenges. Adv. Eng. Intell. Syst. **3**(03) 123–135 (2024)
25. Vashishth, T.K., Sharma, V., Sharma, K.K., Kumar, B., Chaudhary, S., Panwar, R.: Virtual Reality (VR) and Augmented Reality (AR) transforming medical applications. In: AI and IoT-based technologies for precision medicine, pp. 324–348. IGI Global (2023)
26. Mishra, K.N., Chakraborty, C.: A novel approach towards using big data and IoT for improving the efficiency of m-health systems. Adv. Comput. Intell. Tech. Virtual Real. Healthc. 123–139 (2020)

Post-quantum Cryptography Exploration for Quantum-Resilient and Secure Multimedia Data for IoT System

Omit Kumar, Md. Motiur Rahman, Bharat Bhushan, Mustafa Al-Asadi, and Khursheed Aurangzeb

Abstract In recent years, due to the extensive need for high computation power for Internet of Things (IoT)-based application, attentiveness in the direction of quantum computing is witnessed. The world has changed significantly because of the internet and in the current period, most of the devices are connected via internet thus giving rise to IoT. However, the multimedia IoT devices are associated with notable vulnerabilities that are often exploited by adversaries. Quantum computing, with its superior processing capabilities makes securing of multimedia data over the IoT network significantly challenging. Post-Quantum Cryptography (PQC) is regarded as an emerging field responsible for designing algorithm and cryptosystems resilient to quantum attacks, ensuring robust security for the multimedia IoT infrastructures. In this paper, we highlight the principles of cryptography, its reliability for classical computing, and how its inefficiency in the quantum era with the emerging quantum computers gave rise to the need for PQC, a novel cryptosystem resilient in the rising quantum environment. It includes different vulnerabilities for multimedia IoT system and attacks that can be performed by exploiting those vulnerabilities. Finally, this paper discusses the recent advancements and proposed ideas about increasing the efficiency of PQC and making the IoT network secure for the multimedia data in the rising quantum environment.

Keywords Cryptography · Multimedia data · Security · Quantum computing · Internet of Things

---

O. Kumar (✉) · Md. Motiur Rahman · B. Bhushan
Department of Computer Science and Engineering, School of Engineering and Technology, Sharda University, Greater Noida, India
e-mail: omit.kr32@gmail.com

M. Al-Asadi
The University of Mashreq, Baghdad, Iraq
e-mail: mustafa.alasadi@uom.edu.iq

K. Aurangzeb
Department of Computer Engineering, College of Computer and Information Sciences, King Saud University, Riyadh, Saudi Arabia

R. Sharma et al. (eds.), *Multimedia Technologies in the Internet of Things Environment, Volume 4*, Studies in Big Data 173, https://doi.org/10.1007/978-981-96-4356-1_8

1 Introduction

Cryptography is described as the cornerstone of securing information and communication in the digital era. It involves the process of encryption as well as decryption of the plaintext data such that only the authorized party can access those data and thus this process is vital for maintaining authenticity, integrity, confidentiality, and accessibility in communication systems. Symmetric and asymmetric cryptography are the two subcategories of classical cryptography, each having its own encryption and decryption techniques [1]. However, because the rise of Shor's and Grover's like quantum algorithm which have the ability to break these systems, highlights the serious danger that quantum computing poses to commonly used cryptographic techniques, especially asymmetric cryptography [2, 3]. This gives rise to growing need for quantum-resistant encryption mechanism as existing cryptographic standards like RSA and ECC are vulnerable to quantum decryption, making Post-Quantum Cryptography (PQC) essential for future-proof digital security system [4].

PQC represents a significant change in cryptographic approaches, addressing the potential threats associated with quantum computing [5]. PQC departs from conventional security mechanisms that depend on problems like integer factorization (IF) and discrete logarithms which are susceptible to quantum attacks [6]. Rather, it presents novel cryptographic systems and its types based on problems that remain problematic to unravel by a quantum computer. Following these cryptosystems are the leading algorithms and those algorithms are discussed in Sect. 3 that are emerging from the NIST's second-round candidates, offering a promising balance between efficiency and security, making them ideal for resource-constrained (i.e. limited resources for computation) and low-powered application like Internet of Things (IoT).

The Internet has greatly changed the planet, and as of now, it would be difficult to imagine a living without the internet. We live in a period whereby most things around the world are connected to the internet thereby forming what is called the IoT [7]. The nature of IoT makes the support of multimedia applications possible. Multimedia applications supported by internet connectivity and object interaction without intervention of humans lead to a plethora of openings for the advancement in daily life. Multimedia being an information-rich data, features like retina scanning for crime detection and authorization, object detection, motion detection, face recognition can be elevated leveraging multiple data extraction, analytics, and aggregation tools. Multimedia IoT systems, however, come with significant vulnerabilities that are often overlooked by the manufacturers, thus creating devices that are vulnerable to exploitation. As a matter of fact, 75% of devices connected to the internet have been proven to be vulnerable against various cyber threats [8]. As per the researches, the compromised multimedia IoT devices will be the root cause for 25% of the industrial attacks. There are some notable cases, like the worldwide phenomenon Mirai, BrikerBot, and Hajime, which really underlined the severity [9]. The notion of privacy will be in jeopardy as adversaries can use the exploitation of IoT baby monitors and toys as a means to steal sensitive data, including voice recordings, passwords, emails, and video streams. The ease of reprogrammability of multimedia IoT device's firmware further helps the attackers. It even presents life-threatening risks to human, as the US FDA has confirmed the dangers associated with the unauthorized access to implantable devices. These issues call for immediate attention if IoT is ever to have a secure and private future.

The quantum computing's rise has increased the danger to security and privacy concerns for the resource-constrained and low-powered IoT devices, leading to an increasing requirement towards a shift from classical cryptography toward quantum-secure cryptography. The major contribution that this work explores is briefly described below.

- This work explores the field of cryptography, how it works, its different types and different algorithms, and its performance against quantum computing.
- This work highlights the significance of a new cryptosystem that is resilient towards the attacks in the upcoming quantum era.
- This work provides a detailed overview about PQC and its five types based on the problems that are problematic to solve for quantum computers.
- This work provides a comparative analysis of the promising 17 NIST second-round candidate post-quantum algorithms and it different levels.
- This work presents the different vulnerabilities associated with multimedia IoT devices and also states the threats and attacks that are performed by exploiting those vulnerabilities.
- This work illuminates the foremost recent advancements towards the field of PQC for multimedia IoT devices focusing on enhancements in multiple areas like increment in performance and security, compatibility with resource-constrained and battery-dependent IoT devices, and resilience to different quantum attacks.
- This work also highlights several future research directions regarding the improvement in performance, compatibility, and resilience of the quantum-resilient cryptosystems for multimedia IoT devices.

The remainder of this work is structured as follows: Sect. 2 presents the principles of cryptography and its types, quantum computing its rise and emergence, difference between classical computers and quantum computer, threats to cryptographic systems by the quantum computers, and need for a PQC. Section 3 comprises of PQC, its types, and algorithms. Section 4 addresses the vulnerabilities associated with the multimedia IoT devices and different attacks performed by exploiting those vulnerabilities, and Sect. 5 regards the recent advancements.

2 Cryptography and Quantum Computing

Cryptography is the pillar for providing security in a communication, ensuring data confidentiality, authenticity, and integrity. However, the arrival of quantum computers threatens the less secure classical cryptographic system. To understand the implication of this challenge, it is crucial to first explore the principle of cryptography which forms the bedrock of modern-day digital security systems.

2.1 *Principles of Cryptography*

Cryptography is the technique of securing information and communication over unsecured channels. Its main purpose is encrypting plaintext data into ciphertext via some methods and algorithms such that only the relevant parties can decipher it into its original form. The process of transforming plaintext to encoded ciphertext is known as encryption while gaining plaintext message from the received encoded ciphertext is known as the process of decryption. Cryptography provides authenticity, accessibility, integrity, and confidentiality [10]. Authenticity ensures that the parties communicating can authenticate their identities among themselves. Accessibility makes certain that only the authorized entities can access the information resources. Further, integrity prevents the message content from being altered by any unauthorized entities. Finally, confidentiality states that the decryption of the encoded message can only be done by the authorized parties.

Cryptography is presently branched into two main groups, i.e. symmetric cryptography and asymmetric cryptography. These branches of cryptography are discussed in the subsections below.

2.1.1 Symmetric Cryptography

In symmetric cryptography, the key used for encryption as well as decryption are identical and must be exchanged among the two parties at the beginning of the communication. At the sender's end, this secret key is utilized for encryption of the plaintext message before transmitting while at the receiver's end, the same key is used for decryption of the received encoded ciphertext message to get the original plaintext message as stated in the Fig. 1. Let us assume that Alice needs to send a message m to Bob. Then by using a secret key k Alice will encrypt the plaintext message into an encoded ciphertext message before transmitting.

$$C = E_k(m) \tag{1}$$

Here, C represents the encoded ciphertext message, which is obtained after encoding the plaintext message m using the encryption function E () with the help

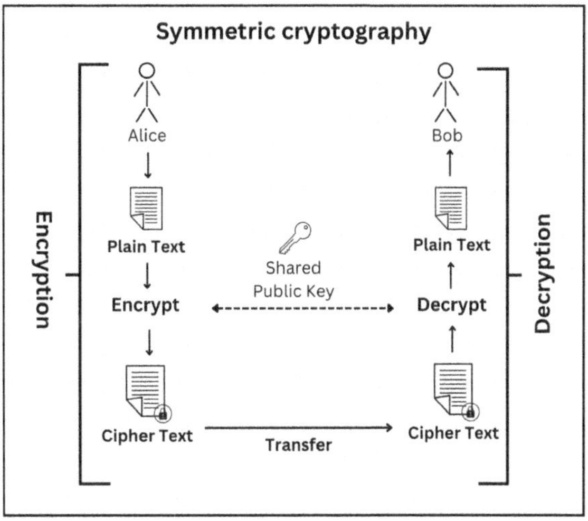

Fig. 1 Working of symmetric cryptography

of a secret key k. For decryption of the encoded message C, Bob will decrypt the received ciphertext C using the same secret key k to get the original message m.

$$D_k(C) = D_k(E_k(m)) = m \tag{2}$$

Here, m is the original message obtained after decryption of the received encrypted message $E_k(m)$ using the decryption function $D_k(C)$ using the same secret key k.

2.1.2 Asymmetric Cryptography

Asymmetric cryptography, also known as public key encryption, was initially introduced in an article by M. Hellman and W. Diffie named New Direction in Cryptography in 1976 [11]. In asymmetric cryptography, every party possesses dual keys, i.e. a private key that is unique for every entity and is securely stored via each communicating party that is utilized for decryption of the encoded message and a public key that is common and is shared publicly to every party involved in the communication and is used for encryption of the plaintext before sending. The sender encrypts the message into a ciphertext using the shared public key, which can then be decrypted into the original plaintext via the private key equipped at the receiver end as shown in Fig. 2.

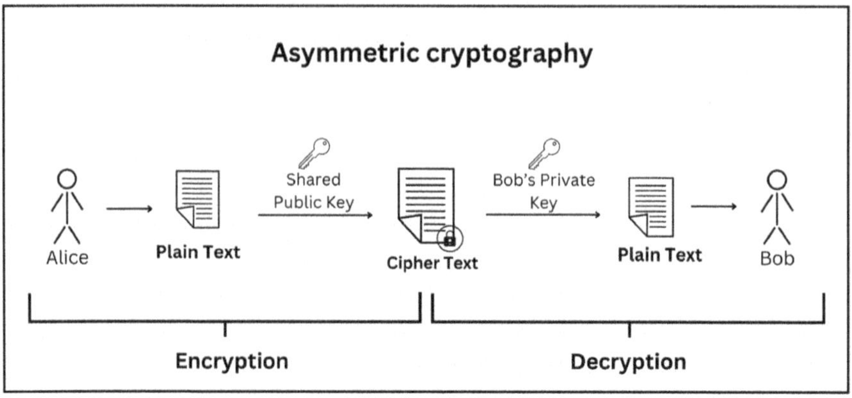

Fig. 2 Working of asymmetric cryptography

Similar to symmetric cryptography, Alice can send encrypted messages to Bob via asymmetric cryptography, but in this cryptography, there are two keys. Let *pk* be the Bob's private key and *sk* be the shared public key. Suppose a message m is sent from Alice to Bob. Alice uses the public key sk shared by the communicator to encrypt the plaintext message m into ciphertext.

$$C = E_{sk}(m) \tag{3}$$

The ciphertext *C* received by Bob is then decrypted using the corresponding decryption algorithm and his private key *pk*.

$$D_{pk}(C) = D_{sk}(E_{sk}(m)) = m \tag{4}$$

2.2 *Quantum Computing's Rise and Emergence*

Quantum Computing refers to a new paradigm of computing that has emerged over past decades. It uses the quantum phenomena, i.e. superposition and entanglement for the purpose of achieving enhanced computation, which can be applied to conventional computational problems. Quantum computing was first introduced in 1980, which became well-known following the article "Simulating physics with Computers" published by the American theoretical physicist Feynman [12]. Feynman pointed out that each quantum state can be in a state of superposition. Quantum Computing uses quantum bits (also known as qubits) for performing operations similar to digital bits used by classical computers. A classical bit used can either be 0 or 1 at a time, while a qubit can be 0 and 1 simultaneously (i.e. it can be described as a coherent superposition of 0 and 1) [13].

Classical computers would not be able to handle the quantum states as the computations would require computers with exponentially large resources. Therefore, the numerical simulation of the quantum system, which can include up to a few hundred or even thousands of qubits, is impractical for conventional devices. In essence, superposition refers to the ability of quantum computers being able to exist in multiple states concurrently.

The quantum system states are provided by the transformation of the matrices and vectors, which are referred to as the main components of quantum computing. The quantum states can be usually described using the Bra-ket notation. The column K-vectors are called ket vectors in this notation, and the row vectors that represent their conjugation and transpositions are called bra vectors. The combination is determined using the scalar product of these vectors.

2.3 Need for Post-quantum Cryptography

Both symmetric and asymmetric cryptography are at a risk from quantum computing although the threat to asymmetric cryptography is more significant. For the post-quantum era, the hash function and symmetric algorithm still prove to be effective. It is assumed that the only thing required by these algorithms is to increase security by increasing their key size/output [14]. The security strength of the existing symmetric cryptosystems can decrease potentially by Grover's algorithm by offering a quadratic speed-up for exhaustive key searching. Since in Grover's algorithm, the queries are executed sequentially, there is an ongoing debate about the practical implication of this attack. However, researchers have estimated that Grover's search against the Advanced Encryption Standard (AES) may be performed by a quantum architecture with 2953, 4449, 6681 qubits, respectively, with 128-, 192-, and 256-bit security [15].

The most popular asymmetric cryptosystems are threatened by quantum computing [16]. Well-known encryption schemes consisting Elliptic-Curve Diffie-Hellman (ECDH), digital signature algorithm (DSA), Elliptic-Curve DSA (ECDSA), RSA are affected by the quantum attacks.

Table 1 shows some of the popular cryptographic algorithms, its current state of usage, and if its quantum-secure or not.

As shown, asymmetric cryptography, including ECDH, ECDSA, and RSA, is not quantum-secure. The security of RSA depends upon how hard solving an IF problem is, which is the difficulty of factoring large bi prime numbers. Elliptic Curve Cryptography (ECC), on the other hand, relies on the assumption that EC Discrete Logarithm (ECDL) problems are difficult to solve. These ECDL problems and IF problems are difficult to solve by a classical computer, but for quantum computers large enough to execute Shor's algorithm, solving these problems becomes as fast as in polynomial times.

Since theoretical estimation suggests that in order to break the existing cryptography, a numerous thousands of qubits are required in a quantum computer. In

Table 1 Cryptosystems and their quantum-secure capability

Cryptography schemes	Current state	Problems on which these cryptosystems are based on	Secure against quantum computing
DSA	Outdated	Discrete logarithm	No
DH			
Symmetric ciphers	In use	Collision-finding and exclusive search for large key space	
ECDH		Elliptic curves discrete logarithm	
ECDSA		Discrete logarithm	
RSA	In use	IF	Yes
PQC	Developing	Isogeny graph, hash tree, multivariate system, error-correction codes, lattice-based problems	

order to reach this goal, technical giants around the globe are surely making efforts. Currently, IBM is leading the way in quantum computing processors, with 1,121-qubits processor and plans on building a 4,158-qubit processor by 2025 by connecting three Kookaburra processors (Kookaburra is a 1,368-qubit processor) and aims to reach 10,000 qubits processors by 2026 [17]. The roadmap of development of qubit processors by IBM is provided in Fig. 3 and they are on track in direction of this development. The prediction has been made that there is a great chance of having a scalable quantum computer is significant in the upcoming future specifically in a decade. Seeing this development in the field of quantum computers, there arise a crucial requirement for a robust quantum-secure encryption and signature schemes.

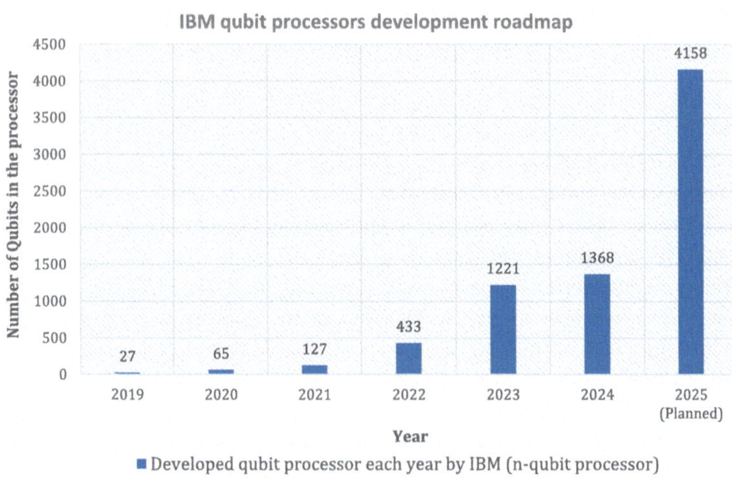

Fig. 3 Development roadmap of IBM in the field of qubit processors

3 Post-quantum Cryptography

PQC is a cryptography scheme for protection assuming that the attackers possess large quantum computers. The primary obstacle of PQC is maintaining cryptographic flexibility and usability without compromising confidentiality. This requires cryptography to stop depending on the discrete logarithmic problem and hardness of IF and begin using problems, which is a challenge to solve even for the quantum computers.

3.1 *Types of PQC*

PQCs are currently categorized into five main types, namely, lattice-based, code-based, hybrid, supersingular elliptic curve isogeny, and multivariate-based cryptosystem as stated in Fig. 4. These cryptosystems are analysed in the subsection given below.

3.1.1 Post-quantum Lattice-Based Cryptosystem

Lattices, which are collections of points in n-dimensional spaces having a periodic structure, form the foundation of lattice-based cryptosystem. The closest vector problem (CVP), shortest vector problem (SVP), and closest vector problem (CVP) are the examples of lattice problems whose alleged difficulty is relevant to the use of lattices in cryptography [18]. Strong security proofs are presented by the lattice-based cryptosystem, and they are fast, easy and relatively efficient to implement. Future lattice-based PQC for IoT devices however ought to effectively handle the storage of big keys and overheads associated with ciphertext.

Fig. 4 Classification of PQC

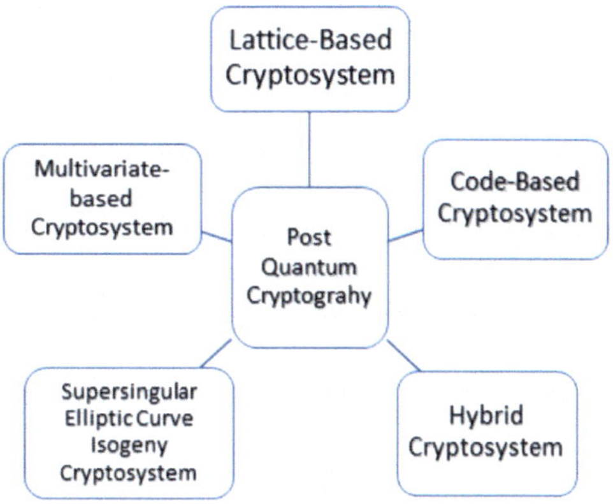

3.1.2 Post-quantum Code-Based Cryptosystem

The theory underlying the error-correction codes which have long since provided redundancy to digital communication forms the foundation for the code-based cryptosystems. McEliece is a code-based relevant and reliable cryptosystem whose performance proved to be reasonably fast for encryption and decryption [19]. Its security is based on syndrome decoding problem and is built on binary Goppa code.

3.1.3 Post-quantum Hybrid Cryptosystem

For providing double security to the transmitted and received data, PQC hybrid cryptosystem merges both the post and pre-quantum cryptosystem. Prior to the full PQC, these post-quantum hybrid cryptosystems seem to be the next step, and additionally, these cryptosystems have already been tested in a practical environment by Tor project and Google.

3.1.4 Post-quantum Supersingular Elliptic Curve Isogeny Cryptosystem

These cryptosystems are derived from the isogeny protocol for ordinary elliptic curves. However, the supersingular curves are required to be non-commutative for prevention against the quantum attacks. The key size is speculated to be a few thousand bits, so the study of applications of optimizations and compression techniques is necessary for IoT development [20].

3.1.5 Multivariate-Based PQC

The foundation of multivariate-based PQC is based on how difficult solving the multivariate equations system is. These equations systems have been proven to be either NP-complete or NP-hard [21]. Currently, the multivariate signature and encryption schemes must be analysed and put into practice in order to tackle the primary shortcomings for IoT-based applications: their typical inefficiency in decryption on devices with limited resources, their size of the key is frequently large causing a potential increase in consumption of energy, and the large ciphertext overhead.

3.2 Algorithms of PQC

The most pertinent and promising PQC algorithms, specifically NIST candidate for PKE and key-establishment algorithms along with a comparative analysis of these algorithms, are provided in the subsections below.

3.2.1 NewHope

NewHope is a PQC based on lattice-based scheme and relies on Ring-LWE problems. Solving a random learning with error (LWE) problems is as difficult as solving some worst-case situations of some specific lattice-based problems under a quantum reduction, as demonstrated by Regev, and popularized the LWE problems [22].

Various distinct instantiations were proposed by the NewHope submission in the NIST, which were NewHope512-CCA-KEM and NewHope1024-CCA-KEM, targeting the level 1 and level 5, respectively, in the call for proposal by NIST (meeting or surpassing the AES-128 and AES-256 brute-force security, respectively).

3.2.2 ROLLO

ROLLO (Rank-Ouroboros, LAKE, and LOCKER) is a PQC based on code-based cryptosystem designed to defend against attacks from conventional commuters as well as quantum computers. ROLLO is the combination of three NIST's competition candidates for PQC standardization (Rank-Ouroboros, LAKE, and LOCKER) [24]. This cryptosystem leverages the same decryption technique for LRPC codes and is based on rank metric codes.

LOCKER (renamed as ROLLO-I) that is in PQ PKE category with security level of IND-CPA2 while LAKE (renamed as ROLLO-II) along with Rank-Ouroboros (renamed as ROLLO-III) comes under PQ KEM category with a security level of IND-CPA [24].

3.2.3 FALCON

FALCON is a digital signature algorithm based on lattice-based cryptography. FALCON is constructed from the security of the shortest integer solution problem in NTRU lattices, using a new trapdoor sampling scheme called Fast Fourier Sampling. It combines NTRU lattice structures with the GPV framework—an approach for hash-and-sign lattice-based signatures. FALCON is viewed as the strong contender, with CRYSTALS-DILITHIUM a candidate being submitted to the standardization process under NIST post-quantum cryptography. Its implementation of floating-point operations and side-channel attacks against it needs further scrutiny. The same can be said for sampling mechanism. Efficiency in its key generation, using less than 30 KB of RAM, makes it perform realistically within any constrained environment [25].

3.2.4 NTRUEncrypt

NTRUEncrypt, also known as NTRU encryption algorithm, is a lattice-based post-quantum cryptosystem and is considered to be one of the most promising candidates in the NIST call for proposal for PQC. It is a substitute for ECC and RSA that relies on a lattice-based shortest vector problem, which is known to be quantum-secure.

In particular, NTRU processes are based on items inside a trunctured polynomial ring, denoted as.

$$R = Z[X]/(X^N - 1) \tag{5}$$

where all polynomials include integer coefficient and maximum degree of $N - 1$ [26].

NTRU contains three parameters (N, p, q), where N denotes polynomial degree bound, p denotes small modulus, and q denotes large modulus. Here, N is assumed to be a prime number, p and q both are coprime numbers and q is much larger than p.

NTRUEncrypt proposed various instances during the submission to the call for proposal of PQC by NIST, namely, NTRUEncrypt-1024, NTRUEncrypt-743, NTRUEncrypt-443.

3.2.5 Classic McEliece

Classic McEliece is a KEM developed to be a one-way under a chosen plaintext attack (OW-CPA), which means that any adversary will be unable to find the codeword efficiently from the public key or the cyphertext, if the chosen codeword is at random. Over the course of 40 years, the security level of the McEliece system has still remained remarkably steady in spite of several attack papers. In spite of the original McEliece parameters, which were intended for 2^{64} security, the system can be readily enhanced to "overkill" numbers, thus providing substantial security against future advancements in computer technologies, including quantum computing [27].

Classic McEliece is built for IND-CCA2 to withstand quantum attacks and provide extremely high security level. This KEM is cautiously constructed by leveraging Goppa codes in the Niederreiter's dual version of McEliece's PKE.

3.2.6 BIKE (Bit Flipping Key Encapsulation)

BIKE is a PQC KEM, code-based cryptosystem, that depends on a variation of the McEliece cryptosystem, i.e. quasi-cyclic moderate-density parity-check (QC-MDPC) codes. This PQC algorithm was submitted to NIST with three instances, which are BIKE-1, BIKE-2, BIKE-3, and each subdivided into three levels, which are 1, 3, and 5. The different levels 1, 3, and 5 of the BIKE algorithms denote the claimed NIST level of that cryptosystem (i.e. Bike level 1 algorithm denoting claimed NIST level 1).

The architectures supported by BIKE were 64-bit little-endian and x84_64 and the security model for BIKE algorithm is IND-CPA security which states that this level of security protects against passive adversaries who only have access to the ciphertext ensuring that even with the knowledge of methodology of converting the plaintext into ciphertext, the encryption remains secure [28, 29].

3.2.7 CRYSTAL-Kyber

CRYSTAL-Kyber, a KEM having a security level of IND-CCA2, is a cryptographic suite that is made up of two primitives: a DSA (Dilithium) and a KEM (Kyber). The security and reliability of this cryptosystem depend on how difficult it is to crack the LWE problem over module lattice. Kyber is currently being incorporated into systems and libraries such as Cloudfare (Integrated Kyber with other PQC forming Cloudfare Interoperable, Reusable Cryptographic Library also known CIRCL), Amazon (Integrated Kyber into AWS Key Management Services), and in IBM (in 2019 advertising "World's First Quantum Computing Safe Tape Drive" leveraging Kyber with Dilithium) [30].

CRYSTAl-Kyber, a finalist at NIST PQC project, submits a list of unique parameter sets, where each set targets a separate level of security. Kyber-512 targeting AES-128 security level, Kyber-768 targeting AES-192, and Kyber-1024 targeting AES-256 [31].

3.2.8 Rainbow

Rainbow is a multivariate scheme, and it can be based on the adaptation of the Unbalanced Oil and Vinegar (UOV) structure from 2005. Rainbow advantages: its small signatures have rapid signing and verification times and use a layered design of UOV provably resistant against side-channel attacks, which render Kipnis-Shamir-type attacks, applicable to the traditional UOV schemes, not workable. While its complex structure has revealed potential vulnerabilities, Rainbow suffered less than any other attack until very recently. Today, the Rainbow family aligns with NIST security levels 1, 3, and 5 and is, therefore, NP hard. However, two attacks brought forward after the NIST Round 3 submission by Ward Buellens have revealed vulnerabilities in their margins and reduce security by a few bit levels along the parameters of NIST. Due to these results, Rainbow's parameter settings are unlikely to be viewed as compliant with NIST's requirement and its developers should study changes. One weakness for Rainbow is that its public and private key sizes are nearly 2,000 times larger, up to 1.8 MB [32].

3.2.9 SPHINCS+

SPHINCS+ is a stateless, hash-based signature scheme that can resist attacks from quantum computers. Even though lattice and code-based schemes are not dependent upon purely cryptographic hash functions, which are extremely resistant to classical and quantum attacks, the security of SPHINCS+ depends solely upon purely cryptographic hash functions. SPHINCS+ is built using a hash tree that is also termed as Merkle tree along with the few-time signature scheme, FTS, so that signatures are sound as well as unique without the assumption of number theory [33].

SPHINCS+ has the major advantage that it resists attacks based on complex mathematical problems due to being based on a hash function whose security has already been established. Its disadvantage is that signature size and signing time are bigger compared to lattice-based ones such as CRYSTALS-DILITHIUM, but the type of security assurance is different in the case of SPHINCS+. No state management is needed. Thus, state misuse cannot be a threat here, and this makes SPHINCS+ the better choice for applications with demanding long-term security requirements [34].

3.2.10 CRYSTALS-DILITHIUM

CRYSTALS-DILITHIUM is a digital signature scheme in the Cryptographic Suite for Algebraic Lattices, based on the hardness of the Short Integer Solutions problem and its module analogue, multiplicative LWE (MLWE). Lattice-based signature scheme whose public key size and signature size are smaller than those of most other schemes belonging to the same category, including those based on uniform sampling. While it cannot meet the stringent NIST Level 5 security requirement, DILITHIUM's core-SVP offers desired performance in many realistic scenarios and flexibility because it uses parameter selection. Such an algorithm was shown experimentally to have impressively high performance in terms of both security and efficiency. NIST picked DILITHIUM for further studies in the development process of standardization of post-quantum cryptography, which involves benchmarking the performance, security, and suitability as a standardized digital [35].

3.2.11 FrodoKEM

FrodoKEM designed to be practical yet conservative and resilience against quantum attacks is a family of KEM [36]. The security and reliability of the cryptosystem depend on cautious parametrization of the popular LWE problems, which is closely related to the general algebraically unstructured lattices. FrodoKEM was developed in parts by Google, NXP, and Microsoft researchers.

In particular, FrodoKEM is built for IND-CCA security on three:

- Targeting Level 1 in NIST call for proposal, meeting or surpassing AES-128's brute-force security is FrodoKEM-640 [37].
- Targeting Level 1 in NIST call for proposal, meeting or surpassing AES-192's brute-force security is FrodoKEM-976 [37].
- Targeting Level 1 in NIST call for proposal, meeting or surpassing AES-256's brute-force security is FrodoKEM-1344 [37].

3.2.12 Three Bears

Three Bears is a lattice-based PQC that utilizes Kyber and NewHope algorithms. This PQC scheme is based on Integer Module Learning with Errors (I-MLWE) problems. ThreeBears got its name from the fact that its modulus has the same "golden-ration-Solinas" shape as Ed448-Goldilocks and also that some of the implementation's arithmetic codes are devised from Goldilocks' arithmetic code [38].

Three Bears algorithm was submitted to NIST's PQC standardization process with its three instances, namely, Three Bears (PapaBear CCA), Three Bears (MamaBear CCA), and Three Bears (BabyBear CCA).

3.2.13 SABER

SABER is a KEM-based PQC scheme with a security level of IND-CCA2 that depends on hardness of MLWR problems making it secure even against quantum systems. Three security levels are offered by SABER suite, namely,

- FireSABER: Similar to the security level of AES-128 [39].
- SABER: Similar to the security level of AES-192 [39].
- LightSABER: Similar to the security level of AES-256 [39].

Table 2 provides a comparative analysis of multiple variants of the NIST candidate algorithms that were discussed in the previous subsections. These algorithms offer varying degrees of security and, given their necessary key sizes and computational complexities, now appear more promising for implementation in IoT devices.

4 Vulnerabilities in Multimedia IOT

Vulnerability is somewhat like a flaw or a hole within a system which if remain unchecked, can lead to a significant issue to the complete system. The security of multimedia data for the IoT network is in critical phase because of the cyber-attacks and breaching vulnerability [40]. There have been numerous reported unexpected vulnerabilities and exploits that were created in order to take advantage of security holes in the multimedia IoT systems. Some of these vulnerabilities related to multimedia IoT are described in the provided subsections below.

4.1 Insufficient Audit Mechanism

Currently, there isn't an existing system to keep track of user activities on IoT devices and periodically review them to guarantee security of these IoT devices. Occurrences

Table 2 Comparative analysis of 13 NIST second-round candidate algorithms

PQC algorithms		Public key size (bits)	Private key size (bits)	Based on
NewHope-512 (CCA) [23]		7,424	15,104	Based on ring-LWE problems
NewHope-1024 (CCA) [23]		14,592	29,440	
ROLLO-II 256 [24]		19,944	320	Based on rank metric codes with LRPC codes
ROLLO-II 192 [24]		16,160	320	
ROLLO-II 128 [24]		12,368	320	
FALCON-512 [25]		7,176	10,240	Based on the shortest integer problem in NTRU lattices
FALCON-1024 [25]		14,344	18,432	
NTRUEncrypt-1024 [26]		32,776	65,552	Based on solving LWE/ Ring-LWE
NTRUEncrypt-743 [26]		8,184	9,384	
NTRUEncrypt-443 [26]		4,888	5,608	
Classic McEliece CPU based [27]		10,862,592	112,640	Based on dual version of Niederreiter using Goppa codes
BIKE-1 [28, 29]	Level 1	20,334	2,130	QC-MDPC McEliece
	Level 3	39,706	3,090	
	Level 5	65,498	4,384	
BIKE-2 [28, 29]	Level 1	10,164	2,130	QC-MDPC Niederreiter
	Level 3	19,853	3,090	
	Level 5	32,750	4,384	
BIKE-3 [28, 29]	Level 1	22,058	2,010	QC-MDPC Ouroboros
	Level 3	43,036	2,970	QC-MDPC Niederreiter
	Level 5	72,262	4,256	
Crystal Kyber-512 [30]		6,400	13,056 (256)	Based on LWE problem with module lattice
Crystal Kyber-512 90s [30]		6,400	13,056 (256)	
Crystal Kyber-768 [30]		9,472	19,200 (256)	
Crystal Kyber-768 90s [30]		9,472	19,200 (256)	
Crystal Kyber-1024 [30]		12,544	25,344 (256)	
Crystal Kyber-1024 90s [30]		12,544	25,344 (256)	
Rainbow [32]	Level 1	1,288,000	824,000	Based on unbalance oil and vinegar (UOV) multivariate scheme, a variant of digital signature algorithm
	Level 3	2,64,000	1,608,000	
	Level 5	14,400,000	12,582,912	

(continued)

Table 2 (continued)

PQC algorithms	Public key size (bits)	Private key size (bits)	Based on
SPHINCS+ 128s [33, 34]	256	512	Based on hash-based signature schemes
SPHINCS+ 192s [33, 34]	384	768	
SPHINCS+ 256s [33, 34]	512	1024	
Crystals-Dilithium-II [35]	10,496	20,224	Based on short integer solutions and MLWE problems
Crystals-Dilithium-III [35]	15,616	32,000	
Crystals-Dilithium-V [35]	20,736	38,912	
FrodoKEM-640 [36]	76,928	159,104	Based on LWE problems with general "algebraically unstructured" lattices
FrodoKEM-976 [36]	125,056	250,368	
FrodoKEM-1344 [36]	172,160	344,704	
Three Bears (PapaBear) [37]	12,672	320	Based on I-MLWE
Three Bears (MamaBear) [37]	9,552	320	
Three Bears (BabyBear) [37]	6,432	320	
SABER KEM (FireSABER) [38, 39]	10,496	24,320	Based on module learning with rounding problem (Mod-LWR)
SABER KEM [38, 39]	7,936	18,432	
SABER KEM (LightSABER) [38, 39]	5,376	24,320	

such as authorization attempts, authentication attempt, application errors, successful or unsuccessful login attempts must always be recorded and in an encrypted file log.

4.2 *Boot Process Vulnerability*

The boot loader, boot process sequence, and firmware are susceptible to get leverage during the boot process. For example, the researchers attacked the Nest Thermostat and the Nike+ fitness tracker in an experimental scenario [41]. Due to this vulnerability, the Google Nest thermostat was compromised. The thermostat was loaded with an onboard kernel argument list, a custom full boot loader and a malevolent initial boot loader. Using the custom boot loader arbitrary payloads such as backdoors later on could be added. They installed their boot loader and configured the processor to boot from the interface of USB or URAT. In addition, they enabled the acceptance of updates from sources other than Nest.

4.3 *Weak Authentication Mechanism*

Strong authentication methods can be implemented to increase the security of various IoT ecosystem interfaces including cloud, web, and mobile. The attacker uses account enumeration and weak credentials to target these insecure interfaces. If any adversary without going through an identity check or bypassing a poor authentication system get access to the IoT nodes then numerous ways gets open by which they can exploit the system [42]. The absence of adequate resources on IoT devices makes it challenging to implement a robust authentication mechanism.

Various attacks that can be done by exploiting the weak authentication mechanism vulnerability are listed below:

4.3.1 Sybil Attack

By changing the identification of compromised nodes, this attack enables a malicious node to pretend like it's a genuine IoT node in the network. The network performance is adversely affected by this attack. The counterfeited device can fabricate the message, eavesdrop on traffic, may flood the channels with bogus packets and much more to prevent the authorized users from accessing the IoT system services [43].

4.3.2 Distributed Denial of Service (DDoS) Attack

It would be simple to utilize IoT nodes as bots to target the DDoS attack. In terms of distributed attacks, the internet is on risk due to this ease. The presence of a large number of 24/7 insecure IoT devices, their barely interactive user interface, and their poor maintenance draw the attacker's attention towards them. Even currently, this exact vulnerability for smart node is exploited by Mirai and its variants [44].

4.3.3 Man-in-the-Middle Attack (MitM)

MitM is described as a well-established technique of breaching network security, which mainly involves creating a separate connection by the adversary with the communication ends, sending and receiving data, and creating an impression that these ends are privately communicating, constituting a direct conversation with one another [45]. However, the attacker is listening to the communication packets, can intercept this communication, and replace or modify data on either ends of the communication. The MitM attack technique is also used in order to get access and modify the communication messages between the application and the cloud platform.

4.3.4 Hello Flood and Homing Attacks

Whenever a new node is added to the network, it sends "HELLO PACKETS" to all the neighbours indicating its arrival to network. This fact is leveraged by the adversaries to perform the hello flood and homing attacks. At this point, every receiving node believes that the sender node is within their communication range. In this attack, the node having a high transmission power is used by the attacker [46].

4.3.5 Dictionary Attack

Dictionary attack is characterized as a kind of brute force attack where the intruder tries every character combination in the dictionary to crack the security to get access to the IoT device or systems. The dictionary attack leads to a highly distributed attack. By compromising the security, it generates millions of infected in order to launch attacks like DDoS, DoS [44].

4.4 Unauthorized Access

The IoT devices having weak, hardcoded, or default credentials are flooded in today's market. It creates an opening for adversary as nobody bothers to modify the default credentials. In the most notable Mirai and DDoS attacks, 60 default credentials were used by the adversaries in order to turn these IoT nodes into malicious ones. It was 1.2 TB per second at its peak [44]. Moreover, hardcoded credentials are the leading cause for unauthorized access in IoT devices and can only be resolved through a software patch included with the shipped baby monitors.

4.5 Improper Patch Management

It has been found that the automated patch-update feature is skipped by the manufacturers in IoT devices. Furthermore, the security patches are also not provided on time by the manufacturers thus lacking integrity assurance. Malicious patches may be found in the unsecure updates that can have an adverse impact on the devices at large [47].

4.6 Improper Encryption

Less reliable wireless communication media like 802.15.4, 802.11. a, LoRa, SigFox, Zigbee, NB-IoT are used in the majority of the IoT devices [48]. Consequently, data leakage attacks are the leading problem for these devices. Additionally, it was discovered that users' personal information was gathered by the majority of the IoT devices for providing the necessary services. Misuse of these personal information in the context of e-health can be fatal and, in certain situations, results in inaccessibility of harsh medical treatment.

4.7 Physical Security of IoT Nodes

The open ports for configuration or maintenance provide a way for the attackers to enter the system. The adversary can acquire direct access to the storage devices by exploiting those open ports, which will give him control over the operating system and allow him to obtain sensitive information such as embedded password. Thus, the lack of physical hardening gives rise to attacks like side-channel and node cloning.

4.7.1 Node Cloning

As the IoT nodes are left in unsupervised surroundings, the clone of it can be made easily. Furthermore, there is no standard method for creating IoT devices with tamper-proof hardware. By replicating the IoT nodes, attackers are able to launch multiple attacks and potentially can get access to the network by using the compromised node [49].

4.7.2 Side-Channel Attacks

The Side-Channel attacks aim to obtain information about the devices performing operations regarding cryptography over the side channel. Physical attributes of a machine during certain processes such as its processing speed, noise level, electro-magnetic emission, and power consumption are all included in these data, and these data are then used for reverse engineering the cryptographic system that the device uses [50].

4.8 No Energy Harvesting

Due to the limited resource nature, the IoT nodes are defenseless against resource exhaustion attacks. Furthermore, there is no existing system for harvesting energy from these low-powered IoT nodes [51]. These resource exhaustion attacks are capable of severely disrupting the communication channel and can lead to unauthorized heavy use of IoT resources such as disk space, CPU time, memory, bandwidth.

Attacks like battery drainage attacks and node outage attacks can be launched by exploiting this vulnerability.

4.8.1 Battery Drainage Attack

IoT nodes run on batteries with low power and above that without a recharging mechanism. In this attack, the adversary overloads the nodes with legal request that it leads to exhaustion. By depleting a node's energy numerous attacks can be conducted [52].

4.8.2 Node Outage

This attack prevents the edge nodes from carrying out their intended system role. Factors like sleep deprivation, unauthorized access, code injection, battery drainage are the major cause of this attack.

4.9 Open Debugging Ports

Through open ports, potential attackers exploit the vulnerabilities in the network services through open ports that operate on the target devices. The majority of the IoT devices are shipped by manufacturers with their debug port disabled and thus providing a vulnerable port that can be exploited in order to gain full access of the system. By exploiting these ports, the adversary is able to spy, steal information, bypass the security, modify the firmware, and inject malicious code. Thus, by exploiting these open ports, a plethora of attacks can be launched such as botnets like Bricker Bot, Mirai which exploits the telnet port [53].

4.10 Broken Access Control

The vulnerability in IoT system named broken access control is a vulnerability primarily related to the risks such as data hijacking and privacy leakage [54]. These vulnerabilities arise from vendor noncompliance with the communication model specification during implementation or from design defect in the interaction among parties while performing a particular function.

The above-stated vulnerabilities and threats associated with those vulnerabilities are summarized in Table 3.

Table 3 IoT vulnerabilities and threats due to those vulnerability

Serial number	Vulnerabilities	Attacks and threats	Description of threats
1	Insufficient audit mechanism	System compromise	Unauthorized access often leading to data theft, manipulation, or disruption
		Data corruption and loss	Loss of data often leading to service disruption and loss of information
2	Boot process vulnerability [41]	Complete device compromise	Attackers gaining full access of the system for manipulating functionalities and extracting sensitive data
		Data theft and exfiltration	Unauthorized access and removal of data from system for malicious purpose
3	Weak authentication mechanism [42]	Sybil attack [43]	Malicious node impersonating a genuine node for malicious activities affecting the network performance
		Distributed denial of service attack [44]	Flooding the target service, server or network with internet traffic making it slow or unavailable
		Man-in-the-middle attack [45]	Breaching network and creating a separate connection to the communication channel for listening or modifying communication packets on either end
		Hello flood attack [46]	Flooding the network with hello packets, overwhelming nodes pretending to be a neighbour and depleting their resources
		Homing attack [46]	Analysing the network for identifying and targeting specific nodes or communication path
		Dictionary attack [44]	A brute force attack for getting access to the system by trying every character in the dictionary
4	Unauthorized access [44]	Data leakage attacks and data theft	Unauthorized access and disclosure of sensitive data often resulting in financial loss or reputation damage
5	Improper patch management [47]	Firmware tampering	Modification of device firmware to introduce vulnerability or malicious code
		Data theft	Unauthorized attacks and removal of data from system
6	Improper encryption [48]	Data leakage attacks	Accidental or intentional disclosure of sensitive data
7	Physical security of IoT nodes	Node cloning attack [49]	Cloning and using the compromised node in getting access to the network

(continued)

Table 3 (continued)

Serial number	Vulnerabilities	Attacks and threats	Description of threats
		Side channel attack [50]	Exploiting physical attributes of a device over a side channel during cryptographic operations to extract secret information
8	No energy harvesting [51]	Battery drainage attack [52]	Overloading the system with legal requests thus depleting its energy and making it vulnerable for various attacks
		Node outage attack	Preventing the edge node from carrying out its role caused due to multiple factors
9	Open debugging ports [53]	Malicious code injection	Introducing malicious code into a system to hamper its security and functionality
		Firmware modification	Altering the device's firmware to introduce vulnerabilities or malicious code
		Security bypass	Attackers evading security measures to gain unauthorized access or control
10	Broken access control [54]	Data hijacking	Intercepting and controlling data transmission to get unauthorized access or modify information
		Privacy leakage	Disclosure of private or sensitive information

5 Recent Advancements

Recent years have seen a rapid emergence in the field of PQC, specifically in the application of IoT network. These developments aim to safeguard this network against the emerging quantum attacks, providing enhanced security. The following section explores the latest advancements that are shaping a better future for secure IoT communication in the quantum environment.

Ebrahimi et al. [55] presented InvRBLWE that is an optimized modification of Ring-LWE scheme for hardware implementations that is highly efficient and is proven to be quantum secure. Lee et al. [56] presented RLizard, a KEM which enhances the efficiency by leveraging the RLWE and rounding problems outperforming the original Lizard algorithm in terms of key generation speed and size. Phoon et al. [57] presented an optimized implementation of QC-MDPC KEM for IoT application. Further, this work provides three contribution implementations in FGPA consisting of the fastest QC-MDPC McEliece decryption, first QC-MDPC KEM implementation, and first count attack-resistant decoder iteration. Zhang et al. [58] constructed an effective countermeasure to the identified vulnerabilities associated with three main operations in every binary search state based on randomization at 58.4% extra slices cost. Althobaiti et al. [59] presented a novel cryptographic system where the geographic location is used as an individual's identity and credential that ensures PQ IoT security. Islam et al. [60] presented a new three-party authentication key agreement (PB-3PAKA) protocol using ideal lattices for mobile users. Further, this work provides the analysis that the devised protocol is secure in the post-quantum environment. Gupta et al. [61] presented the first implemented GPU for the acceleration of PQC algorithms, Kyber-1024, NewHope-1024, and FrodoKEM-976 belonging to different classes: Module-LWE, Ring-LWE, and LWE. Lee et al. [62] presented Kyber, an efficient and novel implementation technique for accelerating a PQ KEM on a GPU (Graphics Processing Unit). Further, this work also presents fully parallel implementation of number theoretic transform (NTT) with combined levels which makes it 2.6 times faster than that of the best available result so far by leveraging a GPU.

In another work, Xu et al. [63] proposed a quantum-resistant and high-performance encryption scheme Ring-ExpLWE where the sampling of the error vector is done using exponential distribution. Kumari et al. [64] presented a lightweight and robust PQ code-based hybrid encryption scheme and lattice-based authentication for IoT-based applications. Furthermore, this work also introduces Bernstein reconstruction for minimizing the computation cost. Gupta et al. [65] designed a new certificateless data authentication protocol in order to provide security characteristics in wireless open communication in the Internet of Vehicles (IoV). Lee et al. [66] presented a novel implementation of Saber algorithms on GPU achieving a result of 31,658 with T4, 120,463 with V100, and 123,418 with RTX3080 GPUs. Hadayeghparast et al. [67] presented an efficient implementation of Ring-BinLWE over the RISC-V platform, achieving a huge 51% increased improvement in speed of CPU cycles while having a low complexity and adding only 6–9% overhead to original core. Kim et al. [68] presented an efficient implementation of Falcon (one of the best digital signature techniques in the current PQC standardization finalist held by NIST) over the ARMv8 environment including Cortex-A70, 50, and 30 series which are widely used for IoT applications.

Additionally, Dharminder et al. [69] designed an effective lattice-based authenticated KEP for IoT-based smart applications by employing ring-based version of learning with errors assumption. Hafeez et al. [70] presented two new methods to accelerate polynomial convolution on GPUs by leveraging tensor cores for improved key encapsulation performance revealing a notable speed gain, especially in the Florete and Sable algorithms. Zhang et al. [71] proposed an efficient and secure authentication mechanism for two-party handover for mobile devices without home agent which is based on NTRU. Furthermore, BAN logic is employed for the validation of the correctness of the presented protocol. Shahidinejad et al. [72] proposed a new decentralized KEM and authentication mechanism for IoT communication from one device to another. Rewal et al. [73] proposed three-party authentication protocol with improved efficiency and security by making the protocol immune to password-guessing attacks.

Similarly, Xu et al. [74] presented a group authentication methodology where identity-based encryption (IBE) and lattice-based aggregate signature algorithm are harnessed in order to achieve concurrent authentication of accessed devices. Gupta et al. [75] designed a novel and lightweight identity-based authenticated KEP by employing lattice cryptography providing multiple security features for open wireless communication for IIoT (Industrial IoT). Further, this work also states that because of the hard assumptions of the lattices, the proposed design is quantum-secure. Mohanty et al. [76] proposed the first Threshold Private Set Intersection (TPSI) protocol based on QC that directly addresses ride-sharing privacy preservation. He further stated this work achieves long-term security and is quantum secure as well because of the application QC. Prateek et al. [77] introduced the Q-Secure-P2-SMA, a privacy preserving protocol for smart meter authentication that combines a hash function with semi-quantum key distribution mechanism to enable realistic implementation for quantum-resistant, privacy-focused metering data transfer. Table 4 summarises these recent advancements in detail, providing a comprehensive overview of their major contribution.

6 Future Research Directions

Effective post-quantum IoT cryptosystem has the potential to improve the security of numerous related domains whose application highly rely on battery-dependent and resource-constrained IoT devices, such as Industry 4.0 [80, 81], smart home automation [82], precision agriculture [83], and smart cities [84, 85]. Furthermore, the results gained might be helpful for developing and enhancing post-quantum networking architecture that utilizes devices with limited resources. Additionally, domains and applications with vital medium- and long-term security such as public safety and defence, smart healthcare, or mission-critical situations can benefit greatly from novel energy-efficient post-quantum IoT systems. For establishing an effective quantum-secure IoT systems, the ideal contributions that would be required are stated below.

- Most of the key establishment approaches especially based on computational methods are the centre of most researchers' focus and have been successful in developing quantum-resistant methods so far. However, now researchers have to explore further and more deeply into the key establishment techniques based on physics, which are commonly known as quantum key distribution (QKD).
- New methods will have to be devised for the analysis, evaluation, and understanding of the effects of quantum computing on various theoretical aspects related to an IoT network.
- Large amount of data comprising structured and unstructured data are generated by multimedia devices, thus processing these data over depletes a large amount of energy. Therefore, a smart feature extraction methodology is essential for the classification and identification of multimedia data maintaining a minimum consumption of energy.

Table 4 Summary of recent advancements

References	Year	Major contribution
Ebrahimi et al. [55]	2019	InvRBLWE, a modification of Ring-LWE for hardware implementation and is quantum secure
Lee et al. [56]	2019	RLizard, a KEM leveraging RingLWE with rounding problem outperforming the Lizard scheme
Alarifi et al. [78]	2020	A new hybrid cryptosystem by merging Mandelbrot sets, Arnold Chaotic map, and DNA sequences in order to secure the HVEC compresses streams
Phoon et al. [57]	2020	QC-MDPC KEM, a key encapsulation mechanism for IoT providing fastest decryption, first implementation in FGPA and first iteration count attack-resistant decoder
Zhang et al. [58]	2020	An effective countermeasure for the identified vulnerabilities for the binary search state at the cost of 58.4% extra slices
Althobaiti et al. [59]	2021	Location-based cryptosystem that guarantees post-quantum IoT security
Islam et al. [60]	2021	PB-3PAKA protocol leveraging ideal lattice-based cryptography for mobile devices
Gupta et al. [61]	2021	First incorporation of GPU for acceleration of PQC algorithms, namely, Kyber-1024, NewHope-1024, and FrodoKEM-976 belonging to different classes
Aribilola et al. [79]	2022	A novel 8-bit non-linear S-box design providing a robust security when embedded with an IoT stream Cipher, Chacha20 in order to improve the protection on visual multimedia data transmission
Lee et al. [62]	2022	A novel efficient implementation methodology for accelerating kyber (a post-quantum KEM) using a GPU
Xu et al. [63]	2022	Ring-ExpLWE, a high-performance encryption scheme for quantum-cryptography
Kumari et al. [64]	2022	Robust & lightweight PQ authentication and hybrid encryption scheme for IoT devices
Gupta et al. [65]	2022	A certificateless data authentication protocol providing security for the wireless open communication in IoV
Lee et al. [66]	2022	Enhanced Saber algorithm by leveraging T4, V100, and RTX3080 GPUs
Hadayeghparast et al. [67]	2022	An enhanced implementation of Ring-BinLWE over RISC-V platform with 51% increased efficiency and only 6–9% overhead
Kim et al. [68]	2022	Enhanced implementation of Falcon over ARMv8 environment including Cortex-A70, 50 and 30 series for IoT
Dharminder et al. [69]	2023	RingLWE and proposed lattice-based cryptosystem for KEM for IoT-enabled smart devices
Hafeez et al. [70]	2023	Leveraged tensor core for enhancing key encapsulation performance resulting in a notable speed gain in the Florete and Sable schemes
Zhang et al. [71]	2023	Two-party NTRU-based authentication mechanism for mobile devices employing BAN logic

(continued)

Table 4 (continued)

References	Year	Major contribution
Shahidinejad et al. [72]	2023	Key exchange and authentication mechanism for IoT D2D communication
Rewal et al. [73]	2023	Saber algorithm's implementation on GPU with results enhancing resilience to password-guessing attacks
Xu et al. [74]	2024	Group authentication system harnessing IBE and lattice-based aggregate signature algorithm to achieve authentication of concurrent accessed device at once
Gupta et al. [75]	2024	A novel identity-based cryptosystem employing lattice cryptography enabling wireless communication in IIoT and making it quantum resistant
Mohanty et al. [76]	2024	TPSI protocol based on quantum cryptography directly addressing ride-sharing privacy preservation
Prateek et al. [77]	2024	Quantum-Secure-P2-SMA protocol for smart meter authentication mechanism by combining hash function with semi-quantum key distribution mechanism

- For further enhancement of the computation speed, incorporation of parallel diffusion and permutation concept can be done to the specified cryptosystems. Furthermore, exploration in the field of multilevel security for the HVEC communication by leveraging the steganography and watermarking algorithms is required.
- For further enhancement of the cryptosystems for the multimedia IoT, permutation and parallel diffusion concepts can also be incorporated.
- Validation and evaluation techniques of energy-efficient as well as quantum-resistant schemes for IoT-based applications will be greatly required.
- For the development of code-based PQC for multimedia IoT, novel coding methodologies must be explored, designed, developed, and optimized.
- PQ IoT designs that are energy efficient are required to be developed and put into practice for small-scale, medium-scale, and large-scale scenarios.

7 Conclusion

IoT is described as a network of devices connected with each other for the purpose of communication and transmission of data. The need towards the IoT-based applications has extensively increased, thus increasing the focus towards quantum computing in this direction. Despite the IoT capabilities, multimedia IoT devices come with some significant vulnerabilities leading to breaches and attacks. The supremacy in the processing potential quantum computers also poses a significant risk for these multimedia IoT devices. The classical cryptosystem is not capable for securing against these quantum computers thus, a rise towards the need for a PQC, a cryptosystem resilient against quantum computers is witnessed. PQC is an emerging field that develops cryptosystem and algorithms to secure against quantum attacks.

In this paper, we explore cryptography and it's working. Furthermore, we discuss its different types and algorithms and its performance analysis relative to quantum computing. The significance of a new cryptosystem is discussed, which can be resistant to quantum computing attacks. We also present PQC's overview and its types and analysis of the 17 NIST second-round candidate PQ algorithms with their different levels. This work also presents multimedia IoT-associated vulnerabilities with their threats and attacks, respectively. This work also lists out the recent advancements in the aspect of multimedia IoT security using PQC for enhancing compatibility with battery-dependent and resource-constrained devices. Finally, we discuss future research directions for further improvement and advancement of the topic.

References

1. Halak, B., Yilmaz, Y., Shiu, D.: Comparative analysis of energy costs of asymmetric vs symmetric encryption-based security applications. IEEE Access **10**, 76707–76719 (2022). https://doi.org/10.1109/access.2022.3192970
2. Oonishi, K., Kunihiro, N.: Shor's algorithm using efficient approximate quantum Fourier transform. IEEE Trans. Quant. Eng. **4**, 1–16 (2023). https://doi.org/10.1109/tqe.2023.3319044
3. Lee, S., Nam, S.Y.: Implementation of grover's iterator for quantum searching with an arbitrary number of qubits. IEEE Access **12**, 43027–43038 (2024). https://doi.org/10.1109/access.2024.3380198
4. Park, J.-Y., Moon, Y.-H., Lee, W., Kim, S.-H., Sakurai, K.: A survey of polynomial multiplication with RSA-ECC coprocessors and implementations of NIST PQC round3 KEM algorithms in EXYNOS2100. IEEE Access **10**, 2546–2563 (2022). https://doi.org/10.1109/access.2021.3138807
5. Borges, F., Reis, P.R., Pereira, D.: A comparison of security and its performance for key agreements in post-quantum cryptography. IEEE Access **8**, 142413–142422 (2020). https://doi.org/10.1109/access.2020.3013250
6. Galbraith, S.D., Gaudry, P.: Recent progress on the elliptic curve discrete logarithm problem. Des. Codes Crypt. **78**(1), 51–72 (2015). https://doi.org/10.1007/s10623-015-0146-7
7. Gupta, B.B., Quamara, M.: An overview of Internet of Things (IOT): architectural aspects, challenges, and protocols. Concurr. Comput. Pract. Exp. **32**(21). https://doi.org/10.1002/cpe.4946
8. Chen, S., Xu, H., Liu, D., Hu, B., Wang, H.: A Vision of IOT: applications, challenges, and opportunities with China perspective. IEEE Internet Things J. **1**(4), 349–359 (2014). https://doi.org/10.1109/jiot.2014.2337336
9. Nakamura, Y., Anh, S.P.: A Baseline Investigation into the Evolution and Prevalence of Mirai and Hajime Utilizing a Network TELESCOPE_SUPP1–3434555.PDF (2023). https://doi.org/10.1109/access.2024.3434555/mm2
10. Ricci, S., Dobias, P., Malina, L., Hajny, J., Jedlicka, P.: Hybrid keys in practice: combining classical, quantum and post-quantum cryptography. IEEE Access **12**, 23206–23219 (2024). https://doi.org/10.1109/access.2024.3364520
11. Diffie, W., Hellman, M.: New directions in cryptography. IEEE Trans. Inf. Theory **22**(6), 644–654 (1976). https://doi.org/10.1109/tit.1976.1055638
12. Feynman, R.P.: Simulating physics with computers. Feynman Comput. 133–153 (n.d.). https://doi.org/10.1201/9780429500459-11
13. Vlachos, P., Karafyllidis, I.G.: Quantum game simulator, using the circuit model of quantum computation. Comput. Phys. Commun. **180**(10), 1990–1998 (2009). https://doi.org/10.1016/j.cpc.2009.05.007

14. Chen, L., Jordan, S., Liu, Y.-K., Moody, D., Peralta, R., Perlner, R., Smith-Tone, D.: Rep. Post-Quant. Cryptogr. (2016). https://doi.org/10.6028/nist.ir.8105
15. Panchi, L., Shiyong, L.: Grover Quantum searching algorithm based on weighted targets. J. Syst. Eng. Electron. **19**(2), 363–369 (2008). https://doi.org/10.1016/s1004-4132(08)60093-6
16. Yan, S.Y.: Quantum safe cryptography. Cybercryptogr. Appl. Cryptogr. Cyberspace Secur. 399–412 (2018). https://doi.org/10.1007/978-3-319-72536-9_8
17. IBM roadmap to Quantum-centric supercomputers (updated 2024)*: IBM quantum computing blog*. IBM roadmap to quantum-centric supercomputers (Updated 2024) | IBM Quantum Computing Blog. (n.d.). https://www.ibm.com/quantum/blog/ibm-quantum-roadmap-2025
18. Blömer, J., Naewe, S.: Sampling methods for shortest vectors, closest vectors and successive minima. Lect. Notes Comput. Sci. 65–77 (n.d.). https://doi.org/10.1007/978-3-540-73420-8_8
19. Chen, B., Zhang, G.: The number of extended irreducible binary GOPPA codes. IEEE Trans. Inf. Theory **69**(12), 7691–7711 (2023). https://doi.org/10.1109/tit.2023.3317290
20. Loi, K.C., Ko, S.-B.: Scalable elliptic curve cryptosystem FPGA processor for NIST prime curves. IEEE Trans. Very Large Scale Integr. (VLSI) Syst. **23**(11), 2753–2756 (2015). https://doi.org/10.1109/tvlsi.2014.2375640
21. Bernstein, D.J.: Introduction to post-quantum cryptography. Post-Quant. Cryptogr. 1–14. https://doi.org/10.1007/978-3-540-88702-7_1
22. Regev, O.: On lattices, learning with errors, random linear codes, and cryptography. In: Proceedings of the Thirty-Seventh Annual ACM Symposium on Theory of Computing (2005). https://doi.org/10.1145/1060590.1060603
23. Schwabe, P.: NewHope (2017). https://newhopecrypto.org/
24. Rollo. ROLLO. (n.d.). https://pqc-rollo.org/
25. Falcon. (n.d.). https://falcon-sign.info/
26. Blömer, J., Naewe, S.: Sampling methods for shortest vectors, closest vectors and successive minima. Theoret. Comput. Sci. **410**(18), 1648–1665 (2009). https://doi.org/10.1016/j.tcs.2008.12.045
27. Canteaut, A., Sendrier, N.: Cryptanalysis of the original McEliece cryptosystem. Lect. Notes Comput. Sci. **187–199** (2000). https://doi.org/10.1007/3-540-49649-1_16
28. Barreto, P.S., Gueron, S., Güneysu, T., Misoczki, R., Persichetti, E., Sendrier, N., Tillich, J.-P.: Cake: code-based algorithm for key encapsulation. Lect. Notes Comput. Sci. **207–226**,(2017). https://doi.org/10.1007/978-3-319-71045-7_11
29. Bit flipping key encapsulation. BIKE. (n.d.). https://bikesuite.org/
30. Bos, J., Ducas, L., Kiltz, E., Lepoint, T., Lyubashevsky, V., Schanck, J. M., Schwabe, P., Seiler, G., Stehle, D.: Crystals—kyber: a CCA-secure module-lattice-based KEM. In: 2018 IEEE European Symposium on Security and Privacy (EuroS&P) (2018). https://doi.org/10.1109/eurosp.2018.00032
31. Crystals. Kyber. (n.d.). https://pq-crystals.org/kyber/
32. Reshikeshan, S.S., Koh, M.B., Illindala, M.: Rainbow signature scheme to secure goose communications from quantum computer attacks. IEEE Trans. Ind. Appl. **57**(5), 4579–4586 (2021). https://doi.org/10.1109/tia.2021.3093375
33. SPHINCS+. (n.d.). https://sphincs.org/
34. Sun, S., Zhang, R., Ma, H.: Efficient parallelism of post-quantum signature scheme SPHINCS. IEEE Trans. Parallel Distrib. Syst. **31**(11), 2542–2555 (2020). https://doi.org/10.1109/tpds.2020.2995562
35. Nguyen, T.-H., Kieu-Do-Nguyen, B., Pham, C.-K., Hoang, T.-T.: High-speed NTT accelerator for crystal-kyber and crystal-dilithium. IEEE Access **12**, 34918–34930 (2024). https://doi.org/10.1109/access.2024.3371581
36. Bos, J., Costello, C., Ducas, L., Mironov, I., Naehrig, M., Nikolaenko, V., Raghunathan, A., Stebila, D.: Frodo: take off the ring! practical, quantum-secure key exchange from LWE. In: Proceedings of the 2016 ACM SIGSAC Conference on Computer and Communications Security (2016). https://doi.org/10.1145/2976749.2978425
37. Frodokem. FrodoKEM. (n.d.). https://frodokem.org/
38. ThreeBears. SourceForge. (n.d.). https://sourceforge.net/projects/threebears/

39. D'Anvers, J.-P., Karmakar, A., Sinha Roy, S., Vercauteren, F.: Saber: Module-LWR based key exchange, CPA-secure encryption and CCA-secure kem. Lect. Notes Comput. Sci. **282–305**,(2018). https://doi.org/10.1007/978-3-319-89339-6_16
40. Siwakoti, Y.R., Bhurtel, M., Rawat, D.B., Oest, A., Johnson, R.C.: Advances in IoT security: vulnerabilities, enabled criminal services, attacks, and countermeasures. IEEE Internet Things J. **10**(13), 11224–11239 (2023). https://doi.org/10.1109/jiot.2023.3252594
41. Sridevi, & Kamble, M.: Privacy and security in wireless devices for the internet of things. Adv. Inf. Secur. Priv. Ethics **105–130** (2022). https://doi.org/10.4018/978-1-6684-3921-0.ch006
42. Roy, P.K., Bhattacharya, A.: An anonymity-preserving mobile user authentication protocol for global roaming services. Comput. Netw. **221**, 109532 (2023). https://doi.org/10.1016/j.comnet.2022.109532
43. Pu, C.: Sybil attack in RPL-based internet of things: Analysis and defenses. IEEE Internet Things J. **7**(6), 4937–4949 (2020). https://doi.org/10.1109/jiot.2020.2971463
44. Borys, A., Kamruzzaman, A., Thakur, H.N., Brickley, J.C., Ali, M.L., Thakur, K.: An evaluation of IoT ddos cryptojacking malware and Mirai botnet. In: 2022 IEEE World AI IoT Congress (AIIoT) (2022). https://doi.org/10.1109/aiiot54504.2022.9817163
45. Olazabal, A.A., Kaur, J., Yeboah-Ofori, A.: Deploying man-in-the-middle attack on IoT devices connected to Long Range Wide Area Networks (lorawan). In: 2022 IEEE International Smart Cities Conference (ISC2) (2022). https://doi.org/10.1109/isc255366.2022.9922377
46. Koosha, M., Farzaneh, B., Farzaneh, S.: A classification of RPL specific attacks and countermeasures in the internet of things. In: 2022 Sixth International Conference on Smart Cities, Internet of Things and Applications (SCIoT), vol. 21, no. 2, pp. 1–7 (2022). https://doi.org/10.1109/sciot56583.2022.9953631
47. Alsabbagh, W., Langendörfer, P.: A flashback on control logic injection attacks against programmable logic controllers. Automation **3**(4), 596–621 (2022). https://doi.org/10.3390/automation3040030
48. Venkatraman, S., Alazab, M., Vinayakumar, R.: A hybrid deep learning image-based analysis for effective malware detection. J. Inf. Secur. Appl. **47**, 377–389 (2019). https://doi.org/10.1016/j.jisa.2019.06.006
49. Faiz, M., Daniel, A.K.: Threats and challenges for security measures on the internet of things. Law, State Telecommun. Rev. **14**(1), 71–97 (2022). https://doi.org/10.26512/lstr.v14i1.38843
50. Yan, Y., Oswald, E., Vivek, S. (2021). An analytic attack against ARX addition exploiting standard side-channel leakage. In: Proceedings of the 7th International Conference on Information Systems Security and Privacy, pp. 89–97. https://doi.org/10.5220/0010223600890097
51. Trappe, W., Howard, R., Moore, R.S.: Low-energy security: limits and opportunities in the internet of things. IEEE Secur. Priv. **13**(1), 14–21 (2015). https://doi.org/10.1109/msp.2015.7
52. Gosavi, P.S., Patil, B.M.: Vampire attack: draining life from wireless AD–Hoc sensor networks. Int. J. Comput. Appl. **144**(9):1–4. https://doi.org/10.5120/ijca2016910424
53. Sachidananda, V., Siboni, S., Shabtai, A., Toh, J., Bhairav, S., Elovici, Y.: Let the cat out of the bag. In: Proceedings of the 3rd ACM International Workshop on IoT Privacy, Trust, and Security (2017). https://doi.org/10.1145/3055245.3055251
54. Ding, W., Hu, H., Cheng, L.: IoTSafe: enforcing safety and security policy with real IoT physical interaction discovery. In: Proceedings 2021 Network and Distributed System Security Symposium (2021). https://doi.org/10.14722/ndss.2021.24368
55. Ebrahimi, S., Bayat-Sarmadi, S., Mosanaei-Boorani, H.: Post-quantum cryptoprocessors optimized for edge and resource-constrained devices in IOT. IEEE Internet Things J. **6**(3), 5500–5507 (2019). https://doi.org/10.1109/jiot.2019.2903082
56. Lee, J., Kim, D., Lee, H., Lee, Y., Cheon, J.H.: RLIZARD: post-quantum Key encapsulation mechanism for IoT devices. IEEE Access **7**, 2080–2091 (2019). https://doi.org/10.1109/access.2018.2884084
57. Phoon, J.-H., Lee, W.-K., Wong, D.C.-K., Yap, W.-S., Goi, B.-M., Phan, R.C.-W.: Optimized IoT cryptoprocessor based on QC-MPDC key encapsulation mechanism. IEEE Internet Things J. **7**(9), 8513–8524 (2020). https://doi.org/10.1109/jiot.2020.2991334

58. Zhang, C., Liu, Z., Chen, Y., Lu, J., Liu, D.: A flexible and generic gaussian sampler with power side-channel countermeasures for quantum-secure internet of things. IEEE Internet Things J. **7**(9), 8167–8177 (2020). https://doi.org/10.1109/jiot.2020.2981133
59. Althobaiti, O.S., Dohler, M.: Quantum-resistant cryptography for the internet of things based on location-based lattices. IEEE Access **9**, 133185–133203 (2021). https://doi.org/10.1109/access.2021.3115087
60. Islam, S.H., Basu, S.: PB-3PAKA: password-based three-party authenticated key agreement protocol for mobile devices in post-quantum environments. J. Inf. Secur. Appl. (2021). https://www.sciencedirect.com/science/article/abs/pii/S221421262100185X
61. Gupta, N., Jati, A., Chauhan, A.K., Chattopadhyay, A.: PQC acceleration using gpus: FrodoKEM, NewHope, and Kyber. IEEE Trans. Parallel Distrib. Syst. **32**(3), 575–586 (2021). https://doi.org/10.1109/tpds.2020.3025691
62. Lee, W.-K., Seo, H., Hwang, S.O., Achar, R., Karmakar, A., Mera, J.M.: DPCrypto: acceleration of post-quantum cryptography using dot-product instructions on gpus. IEEE Trans. Circuits Syst. I Regul. Pap. **69**(9), 3591–3604 (2022). https://doi.org/10.1109/tcsi.2022.3176966
63. Xu, D., Wang, X., Hao, Y., Zhang, Z., Hao, Q., Jia, H., Dong, H., Zhang, L.: Ring-expLWE: a high-performance and lightweight post-quantum encryption scheme for resource-constrained IoT devices. IEEE Internet Things J. **9**(23), 24122–24134 (2022). https://doi.org/10.1109/jiot.2022.3189210
64. Kumari, S., Singh, M., Singh, R., Tewari, H.: A post-quantum lattice based lightweight authentication and code-based hybrid encryption scheme for IoT devices. Comput. Netw. **217**, 109327 (2022). https://doi.org/10.1016/j.comnet.2022.109327
65. Gupta, D.S., Karati, A., Saad, W., da Costa, D.B.: Quantum-defended blockchain-assisted data authentication protocol for internet of vehicles. IEEE Trans. Veh. Technol. **71**(3), 3255–3266 (2022). https://doi.org/10.1109/tvt.2022.3144785
66. Lee, W.-K., Hwang, S.O.: High throughput implementation of post-quantum key encapsulation and decapsulation on GPU for internet of things applications. In: 2022 IEEE World Congress on Services (SERVICES) (2022). https://doi.org/10.1109/services55459.2022.00026
67. Hadayeghparast, S., Bayat-Sarmadi, S., Ebrahimi, S.: High-speed post-quantum cryptoprocessor based on RISC-V architecture for IOT. IEEE Internet Things J. **9**(17), 15839–15846 (2022). https://doi.org/10.1109/jiot.2022.3152850
68. Kim, Y., Song, J., Seo, S.C.: Accelerating falcon on ARMv8. IEEE Access **10**, 44446–44460 (2022). https://doi.org/10.1109/access.2022.3169784
69. Dharminder, D., Reddy, C.B., Das, A.K., Park, Y., Jamal, S.S.: Post-quantum lattice-based Secure Reconciliation enabled key agreement protocol for IOT. IEEE Internet Things J. **10**(3), 2680–2692 (2023). https://doi.org/10.1109/jiot.2022.3213990
70. Hafeez, M.A., Lee, W.-K., Karmakar, A., Hwang, S.O.: High throughput acceleration of scabbard key exchange and key encapsulation mechanism using tensor core on GPU for IoT applications. IEEE Internet Things J. **10**(22), 19765–19781 (2023). https://doi.org/10.1109/jiot.2023.3282255
71. Zhang, S., Du, X., Liu, X.: A novel and quantum-resistant handover authentication protocol in IoT environment. Wirel. Netw. **29**(6), 2873–2890 (2023). https://doi.org/10.1007/s11276-023-03342-4
72. Shahidinejad, A., Abawajy, J.: Decentralized lattice-based device-to-device authentication for the edge-enabled IOT. IEEE Syst. J. **1–11**,(2023). https://doi.org/10.1109/jsyst.2023.3319280
73. Rewal, P., Singh, M., Mishra, D., Pursharthi, K., Mishra, A.: Quantum-safe three-party lattice based authenticated key agreement protocol for mobile devices. J. Inf. Secur. Appl. (2023). https://www.sciencedirect.com/science/article/abs/pii/S2214212623000893
74. Xu, P., Wu, H., Tao, X., Wang, C., Chen, D., Nan, G.: Anti-quantum certificateless group authentication for massive accessing IoT devices. IEEE Internet Things J. **11**(9), 16561–16577 (2024). https://doi.org/10.1109/jiot.2024.3353807
75. Gupta, D.S.: Pilike: post-quantum identity-based lightweight authenticated key exchange protocol for iiot environments. IEEE Syst. J. **18**(1), 15–23 (2024). https://doi.org/10.1109/jsyst.2023.3335217

76. Mohanty, T., Srivastava, V., Debnath, S.K., Das, A.K., Sikdar, B.: Quantum secure threshold private set intersection protocol for IOT-enabled privacy-preserving ride-sharing application. IEEE Internet Things J. **11**(1), 1761–1772 (2024). https://doi.org/10.1109/jiot.2023.3291132
77. Prateek, K., Das, M., Surve, S., Maity, S., Amin, R.: Q-secure-p2-sma: Quantum-secure privacy-preserving smart meter authentication for Unbreakable Security in smart grid. IEEE Trans. Netw. Serv. Manage. **21**(5), 5149–5163 (2024). https://doi.org/10.1109/tnsm.2024.3357103
78. Alarifi, A., Sankar, S., Altameem, T., Jithin, K.C., Amoon, M., El-Shafai, W.: A novel hybrid cryptosystem for secure streaming of high efficiency H.265 compressed videos in IOT multimedia applications. IEEE Access **8**, 128548–128573 (2020). https://doi.org/10.1109/access.2020.3008644
79. Aribilola, I., Lee, B., Naveed Asghar, M.: Möbius transformation and permutation based S-box to enhance IOT multimedia security. IEEE Access **12**, 140792–140808 (2024). https://doi.org/10.1109/access.2024.3466930
80. Fernandez-Carames, T.M., Fraga-Lamas, P.: A review on human-centered IOT-connected smart labels for the industry 4.0. IEEE Access **6**, 25939–25957 (2018). https://doi.org/10.1109/access.2018.2833501
81. Fraga-Lamas, P., Fernandez-Carames, T.M., Blanco-Novoa, O., Vilar-Montesinos, M.A.: A review on industrial augmented reality systems for the industry 4.0 shipyard. IEEE Access **6**, 13358–13375 (2018). https://doi.org/10.1109/access.2018.2808326
82. Suárez-Albela, M., Fraga-Lamas, P., Fernández-Caramés, T., Dapena, A., González-López, M.: Home automation system based on intelligent transducer enablers. Sensors **16**(10), 1595 (2016). https://doi.org/10.3390/s16101595
83. Pérez-Expósito, J., Fernández-Caramés, T., Fraga-Lamas, P., Castedo, L.: Vinesens: an eco-smart decision-support viticulture system. Sensors **17**(3), 465 (2017). https://doi.org/10.3390/s17030465
84. Hernández-Rojas, D.L., Fernández-Caramés, T.M., Fraga-Lamas, P., Escudero, C.J.: A plug-and-play human-centered virtual TEDS architecture for the web of things. Sensors **18**(7), 2052 (2018). https://doi.org/10.3390/s18072052
85. Alam, F., Mehmood, R., Katib, I., Albogami, N.N., Albeshri, A.: Data fusion and IoT for smart ubiquitous environments: a survey. IEEE Access **5**, 9533–9554 (2017). https://doi.org/10.1109/access.2017.2697839

Empowering Metaverse Using Decentralized Blockchain Technology for Multimedia Applications

Ritika Rajput, Bharat Bhushan, Ahmed J. Obaid, and Khursheed Aurangzeb

Abstract Metaverse technology has been becoming more prevalent recently because of its potential to completely alter how people share and interact with each other worldwide. Central to this growth are two key technologies: blockchain and multimedia. Based on these technologies, features of metaverse recently attract the attention due to its immersive and decentralized nature which include other features also. The inclusion of blockchain technology into facilities of metaverse results in the rise of shared ecosystem and it also offers privacy protection, and security and digital asset monitoring in both real world and in virtual assets. The opportunities and challenges associated with the deployment and development of metaverse technologies, including privacy, scalability, and security issues, are fixed by blockchain integration while multimedia technology is pivotal for creating immersive experiences allowing realistic visuals within the metaverse which support the metaverse to attract the users worldwide. In this paper, an overview of the current state of metaverse technologies are provided with multimedia and blockchain technology. This paper begins by description of blockchain technology following with multimedia technology. Further, the paper discusses various components and characteristics of the metaverse, its different use cases, and associated technologies. Furthermore, the paper illustrates the integration of metaverse with blockchain and discusses about how multimedia technology enhances the metaverse. Finally, this paper highlights various recent advancements such as authentication and identity management, management of data, decentralized governance, system of reputation, extended reality technologies, 3D content creation, and digital avatars.

R. Rajput (✉) · B. Bhushan
Department of Computer Science and Engineering, School of Engineering and Technology, Sharda University, Greater Noida, India
e-mail: rajputritika2706@gmail.com

K. Aurangzeb
Department of Computer Engineering, College of Computer and Information Sciences, King Saud University, Riyadh, Saudi Arabia

A. J. Obaid
Faculty of Computer Science and Mathematics, Department of Computer Science, University of Kufa, Najaf, Iraq
e-mail: ahmedj.aljanaby@uokufa.edu.iq

R. Sharma et al. (eds.), *Multimedia Technologies in the Internet of Things Environment, Volume 4*, Studies in Big Data 173, https://doi.org/10.1007/978-981-96-4356-1_9

Keywords Metaverse · Multimedia technology · Blockchain · Augmented reality · Virtual reality · Immersive · Digital · Decentralize

1 Introduction

The Word Metaverse combines of two words "Meta" which means far as beyond and term "verse" means related to the universe [1]. Metaverse considered as the future of Internet applications in both industry and academia. They attract a massive attention toward virtual 3D environments where digital avatars represented as the humans. Metaverse has enticed vast expenditure, e.g., from Adidas, Meta (Facebook), and Microsoft [1]. Neal Stepheson introduced the concept of metaverse in his novel Snow Crash [2]. Metaverse is a virtual world which represents a digital frontier. It is a complete combination of various technologies, including XR, VR/AR, AI, Digital Twin, Internet of things, and blockchain. The metaverse created, retained, and improved with the help of VR and AR convey an immersive experience of 3D.

Multimedia technologies are crucial technologies used to integrate for a better user experience. This contains augmented reality (AR), virtual reality (VR), and 3D modeling which helps in creating an interactive environment that can be manipulated and explored by the users. Multimedia technologies are the building blocks of the metaverse's interactive design. Blockchain is a technology that functions as peer-to-peer network that operates on network of decentralization [3]. Significant elements of blockchain are persistency, anonymity, auditability, 3D content, and social interactivity. These technologies encourage the generation of individualized immersive experiences that are required for the expansion of platform while improving the sensory experience [4].

As the count of mobile devices and users increases, it affects the wireless communication in reality. These virtual spaces are becoming more important for daily work, which in order comes with more risks [5]. Further metaverse contains several characteristics such as immersive, embodied, and distributed virtual uniqueness. These characteristics, however, dive into how multimedia technology integrates with metaverse and enhances the experience of users [6]. The traditional security frameworks that are based upon perimeter-based defenses and security do not work well as these domains are dynamic and open [7]. Safety and confidentiality are important concerns in the metaverse [8]. For the assurance of the user's privacy and addresses the security challenges within the platform, blockchain comes out with effective solution for improving zero-trust structures based on transparency and decentralized network especially on the principle of immutability [9]. Blockchain is based upon several use cases such as reputation system, authentication system, decentralized governance, and data management [10].

This paper examines how multimedia technology and blockchain enhance the immersive and decentralization characteristics of metaverse technology. Multimedia technology along with audio, graphics, and video is essential for creating the realistic virtual experience and enhances the user interaction. Blockchain technology helps metaverse in providing a secure and transparent environment which manages the digital assets and user identities. By integrating blockchain, users' trust and privacy are promoted and by multimedia integration, experience of virtual environment of customers is encouraged.

The remainder of the paper is organized as follows. Section 2 describes blockchain highlighting its characteristics, types, and protocols. Section 3 describes the multimedia technology, characteristics, and components of multimedia technology. Furthermore, Sect. 4 evaluates metaverse technology and its characteristics, hardware components, software components, and technologies behind metaverse. Section 5 verifies how multimedia technology and blockchain enhance metaverse blockchain's potential from a perspective of system as well as user application for the metaverse. Finally, the paper discusses the recent advancements in metaverse.

2 Blockchain

Blockchain technology is based on decentralized network, and it functions as peer-to-peer network. It is normally related with cryptocurrencies such as Bitcoin [11]. It is a database of information about transactions that are transmitted, and preserved as data via a world computer network because all of the transactions which take place in a decentralized fashion eliminate middlemen which are needed for transaction validity and verification of transactions. Blockchain achieved prominence because of its remarkable ability to create a cooperative economy and developed the structure for the present electronic currency maker. Blockchain has been thought of as a revolutionary method for security and depending privacy. The need for blockchain technology is increasing, and its main objective is to safeguard the digital content that all the owners of metaverse use [12]. The ecosystem of metaverse relies on blockchain for the content accounting and deal to check it out that user privacy, integrity, and reputation. The mechanism of consensus protocols is based upon the methodology itself, and it is more convenient for e-commerce platforms. In the expertise of the metaverse, the appropriate tool which is designed for imposing [13].

2.1 Characteristics of Blockchain

Blockchain is decentralized, reliable, transparent, and immutable. It allows the peer-to-peer transactions, enhances traceability, and fosters trust by ensuring data integrity through cryptographic techniques and consensus mechanisms. Various characteristics of blockchain technology are explored in the sub-sections below.

2.1.1 Decentralization

In traditional centralized system of transactions, the central trustworthy agency (e.g., the central bank) first validates each transaction. Consequently, the main difficulty with decentralization lies with the fact that it involves trust and beliefs. There are more issues which include availability, lift flexibility, and fail over, for which architecture of decentralized blockchain is peer to peer [4]. A decentralized ledger is a copied database that is synchronized and shared by all participants of network. It retains the logs of every interaction between the network of users. In order to enter the transactions in the ledger, the recently validated transaction must initially be grouped into blocks. The records can be manufactured by any participant of blockchain and can be added to the blockchain by accumulating new transactions [14].

2.1.2 Persistency

Blockchain provides an environment for faith evaluation and enables the consumers and producers to verify whether their data is unaltered and legitimate [15]. A transaction cannot be erased or reversed once it has gone through the blockchain. This feature ensures that all deals are permanent and inviolable, providing a first-degree belief and conviction. If data has been changed and modified, it will be firstly detected and identified by the network. For this reason, blockchain is considered as an inflexible distributed ledger and also is almost tamper-proof [16].

2.1.3 Anonymity

The network of blockchain which consists of randomly generated address can be communicated. The user specification in blockchain remains concealed, exclusively the matching units of digital addresses are accessible [3]. Within a blockchain network a customer can have many addresses to avoid the submission of his identity. There is no intermediate authority as it is a decentralized system, and there is no surveillance or recording of users' private information. To share the blockchain around the world, public key of cryptography can be used while maintaining relative inconspicuousness. Blockchain also provides the trustless environment to achieve the certain amount of anonymity [17].

2.2 Types of Blockchain

In last few years, blockchain has progressed in immersive technology and based on its different attributes, it is divided into four main types of blockchain: public, private, hybrid, and consortium. Public blockchains are open and decentralized, private blockchains are restricted to specific users, hybrid is a combination of private and public blockchains, and consortium blockchains are shared among organizations [18].

2.2.1 Public Blockchain

Public blockchains are accessible to the public and several independent users by becoming a node which can be associated in the decision-making process, but user may or may not be promoted for their embarrassment in the decision-making process [20]. There is no one in charge and nobody in the network has ownership of the ledgers and are openly free to participate in the network. Public blockchain is highly auditable. The main challenge is with scalability and inefficiency as all nodes demand to authenticate the transaction.

2.2.2 Private Blockchains

These blockchain types are not reachable to the public. It can be usable to organizations or a group of people and only participated members have access to the ledger. A system administrator monitors crucial duties on a private blockchain and has a power to decline and approve access, and it is often to referred as a permissioned blockchain [21]. The controlling organization dictates the amount of accessibility, security, and permits. It tends to be used by organizations or businesses where only a specific number of people are allowed to take part in the network.

2.2.3 Hybrid Blockchain

These types of blockchain allow the currencies to be transferred from one to another blockchain. These blockchains are also referred to as pegged sidechains [22]. It offers mixed features of public and private and it also allows selective auditability but is complex in managing. A hybrid blockchain is an incorporation of both the private and public blockchains. Users can seamlessly connect to confidential ledger with several public blockchains due to the flexibility of the blockchain's hybrid system [20]. In a hybrid blockchain, an operation in a private network is frequently authenticated within that network. Users are able to maintain who is granted to what blockchain data in a network.

2.2.4 Consortium Blockchain

In this type of blockchain, the participants are already well known, reliable, and trusted. It is a type of blockchain which is partially private and also operated by governed organizations [23]. Consortium blockchain is the one where more than one organization manages the network. More than one organization can perform information exchange, do mining, and also serve as node in this type of blockchain [24]. Table 1 provides a brief comparison between various types of blockchain.

Table 1 Comparison of various types of blockchain with respect to its features

Features	Public [20]	Private [21]	Hybrid [20, 22]	Consortium [23, 24]
Usability	IoT operations and document validations	Real estate, supply chain, and asset ownership	Real estate and healthcare sector	Finance, supply chain, and research work
Accessibility	Nature is independent with complete transparency and trust	Fully controlled access	Controlled access and highly scalable in nature	Controlled access and security
Transparency	High transparency	Low transparency	Variable transparency	Moderate transparency
Security	High in security but can face scalability and privacy issues	Controlled security and efficiency	Flexible balance between public and private	Shared security governance
Auditability	Highly auditable	Controlled auditability	Selective auditable	Shared auditability
Immutability	Cannot be tampered	Could be tampered	Could be tampered	Could be tampered
Complexity	Highly complex	Less complex	Complex in terms of architecture and integration	Complex in terms of governance and coordination
Decentralized	Highly decentralized	Partially decentralized	Partially decentralized—hybrid between private and public blockchains	Partially decentralized—between private and public blockchains
Centralized	No	Yes	Partial	Partial
Consensus algorithm	PoW, PoS, DOoS	Company internal	R3, B3I, EWF	Hyperledger, Ripple
Examples	Bitcoin, Ethereum	Hyperledger fabric, Corda	Dragonchain, Quras	R3 Corda, Hyperledger Sawtooth

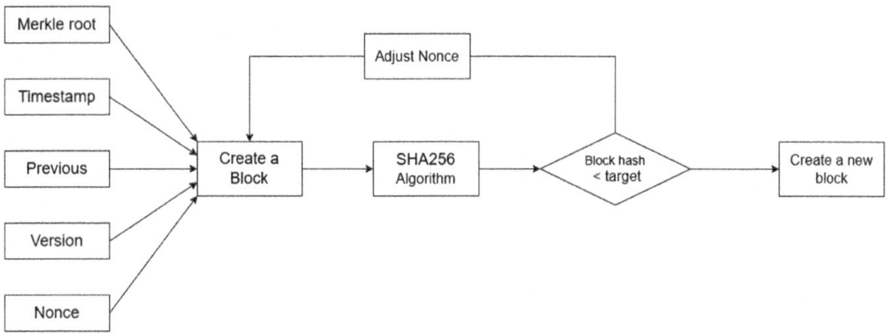

Fig. 1 Proof of work flow

2.3 Consensus Protocols of Blockchain

The backbone of the blockchain is consensus and other technologies are distributed and decentralized. We know that blockchain provides privacy, immutability, transparency, and security. Yet every transaction in the blockchain is thought to be completely verified and secured [25]. Only the process of consensus, which is a crucial part of all blockchain networks, made this happen. We will discuss various consensus algorithms and how they work.

2.3.1 Proof of Work (PoW)

The first time ever functioning blockchain protocol is PoW which is initially implemented as part of the Bitcoin in 2008 though its origins are older than that. Computational power competition node is generated to choose a new block. Here, the solution for cryptographic puzzle is to find the node present in competition [26]. The node which finds the explanation and gets the contingent to create a new block. The creation of block flow in PoW is represented in Fig. 1.The node wants more computational power to solve this tricky puzzle and to get exact answer needed to adjust the nonce value. Within a chain it is easier to overthrow a single block still within the chain correct block emerges, compounded load so a large amount of processing strength is needed to reverse a long chain. It is usually used in Ethereum along with Bitcoin. In Fig. 1, we can see the PoW flow.

2.3.2 Proof of Stake (PoS)

In proof of stake, to form a new block select the path for a node on which the held stake absolutely depends on. The power of computational is not going to be acknowledged [27]. For every time, the nonce is not going to be modified by PoS. For reaching the consensus, the uses of computational should be less. PPcoin, Nxt, ouroboros, and so

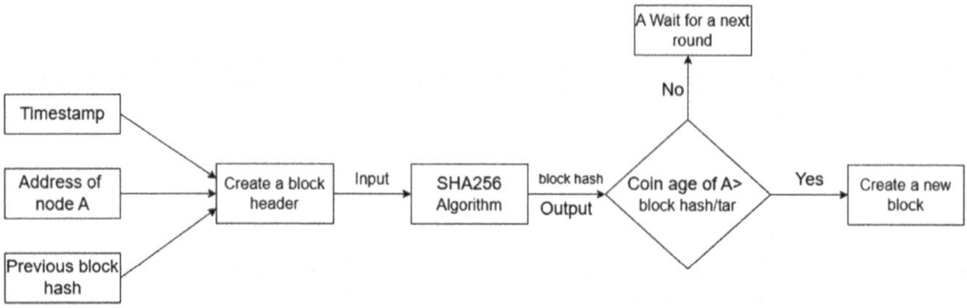

Fig. 2 Proof of stack flow

on are all cryptocurrencies used in PoS consensus protocol [28]. The functioning is shown in Fig. 2.

2.3.3 Delegated Proof of Stake (DPoS)

The DPoS's principle allows that node to pick block producers that can be able to hold the stake vote. They allow that members only whom they assist for granting the rights to assemble the building. Block producers are picked by the node with the most favorable vote, and it can also decrease the computational power [29]. Figure 3 illustrates the flow of DPoS. During the turns if they fail to create block, then the power is removed and goes to the stakeholders for replacing these delegates by identifying the new node. The outstretch of a consensus used the vote of stakeholders in a best way. When we compare the DPoS with the PoW and PoS, they are more efficient and less expensive [30]. DPoS can be clearly explained in Fig. 3.

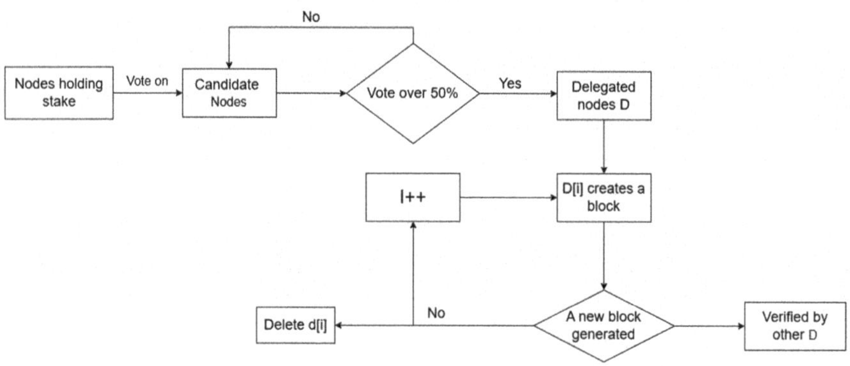

Fig. 3 Delegated proof of stack flow

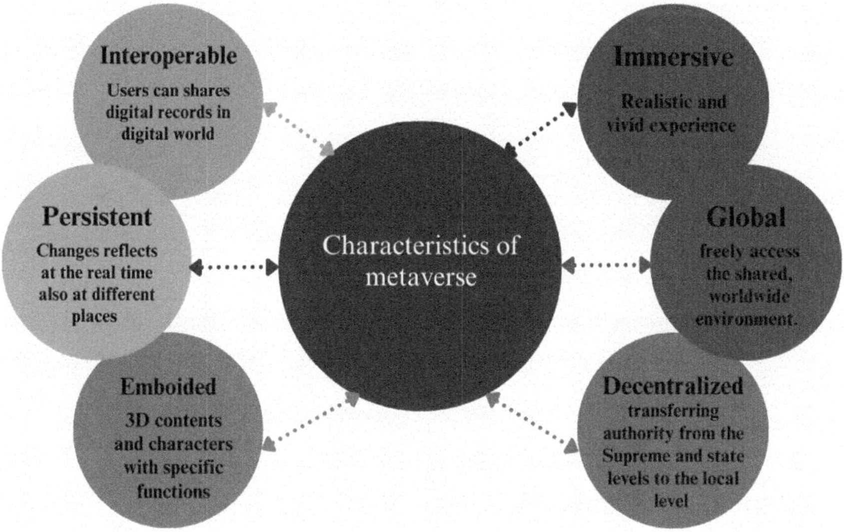

Fig. 4 Characteristics of metaverse

2.3.4 Practical Byzantine Fault Tolerance (PBFT)

For fault tolerance protocol, the practical byzantine tolerance is one of the consensus protocols that can be used. It is more feasible protocol which contains least complexity of the algorithm. The five most crucial phases which follow the order as prepare, request, prepare, reply, and commit. The similar procedure is shown Fig. 4. The messages of a primary node are received by the client [31]. Then these three received nodes can be passed through the five phases that were said. Then to complete the entire round the nodes will give answer to the client. Every node in PBFT maintains the similar state throughout each round of consensus, and makes it a consistent protocol. The Federated Byzantine Agreement is used by the protocol Stellar, which is a PBFT extension [32].

2.3.5 Ripple

Ripple is used for payments as it is one of the protocols of an open source. The nodes are broadcasted throughout the network and the transaction is started by the clients. In ripple, the process will be only carried out when the consensus process validates the nodes. The table of genuine node is maintained by the nodes of ripple called as Unique Node List (UNL). The transactions that support these nodes can be elected on who are on the list. All validating nodes in ripple convey the transaction payload as a suggestion to other nodes which are validated. This proposal's receipt of following the validating node which verifies each of the transactions within it. The suggestions will receive one vote if the local transaction in transaction collection matches with

Table 2 Comparisons between distinct consensus protocols

Consensus protocol	Protocols type	Scalability	Application	Energy utilization
PoW [27]	Probabilistic-finality	More	Public	High
PoS [30, 31]	Probabilistic-finality	More	Public	Low
DPoS [29, 30]	Probabilistic-finality	More	Public	Low
PBFT [31, 32]	Absolute-finality	Less	Permissioned	Negligible
Ripple [1]	Absolute-finality	More	Permissioned	Negligible

the transaction. It proceeds to the following round once it earns 50% of vote [1]. If it hits 80% then it will be documented in the ledger. The comparative description was given in Table 2.

3 Multimedia Technology

Multimedia technology, nowadays, is in huge demand due to advancements in 3D virtual reality. It consisted of components like spatial audio, interactive graphics, and realistic stimulations which are crucial in the creation of fascinating settings and engages the generation toward technology like metaverse by giving these immersive digital experiences [33]. It also allows people to exchange their ideas and information with digital elements. As the 3D virtual reality concept gains prominence, real estate, gaming industries, and education with entertainment are encouraging multimedia to develop virtual worlds, which allow users to interact and explore in 3D spaces with realistic sensation [34]. This makes it more demanding and making multimedia indispensable in today's landscape of digital [35].

3.1 Characteristics of Multimedia Technology

Multimedia technology combines multiple types of information, such as audio, text, images, video, and interactive elements, that creates dynamic experiences. Entertainment, education, marketing, and more, and it presents several characteristics which makes it more effective and versatile [36].

3.1.1 Dynamic Content Delivery

Multimedia delivers the content more dynamically as comparing with the static media. It presents and share the information like updating real time and using transitions through animation, and creating a more fascinating experience for users and gamers. This approach makes the 3D space incredible and enhances user experience by adjusting dynamically to changes in bandwidth, latency, and device requirements in delivering personalized content that ensures optimal quality and seamless interaction [37].

3.1.2 Interactivity

Interactivity in multimedia technology is a core element that adds the amazing transformation of passive media consumption into an engaging user-centered experience. By employing variety of inputs like touch, voice, gestures, or even haptic feedback, it permits users to actively collaborate, control, and affect digital information. Multimedia actively engages the customers with content through elements that interact like sliders, buttons that are clickable, and animations, which make it more personalized and interacting [38].

3.1.3 Immersive

By designing virtual landscapes that resembles the real-world experiences, digital media can enchant users on a deep level. The effects of high-quality graphics, 3D sound, and haptic feedback are combined together to achieve the immersive effect that can respond to user input. Using 3D graphics, VR, and AR, multimedia enhanced simulation and visualization and also visualized complex ideas and stimulated the environments, and multimedia also creates immersion of experiences that is impossible with static and traditional media [39].

3.1.4 Cross-Platform Compatibility

Across various devices, operating systems, and applications, multimedia content critically ensures seamless access and functionality regardless of the platform they use—web, virtual reality, mobile, and many more. It also guarantees the customers to experience consistent quality in photos, videos, movies, audio, and interactive media. It also maintains accessibility across and usability of devices, which are often designed to run on different technologies [40].

3.1.5 Real-Time Interactivity

Real-time processing is very important in applications of multimedia like virtual reality and gaming, where the responses and reactions of users play crucial role to create seamless, real-life-like experience. This technology opens doors to innovative applications, from interactive storytelling and simulations to immersive educational experiences, as it provides users with a heightened sense of control, engagement, and immersion within digital environments [41].

3.1.6 User Control

People are more engaged with customization and manage the media control in multimedia technology that also empowers the personalized content with their tastes and own requirements. It can only become possible with the help of multimedia technology. The awareness that different multimedia systems allow users to manage the type, order of the media, and speed they absorb promotes a more enhanced level of customization [42].

3.1.7 Hypermedia and Linking

Hypermedia and linking are foundational concepts in multimedia technology, allowing users to navigate seamlessly between different types of media and information. Hypermedia expands on traditional hypertext by integrating not only text but also images, audio, video, and animations, creating rich, interconnected digital experiences. With the aim to improve flow of information and accessibility, multimedia may utilize interactive routes and hyperlinks that help the users to traverse the content with respect to the areas of interest and needs [43].

3.2 Components of Multimedia

Multimedia technology consists of various components that work together to create dynamic, interactive, and immersive experiences. Key components include text, which provides essential information and context; graphics, encompassing static images, drawings, and illustrations that add visual appeal; and audio, which includes music, sound effects, and spoken words to enhance engagement and create ambiance. Video integrates moving visuals to capture real-life motion and expression, often used for storytelling or conveying complex ideas visually. Lastly, animation and interactivity. These components were discussed below [5]:

- **Graphics**: Graphics are the information which is present in non-text format, such as photograph, chart, sketch, and represented digitally. Graphics elevates the appearance of the presentation, it usually added to upgrade the effectiveness and presentation of multimedia application [44].

- **Animations**: Animation is the process that provides visual representation by flipping the images. It is quite popular in multimedia and it attracts users and additionally encourages them to communication with the information. Through movement, it provides static graphics life which allows visual explanations, simulations, and interactive interfaces [45].
- **Text**: Text serves as an extra information which provide extra information and context of the scene to the viewers. It appears in all creations of multimedia of any kind, and it can be in various fonts and sizes which can be changed as user's wish [46].
- **Video and Audio**: The video refers as moving images with the sound effect. A significant role is also played by the sound in enhancing the multimedia experience. This is the most common technology that everyone in the world uses such as in movies and television which includes texts.
- **Interactivity**: It creates the personalized experience with the help of multimedia elements that allow users to engage with content, navigating, manipulating, or responding. It enhances how information is presented together and these components form the foundation of multimedia technology that is absorbed and interacts across digital platforms [47].

4 Metaverse

In 1992, Neil Stephenson first coined the term metaverse in his science fiction novel named Snow Crash. The description of metaverse differs generation to generation as popular metaverse is based on the concept generation Z's social beliefs which contains that online egos and offline ones are similar to each other [48]. Further as the content grows a new structure is needed for the future ones. Metaverse is referred as a three-dimensional virtual world and it is a combined word of universe and transcendence meta where avatars engage in cultural, social, economic, and political activities [49]. As the metaverse is wide and continuously growing scope, different definitions and same concepts exist. Although the metaverse is focused mainly on worlds of virtual (like games), it has recently been presented as an outlet for content-centered social interactions with others, thought sharing, and sharing of interest [50]. Here, we have discussed several characteristics of metaverse given below.

4.1 Characteristics of Metaverse

Eight unique characteristics of the metaverse are outlined as follows. The metaverse features immersive virtual environments, interoperability, user-generated content, and social connectivity. It combines augmented and virtual reality, enabling real-time interactions and experiences that blur the lines between real and digital worlds. The characteristics of metaverse explained with the help of Fig. 4 are given below:

- **Immersive**: The metaverse offers users a realistic and vivid experience that extends beyond the current 2D interaction so that users can feel emotionally and psychologically immersed in the digital space [51]. A moment so immersive could be achieved by integrating touch, sound, and sights in a similar temperature and effects of environments.
- **Global**: Regardless of where they reside or which nation they are from, everyone must be capable of freely accessing the shared, worldwide environment that is the metaverse.
- **Decentralized**: A particular organization could not be able to control the metaverse. It should be a free non-realistic space where every individual independently owns their assets and has a say in concerning how the platform grows further ahead [51]
- **Embodied**: Users engage in the virtual world as if they have got a character with a specific function, instead of simply observing its 3D contents [52].
- **Persistent**: There must always be accessibility to the metaverse. It continues to operate regardless of what, even if users abandon the platform [54].
- **Interoperable**: Users can share digital records throughout the digital world while keeping a distinctive profile across platforms. Moreover, the ultimate metaverse generation might even interface with real world.
- **Sustainable**: The virtual world and physical world coexist and sync. The digital realm can reflect any changes made to the physical environment and vice versa [55].

Characteristics of metaverse are explained in Fig. 4

4.2 Components of Metaverse

In this subsection, components of metaverse were given below. They are divided into two sub-sections: hardware components which include head-mounted displays, input methods based on hand, and input based on non-handed methods and motion-head inputs and software components which include object and scene recognition, sound and speech recognition, etc.

4.2.1 Hardware Components

In the metaverse, hardware is not only thing that plays essential role in immersive insight but it also has barriers in technical field. Hardware affects the technological advancement but it should improve to reach the height of real experiences. Head-mounted display (HMD) is the essential part of hardware which enables and opaques the view of immersive involvement [56].

- **Head-Mounted Displays**: The core input tool of metaverse is HMD which is contrary to optical which can be seen through HMD, and video-see-through HMD and non-see-through HMD. Optical-see-through HMD is a method of overlaying the non-real world and for the process of overlaying high hardware specifications are required. Video-see-through HMD is a method used for complement. The issues like minimum battery life of headset are present in HMD with bulky, expensive issues. It reflects both real-world and artificial images to see through the display [57].

- **Input Methods Based on Hands**: Detailed user data modeling is used to feel the material with tactile. The texture to real objects is given by haptic. While the passive haptic technique is the manner that delivers the ability to understand the situation, and at the present time it plays the role of giving the feeling of presence to the target, the active haptic is more efficient going by the way it is used in adjusting between the two and delivering on user feedback. Metaverse through a head-mounted display is special as it provides an egocentric perspective on the world. Third person view data is not available directly in the metaverse, since the viewpoints are inconsistent, and they rotate in many different directions; thus, it is necessary to perform some kind of transformation on the input model to recognize the egocentric video model [58].
- **Input Methods not Based on Hands**: The study of raw wrist movement even without anything to attach directly to your hand is also on the rise as we search for better input devices. As the interest is also emerging rapidly for wearables worn on wrist that ought to provide vibrotactile feedback, we employed a simple model actuator as an input device (a single stand-alone tactor with controllable signal parameters such as frequency, amplitude, and modulation) in order to systematically explore this variation space within inter-subject information transfer. Foveating large bandwidth high spatiotemporal 360 videos with logarithmically decreasing radius in each eccentricity for maximizing the spatial resolution [59].
- **Input Device Based on Motion**: Using treadmills and body tracking, valid motion information (in physical sense of gravity or space) are displayed with complementary devices to enhance the virtual experience (haptic or non-haptic). There are large number of motion input devices, such as those available that can further be classified into passive and active types [60]. Motion capture systems are of two types, namely, active and passive. Passive way is a mode in which generally embraced senses are given to user with pre-defined scenario and active form is the feedback related according to behavior of users. From a fairly crude form of walking, to 360 turns that all are used in every way imaginable for bringing realism on the surface [61]. Body-affecting since the user is essentially risking with their health as well, so it puts a waist-fixing method which has an effect on concerning treadmill.

4.2.2 Software Components

In the physical space, the objective reality is an essential role played by cognitive illusion and users feel subjective reality. There are two different types of cognitives: dynamic cognitive and static cognitive. Sensory balance and body movement are dynamic cognitive and proprioceptive senses are static cognitive. Important features of dynamic cognitive are adaptation, attention, and behavior. Elimination of distortion of detection and recognition is very important in metaverse. Object includes faces, poses, gazes, and gestures for object recognition. These object recognitions go through several processes like recording, tracking, and sensing. There are two types of stimulations and tracking [62].

- **Scene and Object Recognition**: According to distance, the procedure of recognizing the shape, dimensions, place, brightness, and colors of objects is known as object recognition. A good recognition of what components and configuration it has and what state the initial scene is. Scene graphs are an advantageous means to add the explicated traits that were previously recognized as neural network models' shortcomings. Object recognition along with scene recognition is vital, and centered around human scene and non-invasive communication including look, movement, and mental state must be considered [49].
- **Sound and Speech Recognition**: Processing speech and recognizing sounds communicate with other avatars and help understand surroundings. This conversation functions as a direct line of inquiry for directions to be delivered to NPCs (non-player characters) in the metaverse and to fellow avatars. Technology of voice recognition that incorporates the environment seriously while changing the voice recognition is necessary for a realistic world in metaverse. Hearing sounds and receiving speech allow you to perceive the environment around you as well as communicate with other avatars. As for inworld, the direct method of interaction with different avatars is communication and to give orders to NPCs. Since the environment in which metaverse connection is made changes, combined with this and on top of that a technology is needed to separate the noise of surroundings from someone's own voice without consuming background noise. Furthermore, the loudness of sound is a function of distance. To provide the same real-world sense in voice interaction of the metaverse, sound adjustment according to distance fusing environment should be done [63].
- **Scene and Object Generation**: Metaverse objects and techniques used to create environment are split into the creation of an imaginative world and another where the real-world environment gets depicted entail two different things. Time–Space Creation Mechanism (Metaverse): Time–space creation way of the place and object in metaverse is divided into real-world reflection abolition method, to describe a really existing situation or site as well as new void environment generation. One of the practical ways to represent real-world environment is by mimicking known landmarks such as museums, Eiffel Tower and places where we used to daily visit that are home or school in reality. Otherwise, it just becomes an inaccessible locale (underwater, Mars) and fails to give a human feel. Related to People and Things: These are the primary types of object generation. An establishment of object module then produces an avatar and NPC of any human form you want to converse with (e.g., a movie-star or your close family member). Multimodal conversation is fluent in facial expressions and natural joint movements [64].

- **Sound and Speech Synthesis**: Auditory awareness allows you to hear and speak inworld. Metaverse conversation is a type of interaction similar to direct form that allows avatar-to-avatar conversations or ordering an NPC command. Because the metaverse is associated in an infinite number of instances and spaces, we need technology to be able to limp out our voice specifically from everything around us without using a silent booth. The range of sound will also serve as another scalar, along with the volume in terms of distance. You will need to use voice recognition technology in the metaverse that can interact with some ambient and set a volume distance [65].

4.2.3 Contents

Content is worn to provide a mesmerizing experience through user-created events and well-organized stories. It is the indispensable component that maintains the metaverse. Story authenticity, immersive experience, and theoretical coherence are vital elements of content. There are two strategies to be satisfied before building: the paradigm shift technique and the reuse of previously generated information. Content is the foundation of metaverse and serves to create immersive experiences with organized stories that are underworld adventures or user-generated events. Story reality, immersive experience, and logical completeness are crucial for any kind of content. Content creation—there are two kinds of ways, one is a paradigm shift process and other to reuse the existed ones [66].

4.3 Technologies Behind Metaverse

The metaverse is described the 3D world and following generation variant of the global web in the terms of technology. It is a complete combination of various cutting-edge technologies, including VR/AR, XR, AI, DT, 5G/6G, and blockchain. Cloud edge and wireless networks processing are also included. These technologies all support the metaverse in various contexts and they are complementary with each other [67]. The section that follows is a presentation of these technologies given below. The visual representation of these technologies of metaverse can be shown in Fig. 5.

4.3.1 Blockchain

Blockchain is a distributed, decentralized network that employs stored information in block form with digital signature. It builds the economic structure with the help of metaverse with NFT, digital currency and DeFi [68].

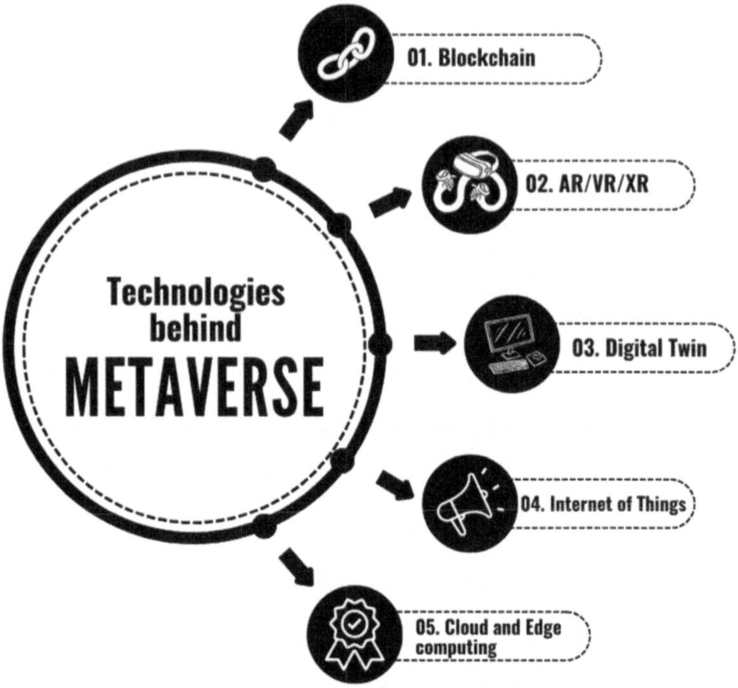

Fig. 5 Technologies of metaverse

4.3.2 Interactivity Technologies (AR/VR/XR)

These innovations in technology are important to the metaverse's capacity to be immersive. Instead of employing standard tools like smartphones and laptops, they offer user to have a immersive tour of the metaverse through devices that are worn like haptic gloves and VR headsets [69].

4.3.3 Digital Twin

DT constructs digital representations of tangible objects based on real-world data. It enables the metaverse's synchronized characteristic feasible allowing the coexistence of physical and virtual realities. Real-life items may show up in the digital world, and any adjustments made to them there will additionally have an effect on real world [70].

4.3.4 Internet of Things

A huge amount of data for the metaverse is rendered available by IoT, which combines many dissimilar kinds of cameras and sensors. Designing the real world into the unreal and vice versa is made easier with the help of digital twins [71].

4.3.5 Cloud and Edge Computing

Despite abrogate technology repeatedly has deadline storage and processing capabilities, the huge amounts of users accessing the platform could ultimately overload the servers handling the metaverse's apps and functionalities [52]. Therefore, processing the huge amount of data in the metaverse while offering a sufficient amount of service relies on an ideal mix and utilization of complex cloud and edge computing [72].

5 Blockchain and Multimedia Technology-Enabled Metaverse

This section analyzes blockchain's potential from a perspective of system as well as user application for metaverse and discussed about how multimedia technology is enhancing the metaverse. Blockchain includes identity management, management of authentication, reputation system, data management of decentralized and governance. These fall under the system perspective. In user application technologies including metaverse virtual services and economic system are gathered in Fig. 6.

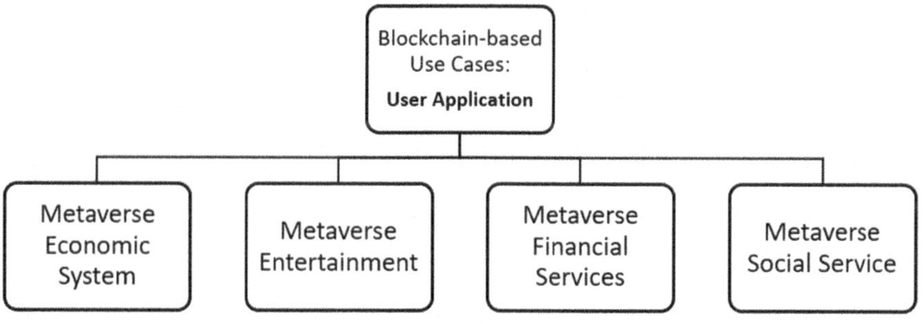

Fig. 6 Blockchain-based use cases

5.1 Blockchain Integrated in the Metaverse

When we imagine metaverse, a range of breathtaking experiences or enjoyable games may come in our mind. In order to provide the customer with the best possible experiences, a large amount of private data are gathered enormously by metaverse. The users will be targeted in the real world if the wrong people discover the information. It ensures the data privacy and security. For making the decisions of stakeholders, the AI models rely on this data. Blockchain provides the quality of data. The metaverse based on AR and VR devices conclude in more immersive and connected globe. It delivers secure and smooth data. The data of metaverse must be accurately and consistently maintained. Blockchain also maintained the data integrity in metaverse [70].

5.2 Integration of Multimedia Technology with Metaverse

After integrating multimedia technology with the metaverse, it plays an important role in establishing a dynamic, immersive, and interactive virtual world that becomes a bridge to fill the gap between real and virtual reality. For engaging the users with rich interactive spaces that enable the creation of realistic environments which can transform the physical world by uplifting the multimedia technology, advanced 3D graphics, augmented reality, and virtual reality. Fully immersive experiences were offered by the VR headsets, while digital elements were blended by AR with real-world technology which make it possible to interact with users with both together. Another layer of reality is brought in by spatial audio, which changes the sound with respect to a user's movements and location [73]. This enhances the overall atmosphere of virtual places and makes interactions appear more genuine. Many algorithms of AI and machine learning are rapidly integrated into metaverse to make it more realistic and provide the experiences which are personalized that are modified based on user preferences and behavior [74]. The content of multimedia was facilitated by metaverse which was shared and created through user-generated platforms, where individuals can create their own imagination into virtual reality with the help of tools, game engines, and collaborative spaces. Devices like motion suits and VR gloves are the advent of haptic feedback that allows users of metaverse to engage through mimicking touch, physical interaction, and advance immersion. Also includes virtual meetings, interactive performance, and concerts which are made possible by multimedia technology that connects people to interact from all over the world and auctioned by creators using NFTs and digital asset and gives them an ownership of virtual environment in a whole new dimension. The revolution in the integration of multimedia technology is how we interact, connect, and collaborate with online assets [34].

5.3 *Blockchain-Based Metaverse: User Application*

According to several applications and features of metaverse, a short brief of some metaverse platforms which discusses blockchain-based use cases for various applications is illustrated in Fig. 6.

5.3.1 Metaverse Economic System

In decentralized metaverse, digital assets involving blockchain-native NFT-based content created by consumers, cryptocurrencies, fungible coins, and online real estate would all be incorporated into the virtual marketplace. The blockchain nature of decentralized metaverse with its own virtual currency of metaverse economic system. Users trade them in marketplaces and produce content (i.e., UGC) to earn cryptocurrency and tokens [75]. For preventing counterfeiting and tokenize UGC, NEFT is used to ensure its distinctiveness. In this use case, metadata is to reduce storage cost to store in off-chain environment. Similarly, in metaverse, virtual real estate as NFT can be tokenized to make it valuable and scarce. In this regard, Roblox creators can get particular Robus, the platform's virtual money, in regard of UGCs like game scenes and products [76].

5.3.2 Metaverse Entertainment

In the entertainment industry, applications for the metaverse are vast: from gaming and virtual events to other experiences. A popular metaverse game is Roblox that is basically a global platform for user-generated games, allowing customer to design their own block-based avatars and play various types of 3D online mini-games. When it comes to metaverse NFT, gaming and blockchain can be used in tokenizing game assets from the inside of the game so they become unique items with real value. In a similar line of thought, the authors in [77] introduced gaming platform that consists of using NFTs as prototypes to represent all kinds and types of assets within the games so it has ability in tracking and managing rewards. Furthermore, blockchain is applied to the profile management and exchange of data between players in game. Conversely, metaverse games related to gambling (like the poker game in particular regions of decentralized land) could be structured to have an advantage for players with a high probability that they will lose their tokens.

5.3.3 Metaverse Financial Services

In the real world, financial activities range from lending to borrowing, saving and stock investing. In like manner, the metaverse may also be a reversion to those services and activities—for which blockchain together with appearing DeFi technologies carves out suitable gears to implement them in truly decentralized fashion. Broadly speaking, there are three varieties of financial activities that can be broken down into the following categories: personal finance, corporate finance, and public (government) finance [78]. The metaverse may also be used for personal finance, such as lending and borrowing funds or saving and investing in the virtual world. These objectives are well served by blockchain technology because it provides the much-needed transparency and verifiability attributes of documents. The BlockCIS is a smart contract-based insurance framework introduced by the authors in [79]. If a customer node is subjected to cyber-attack, uploads logs (firewall and storage log) as evidence of the intrusion into Block CIS system.

5.3.4 Metaverse Social Services

The offering of social services could be modernized and made more convenient and accessible by the metaverse, which is envisioned as a fully virtual society. Hospital physical appearance is no more necessary because of the metaverse's ability to offer telemedicine services, remote monitoring, and virtual consultations [80]. In terms of education, the metaverse can facilitate immersive learning experiences and serve as a platform for virtual classrooms. Geographical restrictions can be lifted, and educational options can be expanded, significantly increasing access to education.

6 Recent Advancements

Recent advancements in the metaverse have expedited the growth of immersive virtual world that brings together the digital settings and people and also include improved interoperability between platforms, realistic AI-driven avatars, enhanced AR integration, and the rise of blockchain for asset ownership. These developments enhance social experiences and enterprise applications, driving greater user engagement and collaboration in virtual environments. Metaverse applications in various field are given below.

6.1 Identity and Authentication Management

Identity and authentication management in the metaverse is essential for ensuring secure and seamless user experiences. As users navigate diverse virtual environments, robust systems are needed to verify identities through multi-factor authentication, biometrics, and decentralized identity solutions. These measures not only protect against unauthorized access but also enable users to maintain ownership of their digital personas and assets across various platforms. With the rise of social interactions, gaming, and commerce within the metaverse, effective identity management fosters trust, enhances privacy, and ensures compliance with regulations, creating a guarded space for users to engage and communicate in immersive digital experiences.

Yang et al. [80] proposed chameleon signature and a two-factor authentication framework which is based on biometric-based authentication and further it aims for impersonation in physical world. Brij et al. [81] suggested combination fuzzy logic and convolutional neural network (CNN) techniques in two-phase security framework that is used for biometric authentication and a lightweight cryptographic protocol for the metaverse security. Zhang et al. [82] aimed to recognize an old-friends an anti-disguise authentication method from physical world draws on the idea of first impression.

6.2 Decentralized Governance

Decentralized governance in the metaverse empowers users by distributing authority of decision-making across a network rather than centralizing it within a single entity. This model fosters community engagement and inclusivity, allowing participants to have a direct say in shaping virtual environments, policies, and economic systems. Utilizing blockchain technology, decentralized governance enables transparent voting and consensus mechanisms, ensuring that all voices are heard and that actions are accountable. By minimizing the risk of censorship and enhancing user autonomy, this approach promotes trust and innovation within the metaverse. As communities grow, decentralized governance will be essential for creating vibrant, democratic ecosystems that reflect the values and interests of their participants.

Goldberg et al. [83] proposed the importance of platform governance. In the context of metaverse, centralized architecture-related problems are discussed. Some of these issues can potentially mitigate the blockchain-based governance and open standards. Martha et al. [84] proposed that over various use cases and its derivatives currently ownership rising and digital assets forward with the acceleration utilization of blockchain. Hunhevicz et al. [85] aimed that combination of blockchain provides initial manufacturing of the paradigm of common pool resource governance, commonly known as "crypto commons," and project delivery as decentralized. Cheng et al. [86] aimed a blockchain-enabled architecture to address issues for metaverse which is adaptive and modular. The solution is also included based on modular blockchain. Elsadig et al. [87] proposed that blockchain helps in managing and creating the platform for metaverse and further intended a reference for practitioners and researchers in the area of blockchain-based metaverse.

6.3 Reputation System

A reputation system in the metaverse plays a crucial role in fostering trust and enhancing user interactions. By assessing users based on their behaviors, contributions, and interactions, these systems provide a transparent metric of reliability and credibility. Users earn reputation points through positive actions, such as creating valuable content, participating in community events, or engaging in fair trading. Conversely, negative behaviors, such as fraud or harassment, can result in penalties. This feedback mechanism encourages users to uphold community standards, facilitating safer environments for socializing, commerce, and collaboration. As the metaverse evolves, effective reputation systems will be vital for maintaining a healthy, engaged community.

Ryu et al. [88] proposed a secure communication system model that transparently manages user identification data using blockchain technology in metaverse. Further proposed a scheme using biometric and ECC mutual authentication which provide security between users and platforms. Awan et al. [89] introduce designed methodology of distributed systems to strengthen the security parameters, which within the metaverse focused exclusively on its applicability. The range of security threats mechanism can effectively counters, including Good/Bad mouthing, the Sybil and on/off attacks. Liu et al. [90] proposed model of reputation, which improves the probability of success and credibility of consensus with concern to not increase the number of participating vehicles. To score the vehicles vehicular edge metaverse maintained, so high reputation vehicles can be selected to participate in practical Byzantine fault tolerant (PBFT).

6.4 Data Management

It emphasizes security, scalability, and real-time processing for immersive to support virtual space. These technologies ensure that the data of users can be efficiently verified, stored, and processed which foster its growth as an interactive digital realm, which is dynamic.

Paik et al. [91] aimed to highlight a board software arrangement as a data store when incorporating blockchain concepts and considerations that can analyze the viewpoint of a developer toward blockchains. The work is to increase the compassionate matching of data storing in blockchain technology. Tao et al. [92] understand a structural identity perspective of the blockchain transaction network which symbolize the entire network structure and expose the relations among multiple entities. Truong et al. [10] proposed to focus on Meta Trade, which offers secure AIGC trade, deals with these critical trust issues, provides a blockchain-powered DAM framework, and management in the setting of the trustless metaverse. With Meta Trade, participants are able to deal with the having to take on assuming the role of an impartial third party. Ooi et al. [93] proposed to appreciate the growing importance of the metaverse and provided five valuable key areas insights from the viewpoint of engineering management: knowledge management, technology management, decision-making, logistics management, and supply chain management.

6.5 Extended Reality Technologies

XR technologies combine real and virtual worlds to create immersive digital experiences which encompass virtual reality (VR), augmented reality, and mixed reality. VR enables the experiences from gaming to virtual travel. Together, these technologies revolutionized the industries from education to healthcare, entertainment, etc.

Javed et al. [94] highlight functions in order to meet the deadline of driven XR applications with strict requirements and has a unique characteristic of mmWave networks to re-think. Due to their ability of inherence for providing mitigate interference and high data rates, mmWave communications are the cellular networks for next generation providing a paying way. Shin et al. [95] proposed a web-based platform that integrates exhibitions based on virtual reality (VR) and augmented reality (AR) called extended reality (XR). It includes an XR cloud which helps an AR client for real-world users and VR clients for users for sharing and managing data in XR spaces.

6.6 3D Content Creation and Digital Avatars

3D content creation and digital avatars enables the users so that they can enjoy the experience of highly personalized and interactive virtual worlds. They both are the key elements of the metaverse. Photorealistic 3D models and environments are rendered by these advancements due to increase in complexity and personalized AI-driven avatars have made it possible to capture expressions of user and their body language.

Cheymol et al. [96] proposed these limitations by discussing and addressing them. These limitations are a methodology for dissimilar avatar characterization, different fields of application, their impacts on customer experience, and at last future research direction. Park et al. [97] suggested a web-based platform that enhances the learning pattern with analyzing the facial expressions of students (Table 3).

Table 3 Recent advancement of metaverse technologies

Metaverse recent advancement	Authors	Year	Contribution
Identity and authentication management	Yang et al. [80]	2023	Develops a secure authentication framework ensuring avatar traceability in the metaverse
	Gupta et al. [81]	2024	Offers a lightweight cryptographic authentication technique for metaverse security that is based on fingerprints and fuzzy logic
	Zhang et al. [82]	2024	Introduces an anti-disguise authentication system based on first impressions of avatars in the metaverse
Decentralized governance	Goldberg et al. [83]	2023	Conducts an empirical analysis of governance and voting within DAOs in the metaverse context
	Martha et al. [84]	2023	Presents a conceptual framework for utilizing DAOs in virtual property transaction governance in the metaverse
	Hunhevicz et al. [85]	2024	Discusses governance structure and future directions for the distributed project management on digital commons
	Cheng et al. [86]	2024	Proposes a modular blockchain-enabled architecture for decentralized metaverse applications
	Elsadig et al. [87]	2024	Explores the roles of blockchain in the metaverse, presenting concepts, taxonomy, advances, and research issues
Reputation system	Ryu et al. [88]	2022	Designs a secure mutual authentication scheme for metaverse environments utilizing blockchain technology
	Awan et al. [89]	2023	Proposes a blockchain-based trust management model for avatar and non-realistic organization interactions
	Liu et al. [90]	2024	Explores reputation management strategies for consensus mechanisms in vehicular edge metaverse applications
Data management	Paik et al. [91]	2019	Analyzes data management in blockchain systems, focusing on architecture and governance
	Tao et al. [92]	2023	Introduces structural identity representation learning for blockchain-enabled metaverse applications based on complex network analysis
	Truong et al. [10]	2024	Develops blockchain buildings stripped of trust for the metaverse's business and regulation of AI-generated content
	Ooi et al. [93]	2024	Provides an overview of the metaverse in engineering management, discussing opportunities and challenges
Extended reality technologies	Javed et al. [94]	2023	Enhanced XR application performance in multi-connectivity mmWave networks
	Shin et al. [95]	2024	Constructed an environment for metaverse presentations using extended reality

(continued)

Table 3 (continued)

Metaverse recent advancement	Authors	Year	Contribution
3D content creation and digital avatars	Cheymol et al. [96]	2023	Outlined the impacts and applications of "dissimilar" avatars in virtual reality
	Park et al. [97]	2023	Enhanced the learning experiences by interacting AR and human facial recognition technology

7 Conclusion

In summary, the metaverse concept has been increasingly popular over recent years and with many technologies burgeoning, such as AI and blockchain including accompanying AR/VR/XR/IoT on a similar trajectory makes, it is more plausible of becoming reality. This research paper is an intent to make a masterpiece on the metaverse technology where it can completely transform the way of living, working, and communicating with people. It also explains the consensus protocols for the secured future directions as well as briefly describe blockchain and multimedia with their characteristics, components, and use.

This paper presents how multimedia and blockchain technology is integrated with metaverse and evaluates their ability to radically alter how virtual worlds are operated and generated in the metaverse. The role that blockchain serves is to establish a decentralized infrastructure that is vital in advancing accessibility and making it simpler for users to conduct transactions with one another and transfer virtual assets. As further the function of multimedia technology in metaverse is very crucial for the immersive and 3D effect in its virtual space which attracts the attention of customers and provide them a realistic environment. This paper also reviewed the latest trends of metaverse approaches (i.e., application, user interaction, and implementation) that were presently amiable and needed in the future and explained the collaboration of multimedia and blockchain with metaverse and finally highlights the recent advancements in multimedia, blockchain, and metaverse technology.

References

1. Ali, O., Jaradat, A., Kulakli, A., Abuhalimeh, A.: A comparative study: blockchain technology utilization benefits, challenges and functionalities. IEEE Access **9**, 12730–12749 (2021). https://doi.org/10.1109/access.2021.3050241
2. Sparkes, S.M.: What is a metaverse. New Sci. **251**(3348), 18 (2021). https://doi.org/10.1016/s0262-4079(21)01450-0
3. Park, S.-M., Kim, Y.-G.: A metaverse: taxonomy, components, applications, and open challenges. IEEE Access **10**, 4209–4251 (2022). https://doi.org/10.1109/access.2021.3140175
4. Suzuki, S., Kanematsu, H., Barry, D.M., Ogawa, N., Yajima, K., Nakahira, K.T., Shirai, T., Kawaguchi, M., Kobayashi, T., Yoshitake, M.: Virtual experiments in metaverse and their

applications to collaborative projects: the Framework and its significance. Proced. Comput. Sci. **176**, 2125–2132 (2020). https://doi.org/10.1016/j.procs.2020.09.249
5. Chung, J.-M.: Cloud computing and edge cloud technologies. Emerging metaverse XR and video multimedia technologies, pp. 279–304 (2022). https://doi.org/10.1007/978-1-4842-8928-0_8
6. Duan, H., Li, J., Fan, S., Lin, Z., Wu, X., Cai, W.: Metaverse for social good. In: Proceedings of the 29th ACM International Conference on Multimedia (2021). https://doi.org/10.1145/3474085.3479238
7. Qi, P., Chen, Z.: The origin, characteristics and prospect of metaverse. Adv. Educ. Human. Soc. Sci. Res. **1**(1), 315 (2022). https://doi.org/10.56028/aehssr.1.1.315
8. Siyaev, A., Jo, G.-S.: Towards aircraft maintenance metaverse using speech interactions with virtual objects in mixed reality. Sensors **21**(6), 2066 (2021). https://doi.org/10.3390/s21062066
9. Shih, J.C., Meier, F., Rai, A.: A framework for online updates to safe sets for uncertain dynamics. IEEE/RSJ Int. Conf. Intell. Robot. Syst. (IROS) **2020**, 5994–6001 (2020). https://doi.org/10.1109/iros45743.2020.9341606
10. Truong, V.T., Le, H.D., Le, L.B.: Trust-free blockchain framework for AI-generated content trading and management in Metaverse. IEEE Access **12**, 41815–41828 (2024). https://doi.org/10.1109/access.2024.3376509
11. Hassija, V., Bansal, G., Chamola, V., Kumar, N., Guizani, M.: Secure lending: Blockchain and prospect theory-based decentralized credit scoring model. IEEE Trans. Netw. Sci. Eng. **7**(4), 2566–2575 (2020). https://doi.org/10.1109/tnse.2020.2982488
12. Zalan, T., Barbesino, P.: Making the metaverse real. Dig. Bus. **3**(2), 100059 (2023). https://doi.org/10.1016/j.digbus.2023.100059
13. Lee, Y., Moon, C., Ko, H., Lee, S.-H., Yoo, B.: Unified representation for XR content and its rendering method. In: The 25th International Conference on 3D Web Technology (2020). https://doi.org/10.1145/3424616.3424695
14. Alangot, B., Reijsbergen, D., Venugopalan, S., Szalachowski, P., Yeo, K.S.: Decentralized and lightweight approach to detect eclipse attacks on PoWblockchains. IEEE Trans. Netw. Serv. Manage. **18**(2), 1659–1672 (2021). https://doi.org/10.1109/tnsm.2021.3069502
15. Stokel-Walker, C.: Welcome to the metaverse. New Sci. **253**(3368), 39–43 (2022). https://doi.org/10.1016/s0262-4079(22)00018-5
16. Cali, U., Kuzlu, M., Karaarslan, E., & Jovanovic, V. (2022a). Opportunities and challenges in metaverse for industry 4.0 and beyond applications. In: 2022 IEEE 1st Global Emerging Technology Blockchain Forum: Blockchain & Beyond (iGETblockchain). https://doi.org/10.1109/igetblockchain56591.2022.10087104
17. Shah, M.A., Uddin, F.: Exploring security and privacy challenges for banks in the evolving landscape of the metaverse. Advances in Information Security, Privacy, and Ethics, pp. 280–303 (2024). https://doi.org/10.4018/979-8-3693-3824-7.ch011
18. Suhail, S., Hussain, R., Jurdak, R., Hong, C.S.: Trustworthy digital twins in the industrial internet of things with Blockchain. IEEE Intern. Comput. **26**(3), 58–67 (2022). https://doi.org/10.1109/mic.2021.3059320
19. Liu, H., Luo, X., Liu, H., Xia, X. (2021). Merkle tree: a fundamental component of blockchains. In: 2021 International Conference on Electronic Information Engineering and Computer Science (EIECS), **2**, 556–561. https://doi.org/10.1109/eiecs53707.2021.9588047
20. Lin, C., He, D., Kumar, N., Huang, X., Vijayakumar, P., Choo, K.-K.R.: HomeChain: a blockchain-based secure mutual authentication system for smart homes. IEEE Internet Things J. **7**(2), 818–829 (2020). https://doi.org/10.1109/jiot.2019.2944400
21. Panda, S.S., Jena, D., Mohanta, B.K., Ramasubbareddy, S., Daneshmand, M., Gandomi, A.H.: Authentication and key management in distributed IOT using Blockchain technology. IEEE Internet Things J. **8**(16), 12947–12954 (2021). https://doi.org/10.1109/jiot.2021.3063806
22. Cui, Z., Xue, F., Zhang, S., Cai, X., Cao, Y., Zhang, W., Chen, J.: A hybrid blockchain-based identity authentication scheme for Multi-WSN. IEEE Transactions on Services Computing, pp. 1–1 (2020). https://doi.org/10.1109/tsc.2020.2964537

23. Al-Shaibani, H., Lasla, N., Abdallah, M.: Consortium blockchain-based decentralized stock exchange platform. IEEE Access **8**, 123711–123725 (2020). https://doi.org/10.1109/access.2020.3005663
24. Wang, J., Wu, L., Choo, K.-K.R., He, D.: Blockchain-based anonymous authentication with key management for smart grid edge computing infrastructure. IEEE Trans. Industr. Inf. **16**(3), 1984–1992 (2020). https://doi.org/10.1109/tii.2019.2936278
25. Manolache, M.A., Manolache, S., Tapus, N.: Decision making using the blockchain proof of authority consensus. Proced. Comput. Sci. **199**, 580–588 (2022). https://doi.org/10.1016/j.procs.2022.01.071
26. Zhang, S., Lee, J.: Analysis of the main consensus protocols of blockchain. ICT Express **6**(2), 93–97 (2019). https://doi.org/10.1016/j.icte.2019.08.001
27. Yang, F., Zhou, W., Wu, Q., Long, R., Xiong, N.N., Zhou, M.: Delegated proof of stake with downgrade: a secure and efficient blockchain consensus algorithm with Downgrade mechanism. IEEE Access **7**, 118541–118555 (2019). https://doi.org/10.1109/access.2019.2935149
28. Johar, S., Ahmad, N., Durrani, A., Ali, G.: Proof of pseudonym: blockchain-based privacy preserving protocol for intelligent transport system. IEEE Access **9**, 163625–163639 (2021). https://doi.org/10.1109/access.2021.3133423
29. Xu, G., Liu, Y., Khan, P.W.: Improvement of the DPoS consensus mechanism in blockchain based on vague sets. IEEE Trans. Industr. Inf. **16**(6), 4252–4259 (2020). https://doi.org/10.1109/tii.2019.2955719
30. Santiago, C., Ren, S., Lee, C., Ryu, M.: Concordia: a streamlined consensus protocol for blockchain networks. IEEE Access **9**, 13173–13185 (2021). https://doi.org/10.1109/access.2021.3051796
31. Kumar Jha, A.: Hybrid consensus mechanism (HCM): achieving efficient and secure consensus in blockchain networks. SSRN Electron. J. (2023). https://doi.org/10.2139/ssrn.4413290
32. Murovec, B., Spaniol, J., Campos, J.L., Keshavarz, B.: Multisensory effects on illusory self-motion (Vection): the role of visual, auditory, and tactile cues. Multisens. Res. **34**(8), 869–890 (2021). https://doi.org/10.1163/22134808-bja10058
33. Xu, G., Liu, Y., Khan, P.W.: When internet of things meets metaverse: convergence of physical and cyber worlds. IEEE Trans. Industr. Inf. **16**(6), 4252–4259 (2020). https://doi.org/10.1109/tii.2019.2955719
34. Joshi, J.B.D., Li, Z.K., Fahmi, H., Shafiq, B., Ghafoor, A.: A model for secure multimedia document database system in a distributed environment. IEEE Trans. Multimed. **4**(2), 215–234 (2002). https://doi.org/10.1109/tmm.2002.1017735
35. Wang, X., Zhu, W., Tian, Y., Gao, W.: Multimedia intelligence: when multimedia meets artificial intelligence. In: Proceedings of the 28th ACM International Conference on Multimedia, 4775–4776 (2020). https://doi.org/10.1145/3394171.3418547
36. Lu, Y., Wang, X., Zhang, W., Chen, H., Peng, L., Zhao, W.: Performance analysis of multimedia retrieval workloads running on multicores. IEEE Trans. Parallel Distrib. Syst. **27**(11), 3323–3337 (2016). https://doi.org/10.1109/tpds.2016.2533606
37. Zhang, X., Zheng, G., Lambotharan, S., Nakhai, M.R., Wong, K.-K.: A reinforcement learning-based user-assisted caching strategy for dynamic content library in small cell networks. IEEE Trans. Commun. **68**(6), 3627–3639 (2020). https://doi.org/10.1109/tcomm.2020.2977895
38. Armbruster, H., Wimmer, K.: Broadband multimedia applications using ATM networks: high-performance computing, high-capacity storage, and high-speed communication. IEEE J. Sel. Areas Commun. **10**(9), 1382–1396 (1992). https://doi.org/10.1109/49.184868
39. Li, Q.: The definition of immersive communication. Immersive Communication, pp. 43–55 (2019). https://doi.org/10.4324/9780429355530-2
40. Xu, T., Wang, D., Liu, G.: Banian: a cross-platform interactive query system for structured big data. Tsinghua Sci. Technol. **20**(1), 62–71 (2015). https://doi.org/10.1109/tst.2015.7040514
41. Tian, Y., Bai, H., Zhao, S., Fu, C.-W., Yu, C., Qin, H., Wang, Q., Heng, P.-A.: Kine-appendage: enhancing freehand VR interaction through transformations of virtual appendages. IEEE Trans. Visual Comput. Gr. **30**(7), 3298–3313 (2024). https://doi.org/10.1109/tvcg.2022.3230746

42. Tang, Q., Groba, Á.M., Juárez, E., Sanz, C.: Closed-loop power-control governor for multimedia mobile devices. IEEE Trans. Consum. Electron. **63**(2), 153–161 (2017). https://doi.org/10.1109/tce.2017.014824
43. Royer, J., Nguyen, H., Martinot, O., Preteux, F.: Multimedia interactive services automation based on content indexing. Bell Labs Tech. J. **13**(2), 147–154 (2008). https://doi.org/10.1002/bltj.20310
44. Liang, B.-S., Lee, Y.-C., Yeh, W.-C., Jen, C.-W.: Index rendering: hardware-efficient architecture for 3-D graphics in multimedia system. IEEE Trans. Multimed. **4**(3), 343–360 (2002). https://doi.org/10.1109/tmm.2002.802008
45. Han, Y.: Computer animation in mobile phones using a motion capture database compressed by polynomial curve-fitting techniques. IEEE Trans. Consum. Electron. **54**(3), 1008–1016 (2008). https://doi.org/10.1109/tce.2008.4637580
46. Tsai, Y.-Y., Lin, W.-C., Cheng, K.B., Lee, J., Lee, T.-Y.: Real-time physics-based 3D biped character animation using an inverted pendulum model. IEEE Trans. Visual Comput. Gr. **16**(2), 325–337 (2010). https://doi.org/10.1109/tvcg.2009.76
47. Tan, R., Guan, S.-U.: A dynamic petri net model for iterative and interactive distributed multimedia presentation. IEEE Trans. Multimed. **7**(5), 869–879 (2005). https://doi.org/10.1109/tmm.2005.854377
48. Yousefpour, A., Fung, C., Nguyen, T., Kadiyala, K., Jalali, F., Niakanlahiji, A., Kong, J., Jue, J.P.: All One needs to know about fog computing and related edge computing paradigms: a complete survey. J. Syst. Architect. **98**, 289–330 (2019). https://doi.org/10.1016/j.sysarc.2019.02.009
49. Dwivedi, Y.K., Hughes, L., Baabdullah, A.M., Ribeiro-Navarrete, S., Giannakis, M., Al-Debei, M.M., Dennehy, D., Metri, B., Buhalis, D., Cheung, C.M., Conboy, K., Doyle, R., Dubey, R., Dutot, V., Felix, R., Goyal, D., Gustafsson, A., Hinsch, C., Jebabli, I., Wamba, S.F.: Metaverse beyond the hype: multidisciplinary perspectives on emerging challenges, opportunities, and agenda for research, practice and policy. Int. J. Inf. Manage. **66**, 102542 (2022). https://doi.org/10.1016/j.ijinfomgt.2022.102542
50. Chen, X., Zou, D., Xie, H., Wang, F.L.: Metaverse in education: contributors, cooperations, and research themes. IEEE Trans. Learn. Technol. **16**(6), 1111–1129 (2023). https://doi.org/10.1109/tlt.2023.3277952
51. Porcu, S., Floris, A., Atzori, L.: Quality of experience in the metaverse: an initial analysis on quality dimensions and assessment. In: Quality of Experience in the Metaverse: An Initial Analysis on Quality Dimensions and Assessment (2020). https://doi.org/10.1109/qomex55416.2022.9900897
52. Çallı, B.A., Ediz, Ç.: Top concerns of user experiences in Metaverse games: a text-mining based approach. Entertain. Comput. **46**, 100576 (2023). https://doi.org/10.1016/j.entcom.2023.100576
53. Koul, S., Singh, S., Verma, R.: Decentralised content creation in digital learning: a blockchain concept. In: Smart Innovation, Systems and Technologies (pp. 583–591) (2021). https://doi.org/10.1007/978-981-16-4177-0_57
54. Kim, K., Kwon, O., Yu, J.: Evaluation of an HMD-based multisensory virtual museum experience for enhancing sense of presence. IEEE Access **11**, 100295–100308 (2023). https://doi.org/10.1109/access.2023.3311135
55. Vashistha, N., Hossain, M.M., Shahriar, M.R., Farahmandi, F., Rahman, F., Tehranipoor, M.M.: EChain: a blockchain-enabled ecosystem for electronic device authenticity verification. IEEE Trans. Consum. Electron. **68**(1), 23–37 (2021). https://doi.org/10.1109/tce.2021.3139090
56. Birnie, L., Abhayapala, T., Tourbabin, V., Samarasinghe, P.: Mixed source sound field translation for virtual binaural application with perceptual validation. IEEE/ACM Trans. Audio Speech Lang. Process. **29**, 1188–1203 (2021). https://doi.org/10.1109/taslp.2021.3061939
57. Yan, S., Xiong, Y., Lin, D.: Spatial temporal graph convolutional networks for skeleton-based action recognition. In: Proceedings of the AAAI Conference on Artificial Intelligence, **32**(1) (2018). https://doi.org/10.1609/aaai.v32i1.12328

58. Tataria, H., Shafi, M., Molisch, A.F., Dohler, M., Sjoland, H., Tufvesson, F.: 6G Wireless systems: vision, requirements, challenges, insights, and opportunities. Proc. IEEE **109**(7), 1166–1199 (2021). https://doi.org/10.1109/jproc.2021.3061701
59. Toichoa Eyam, A., Martínez Lastra, J.L.: Metastates: an approach for representing human workers' psychophysiological states in the industrial metaverse. IEEE Access **12**, 81682–81691 (2024). https://doi.org/10.1109/access.2024.3411877
60. Kim, J.-H., Lee, J., Kim, Y.: Motion generation and analyzing the user's arm muscles via leap motion and its data-driven representations. IEEE Access **12**, 47787–47796 (2024). https://doi.org/10.1109/access.2024.3383318
61. Si, Z., Li, S., Wei, G., Lin, X., Zhao, W.: An all-electric neural device and network based on laterally coupled Nanomagnets for binary image recognitions. IEEE Trans. Electron Devices **69**(6), 3130–3134 (2022). https://doi.org/10.1109/ted.2022.3167936
62. Li, Z., Chan, J., Walton, J., Benko, H., Wigdor, D., Glueck, M.: Armstrong: an empirical examination of pointing at non-dominant arm-anchored UIs in virtual reality. In: Armstrong: An Empirical Examination of Pointing at Non-dominant Arm-anchored UIs in Virtual Reality (2021). https://doi.org/10.1145/3411764.3445064
63. Foy, C.R., Dudley, J.J., Gupta, A., Benko, H., Kristensson, P.O.: Understanding, detecting and mitigating the effects of coactivations in Ten-Finger Mid-Air typing in virtual reality. Underst. Detect. Mitig. Eff. Coactivations Ten-finger Mid-air Typ. Virtual Real. (2021). https://doi.org/10.1145/3411764.3445671
64. Ren, Z., Misra, I., Schwing, A.G., Girdhar, R.: 3D spatial recognition without spatially labeled 3D. 3D Spatial Recognition Without Spatially Labeled 3D (2021). https://doi.org/10.1109/cvpr46437.2021.01300
65. Tiomkin, S., Malah, D., Shechtman, S.: Statistical text-to-speech synthesis based on segment-wise representation with a norm constraint. IEEE Trans. Audio Speech Lang. Process. **18**(5), 1077–1082 (2010). https://doi.org/10.1109/tasl.2010.2040795
66. Mehta, D., Sotnychenko, O., Mueller, F., Xu, W., Elgharib, M., Fua, P., Seidel, H., Rhodin, H., Pons-Moll, G., Theobalt, C. (2020). XNect. ACM Trans. Gr. **39**(4). https://doi.org/10.1145/3386569.3392410
67. Xu, M., Ng, W.C., Lim, W.Y., Kang, J., Xiong, Z., Niyato, D., Yang, Q., Shen, X., Miao, C.: A full dive into realizing the edge-enabled metaverse: visions, enabling technologies, and challenges. IEEE Commun. Surv. Tutor. **25**(1), 656–700 (2023). https://doi.org/10.1109/comst.2022.3221119
68. Jeon, H., Youn, H., Ko, S., Kim, T.: Blockchain and AI meet in the metaverse. Blockchain Potential AI (2022). https://doi.org/10.5772/intechopen.99114
69. Adams, H., Stefanucci, J., Creem-Regehr, S., Pointon, G., Thompson, W., Bodenheimer, B.: Shedding light on cast shadows: an investigation of perceived ground contact in AR and VR. IEEE Trans. Visual Comput. Gr. **28**(12), 4624–4639 (2021). https://doi.org/10.1109/tvcg.2021.3097978
70. Far, S.B., Rad, A.I.: Applying digital twins in metaverse: user interface, security and privacy challenges. DergiPark (Istanbul University) (2022). https://dergipark.org.tr/tr/pub/jmv/issue/67967/1072189
71. Li, K., Cui, Y., Li, W., Lv, T., Yuan, X., Li, S., Ni, W., Simsek, M., Dressler, F.: When internet of things meets metaverse: convergence of physical and cyber worlds. IEEE Internet Things J. **10**(5), 4148–4173 (2022). https://doi.org/10.1109/jiot.2022.3232845
72. Kim, W.: Edge computing server deployment technique for cloud VR-based multi-user metaverse content. J. Korea Multimed. Soc. **24**(8), 1090–1100 (2021). https://doi.org/10.9717/kmms.2021.24.8.1090
73. Yu, F., Lin, H., Wang, X., Yassine, A., Hossain, M.S.: Blockchain-empowered secure federated learning system: architecture and applications. Comput. Commun. **196**, 55–65 (2022). https://doi.org/10.1016/j.comcom.2022.09.008
74. Sin, Z.P., Jia, Y., Wu, A.C., Zhao, I.D., Li, R.C., Ng, P.H., Huang, X., Baciu, G., Cao, J., Li, Q.: Toward an edu-metaverse of knowledge: immersive exploration of university courses. IEEE Trans. Learn. Technol. **16**(6), 1096–1110 (2023). https://doi.org/10.1109/tlt.2023.3290814

75. Le Duy, H., Tuan Truong, V., Le Bao, L.: Blockchain-empowered metaverse: decentralized crowdsourcing and marketplace for trading machine learning data and Models. IEEE Access **12**, 68556–68572 (2024). https://doi.org/10.1109/access.2024.3401076
76. Wang, J., Chen, S., Liu, Y., Lau, R.: Intelligent metaverse scene content construction. IEEE Access **11**, 76222–76241 (2023). https://doi.org/10.1109/access.2023.3297873
77. Fernández-Caramés, T.M., Fraga-Lamas, P.: Forging the industrial metaverse for industry 5.0: where extended reality, IIoT, opportunistic edge computing, and digital twins meet. IEEE Access **12**, 95778–95819 (2024). https://doi.org/10.1109/access.2024.3422109
78. Bobko, T., Corsette, M., Wang, M., Springer, E.: Exploring the possibilities of edu-metaverse: a new 3D ecosystem model for innovative learning. IEEE Trans. Learn. Technol. 1–13 (2024). https://doi.org/10.1109/tlt.2024.3364908
79. Wang, F., Qin, R., Wang, X., Hu, B.: MetaSocieties in metaverse: metaeconomics and meta-management for metaenterprises and metacities. IEEE Trans. Comput. Soc. Syst. **9**(1), 2–7 (2022). https://doi.org/10.1109/tcss.2022.3145165
80. Yang, K., Zhang, Z., Youliang, T., Ma, J.: A secure authentication framework to guarantee the traceability of avatars in Metaverse. IEEE Trans. Inf. Forensics Secur. **18**, 3817–3832 (2023). https://doi.org/10.1109/tifs.2023.3288689
81. Gupta, B.B., Gaurav, A., Arya, V.: Fuzzy logic and biometric-based lightweight cryptographic authentication for metaverse security. Appl. Soft Comput. **164**, 111973 (2024). https://doi.org/10.1016/j.asoc.2024.111973
82. Zhang, Z., Yang, K., Tian, Y., Ma, J.: An anti-disguise authentication system using the first impression of avatar in metaverse. IEEE Trans. Inf. Forensics Secur. **19**, 6393–6408 (2024). https://doi.org/10.1109/tifs.2024.3410527
83. Goldberg, M., Schär, F.: Metaverse governance: an empirical analysis of voting within decentralized autonomous organizations. SSRN Electron. J. (2022). https://doi.org/10.2139/ssrn.4310845
84. Martha, G.I., Gaol, F.L., Supangkat, S.H., Ranti, B.: Utilization of decentralized autonomous organization (DAO) for virtual property transaction governance in metaverse: a conceptual framework. In: 2023 3rd International Conference on Intelligent Cybernetics Technology & Applications (ICICyTA), pp. 306–311 (2023). https://doi.org/10.1109/icicyta60173.2023.10429035
85. Hunhevicz, J.J., Hall, D.M., Brasey, P.-A., Bonanomi, M.M.M., Fischer, M.: Decentralized project delivery on the crypto commons: conceptualization, governance mechanisms, and future research directions. Proj. Leadersh. Soc. **5**, 100132 (2024). https://doi.org/10.1016/j.plas.2024.100132
86. Cheng, Y., Guo, Y., Xu, M., Hu, Q., Yu, D., Cheng, X.: An adaptive and modular blockchain enabled architecture for a decentralized metaverse. IEEE J. Sel. Areas Commun. **42**(4), 893–904 (2024). https://doi.org/10.1109/jsac.2023.3345432
87. Elsadig, M., Alohali, M.A., Ibrahim, A.O., Abulfaraj, A.W.: Roles of blockchain in the metaverse: concepts, taxonomy, recent advances, enabling technologies, and open research issues. IEEE Access **12**, 38410–38435 (2024). https://doi.org/10.1109/access.2024.3367014
88. Ryu, J., Son, S., Lee, J., Park, Y., Park, Y.: Design of secure mutual authentication scheme for metaverse environments using blockchain. IEEE Access **10**, 98944–98958 (2022). https://doi.org/10.1109/access.2022.3206457
89. Awan, K.A., Din, I.U., Almogren, A., Seo-Kim, B.: Blockchain-based trust management for virtual entities in the metaverse: a model for avatar and virtual organization interactions. IEEE Access **11**, 136370–136394 (2023). https://doi.org/10.1109/access.2023.3337806
90. Liu, L., Feng, J., Wu, C., Chen, C., Pei, Q.: Reputation management for consensus mechanism in vehicular edge metaverse. IEEE J. Sel. Areas Commun. **42**(4), 919–932 (2024). https://doi.org/10.1109/jsac.2023.3345382
91. Paik, H.-Y., Xu, X., Bandara, H.M., Lee, S.U., Lo, S.K.: Analysis of data management in blockchain-based systems: from architecture to governance. IEEE Access **7**, 186091–186107 (2019). https://doi.org/10.1109/access.2019.2961404

92. Tao, B., Dai, H.-N., Xie, H., Wang, F.L.: Structural identity representation learning for blockchain-enabled metaverse based on complex network analysis. IEEE Trans. Comput. Soc. Syst. **10**(5), 2214–2225 (2023). https://doi.org/10.1109/tcss.2022.3233059
93. Ooi, K.-B., Tan, G.W.-H., Al-Emran, M., Al-Sharafi, M.A., Arpaci, I., Zaidan, A.A., Lee, V.-H., Wong, L.-W., Deveci, M., Iranmanesh, M.: The metaverse in engineering management: overview, opportunities, challenges, and future research agenda. IEEE Trans. Eng. Manage. **71**, 13882–13889 (2024). https://doi.org/10.1109/tem.2023.3307562
94. Javed, M.A., Liu, P., Panwar, S.S.: Enhancing XR application performance in multi-connectivity enabled mmwave networks. IEEE Open J. Commun. Soc. **4**, 2421–2438 (2023). https://doi.org/10.1109/ojcoms.2023.3322383
95. Shin, C., Oh, S., Jeong, H.: Extended reality platform for metaverse exhibition. J. Web Eng. 1055–1074 (2024). https://doi.org/10.13052/jwe1540-9589.2275
96. Cheymol, A., Fribourg, R., Lécuyer, A., Normand, J.-M., Argelaguet, F.: Beyond my real body: characterization, impacts, applications and perspectives of "dissimilar" avatars in virtual reality. IEEE Trans. Visual Comput. Gr. **29**(11), 4426–4437 (2023). https://doi.org/10.1109/tvcg.2023.3320209
97. Park, E.: Development of web content for music education using AR human facial recognition technology. J. Web Eng. (2023). https://doi.org/10.13052/jwe1540-9589.2252

Cloud Computing and Multimedia IoT

Raju Anitha, V. Esther Jyothi, Ramakrishna Regulagadda, Sivudu Macherla, D. Naga Malleswari, G. Siva Nageswara Rao, and Puvvada Nagesh

Abstract Cloud computing and multimedia Internet of Things (IoT) represent two transformative technologies that have reshaped the landscape of computing and communication. Cloud computing offers scalable and on-demand resources, enabling efficient storage, processing, and analysis of vast amounts of data. On the other hand, multimedia IoT involves the integration of multimedia elements into the IoT ecosystem, enriching the sensory experience and enabling new applications. This paper explores the intersection of cloud computing and multimedia IoT, examining the synergies and challenges in leveraging cloud services for managing multimedia data in IoT environments. The study delves into the architectural considerations, security implications, and performance optimizations required to harness the full potential of this convergence. Through a comprehensive analysis, this work contributes valuable insights into the evolving paradigm of cloud computing and multimedia IoT, offering guidance for researchers, practitioners, and policymakers navigating this dynamic landscape.

Keywords Cloud computing · Multimedia · Internet of Things (IoT) · Edge computing · Data management · Security · Scalability · Performance optimization · Sensory computing · Multimedia IoT applications

---

R. Anitha
Department of Computer Science and Engineering, Koneru Lakshmaiah Education Foundation, Vaddeswaram, Andhra Pradesh, India

V. Esther Jyothi
Department of Computer Applications, V R Siddhartha Engineering College, SAHE-A Deemed to be University, Vijayawada, India

R. Regulagadda (✉) · S. Macherla · D. Naga Malleswari · G. Siva Nageswara Rao · P. Nagesh
Department of CSE, Koneru Lakshmaiah Education Foundation, Guntur, Andhra Pradesh, India
e-mail: regulagadda@gmail.com

R. Sharma et al. (eds.), *Multimedia Technologies in the Internet of Things Environment, Volume 4*, Studies in Big Data 173, https://doi.org/10.1007/978-981-96-4356-1_10

1 Introduction

Cloud computing and the Internet of Things (IoT) have individually transformed numerous industries through the provision of scalable resources and interconnected devices. The utilization of cloud computing entails the provision of adaptable, on-demand entry to a communal pool of configurable computing resources, allowing enterprises to improve their operations without substantial initial investments in infrastructure. This fundamental change has enabled entities to efficiently expand their operations, realize cost reductions, and concentrate more on their fundamental capabilities by exploiting cloud-based services. Concurrently, the IoT environment has experienced a significant expansion, linking billions of devices globally. These devices gather and share data, delivering immediate insights and facilitating automation across a variety of sectors including smart cities, healthcare, agriculture, and industrial automation [1]. The surge in IoT devices has resulted in the creation of extensive volumes of data, necessitating robust approaches for data storage, processing, and examination. The integration of multimedia elements such as audio, video, and images into the IoT framework, known as multimedia IoT, enhances data collection and creates opportunities for applications requiring rich media content.

The combination of cloud computing and multimedia IoT provides a promising prospect for the effective management and processing of multimedia data. Through the utilization of cloud resources, businesses can address the storage and computational needs of multimedia IoT applications, ensuring scalability, cost-effectiveness, and improved security. This fusion facilitates the development of innovative applications in various industries, enriching sensory experiences and delivering real-time analytics. Nevertheless, this integration also brings about challenges including latency, bandwidth limitations, data privacy concerns, and interoperability issues that need to be resolved for the full realization of its potential [2]. This section delves into these synergies and challenges, offering valuable insights and guidance for researchers, practitioners, and policymakers navigating this ever-evolving landscape.

Cloud computing is critical within the IoT ecosystem because it provides scalable storage, robust data processing capabilities, and cost-effective solutions. Enterprises can easily manage the massive amounts of data generated by IoT devices by leveraging cloud platforms, eliminating the need for expensive on-premises infrastructure. The combination of sophisticated data analytics and machine learning made possible by cloud services allows for immediate data processing and useful insights, improving operational efficiency and decision-making processes. Furthermore, the cloud's pay-as-you-go pricing structure and embedded maintenance services help companies reduce operational costs, while strict security processes and compliance standards ensure data security. The following figure shows the role of cloud computing in IoT [3].

1.1 Enhanced Data Storage and Processing in Cloud Computing for IoT

The integration of cloud computing with IoT addresses the critical challenges associated with the storage and processing of the vast volumes of data generated by IoT devices. These devices, which range from basic sensors to advanced multimedia systems, are continuously generated by IoT devices. The devices, which range from basic sensors to advanced multimedia systems, continuously generate data, often in real time, which must be stored, processed, and analyzed to extract meaningful insights (Fig. 1).

Cloud computing offers a robust solution to these challenges by providing virtually unlimited storage capacity and extensive processing power. This is achieved through data centers distributed globally, allowing for efficient data storage and retrieval. Cloud platforms can store data in a distributed manner, ensuring that even if one data center experiences an issue, data can still be accessed from another, thus providing high availability and reliability.

Fig. 1 Role of cloud computing in IoT

1.2 Data Management and Scalability

Cloud storage systems are designed to scale effortlessly, accommodating the growing volumes of data generated by IoT devices. This is crucial for industries like healthcare, where devices such as wearable health monitors, smart medical equipment, and patient tracking systems continuously produce data. For example, wearable devices can monitor patients' heart rates, glucose levels, or physical activity in real time [4]. This data can be transmitted to the cloud, where it is stored in large datasets that can be analyzed to identify trends or anomalies, providing invaluable insights into patient health.

Complex data analysis is made possible by the processing capacity made available by cloud computing platforms, which would be difficult to accomplish on conventional systems. IoT data can be analyzed at scale by cloud services using artificial intelligence, machine learning algorithms, and big data analytics. For example, cloud-based analytics may sort through enormous volumes of patient data in the healthcare industry to find trends that may point to possible health problems, such as early symptoms of chronic illnesses or the possibility of prescription side effects. More individualized treatment regimens and speedier diagnoses may result from these real-time analysis capabilities [5].

Another critical aspect is data security and compliance. Healthcare data, for instance, is highly sensitive and subject to stringent regulatory requirements like HIPAA (Health Insurance Portability and Accountability Act) in the United States. Cloud providers implement robust security measures, including encryption, access controls, and regular security audits, to protect data from breaches and unauthorized access. They also offer compliance management services to help organizations adhere to various regulations, ensuring that data handling practices meet the required standards [6].

2 Convergence of Cloud Computing and Multimedia IoT

2.1 Technological Synergy

Multimedia is combined with cloud computing and the Internet of Things which are two strong technologies that complement each other and increase their potential. Cloud computing provides a stable and expandable foundation for handling the enormous volumes of data produced by Internet of Things devices, particularly those that deal with multimedia information like audio, video, and photos. Multimedia Internet of Things devices are those that collect and send multimedia data. Cloud computing provides a solution to this challenge by offering virtually unlimited storage and powerful computing resources that can be accessed on demand [7].

2.2 *Data Storage and Processing*

One of the key benefits of cloud computing in the context of multimedia IoT is its ability to store and process large volumes of data efficiently. IoT devices continuously generate data that needs to be processed in real time or near-real time. Cloud platforms provide scalable storage solutions that can accommodate this data, ensuring that it is readily available for analysis and retrieval. Additionally, cloud computing offers advanced data processing capabilities that can handle complex computations and large-scale data analytics, enabling more sophisticated and efficient processing of multimedia content. For example, in a smart city environment, thousands of cameras may be deployed for surveillance purposes, generating terabytes of video data daily. Processing and analyzing this data locally would require significant computational resources and storage capacity. By leveraging cloud computing, this data can be uploaded to the cloud where it can be processed using powerful algorithms, stored securely, and accessed as needed, and this approach not only ensures that the data is handled efficiently but also allows for advanced analytics, such as facial recognition or behavioral analysis, to be performed at scale [8].

2.3 *Scalability and Flexibility*

Another significant advantage of combining cloud computing with multimedia IoT is the scalability and flexibility it offers. Cloud platforms are designed to scale resources up or down based on demand, which is particularly important for IoT applications where data volumes can vary significantly. This scalability ensures that resources are always available to handle the incoming data, regardless of spikes in usage or unexpected increases in data generation.

For instance, during a large public event in a smart city, the number of active IoT devices and the volume of multimedia data generated can increase dramatically. Cloud computing can dynamically allocate additional resources to accommodate this surge, ensuring that all data is processed and stored without delay. Once the event is over, resources can be scaled back, optimizing costs and maintaining efficiency [9].

3 Cloud Architecture for Multimedia IoT

Data ingestion is the process of gathering and importing information for immediate use of storage in a database, In a multimedia IoT system, data ingestion entails gathering multimedia content (e.g., video, audio, photos) from IoT devices and transmitting it to the cloud. This procedure necessitates robust and scalable technologies to handle the large volume and variety of multimedia data.

Cloud platforms provide a variety of storage options to accommodate multimedia data, including.

Object storage: It is ideal for storing big amounts of unstructured data like films and photographs. Examples include Amazon S3, Google Cloud storage, and Azure Blob storage [10].

File storage: Ideal for applications that require a shared file system. Examples include Amazon EFS (elastic file system) and azure files.

Block storage: It is used in applications that demand low-latency access to raw storage. Examples include Amazon EBS (elastic block store) and Azure managed disks.

3.1 Processing and Analytics

Processing and interpreting multimedia data necessitates tremendous processing resources. Cloud computing provides the resources required to conduct complicated analyses such as video encoding, real-time streaming, picture recognition, natural language processing, and cloud-based processing. Cloud computing provides the resources required to conduct complicated analyses such as video encoding, real-time streaming, picture recognition, and natural language processing. Cloud-based processing can be classified as: real-time processing refers to the processing of data as it enters to provide prompt insights or actions. This is critical for applications requiring live video streaming, real-time surveillance, and interactive gaming. Batch processing involves processing massive amounts of data at predetermined periods. This is important for tasks such as video archiving, historical data analysis, and bulk picture processing (Fig. 2).

4 Cloud Architectures for Multimedia Processing

In today's digital world, managing the enormous volumes of multimedia data generated and consumed requires the use of cloud systems. Cloud solutions are perfect for handling multimedia IoT (Internet of Things) applications because of their scalability and flexibility. These applications process and store a variety of data kinds, including audio, video, and image. The main elements of cloud-based multimedia processing are examined in this part, including content delivery networks, data processing frameworks, storage options, media encoding and transcoding, and real-time data processing [11].

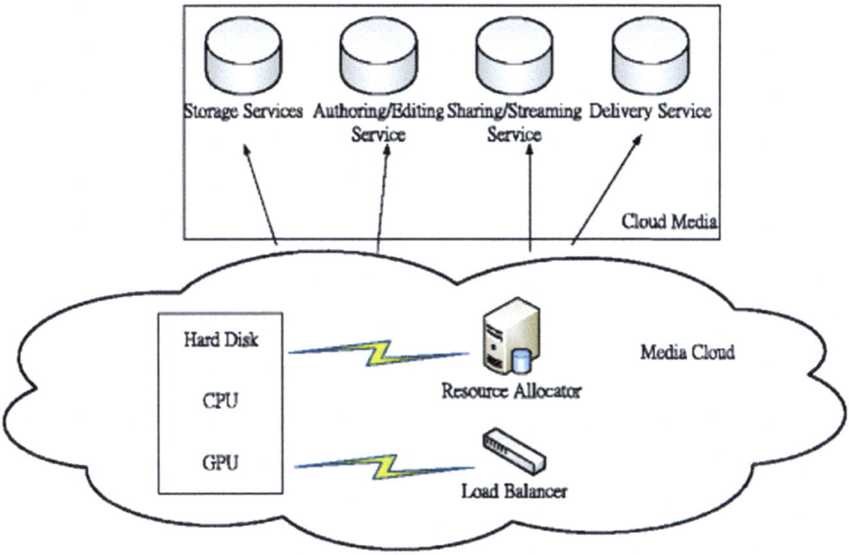

Fig. 2 Cloud architecture in multimedia

4.1 Solution Scalable Storage

Because multimedia data is generally vast and complicated by nature, reliable storage solutions are required. Because they can manage enormous amounts of unstructured data, object storage systems like Amazon S3, Google Cloud Storage, and Azure Blob Storage are frequently used. These systems group data into objects that are made up of the data itself, associated metadata, and a unique identifier. These systems classify data into objects, which are made up of the data itself, metadata, and a unique identifier. The inherent scalability of object storage allows it to grow seamlessly in response to the increasing data needs of multimedia applications. Moreover, object storage offers exceptional durability and resilience against data loss because data is replicated across multiple geographic locations. This redundancy is especially important for multimedia data, which usually requires high availability and reliability [12].

However, file storage options with a more conventional file system interface, such as Amazon EFS and Azure Files, can be helpful for programs that need shared file access. The ability of file storage to preserve a hierarchical directory structure makes handling and navigating huge datasets simpler. Workflows involving collaborative media editing, where several users may require simultaneous access to the same files, benefit from this arrangement. Large files must be accessed quickly for operations like video editing and streaming, which is why file storage systems are made to manage high throughput and deliver reliable performance [13].

Block storage provides high-performance storage options perfect for situations demanding low-latency access to data. Examples of this type of storage are Amazon EBS and Azure Managed Disks. Databases and virtual machines (VMs) used in multimedia processing pipelines benefit greatly from this kind of storage. Databases, for instance, may index data or store metadata that must be readily accessed for media retrieval procedures. Block storage gives customers precise control over storage volumes, enabling them to customize cost and performance attributes to meet their unique requirements. Because of its adaptability, block storage is essential for maximizing the efficiency of multimedia applications [14].

4.2 Frameworks for Data Processing

Multimedia data processing frequently calls for intricate calculations and extensive data transformations. Apache Spark and Hadoop are two key frameworks for effectively completing these tasks. Big data processing has been made possible by Apache Hadoop, which is known for its MapReduce programming model and Hadoop Distributed File System (HDFS). By distributing data among several nodes, HDFS offers parallel processing and fault tolerance. This configuration is useful for batch processing huge amounts of multimedia data, including extracting metadata from photos or analyzing video logs for content indexing [15].

However, Apache Spark has become more well known as real-time processing capabilities are required by multimedia applications. When it comes to data analytics, Spark's in-memory processing powers greatly outpace Hadoop's disk-based processing. Because of this, Spark works especially well in interactive and iterative processes where speed is of the essence, like real-time video analytics. Because of Spark's adaptability, more complex analyses of multimedia content are made possible by its application to advanced analytics, such as machine learning and graph processing. For example, models for automatic video tagging, facial recognition, or sentiment analysis in audio files can be constructed using Spark's MLlib module.

4.3 Content Delivery Networks (CDNs)

The efficient global delivery of multimedia material is largely dependent on content delivery networks or CDNs. Content is cached at strategically placed edge servers by CDNs like Akamai, Cloudflare, and Amazon CloudFront. Because of this geographic dispersion, latency is decreased and consumers don't have to wait long to receive multimedia material. For high-bandwidth applications like video streaming, where buffering can negatively affect user experience, CDNs are extremely important. Content delivery networks (CDNs) can offer quicker download speeds and a more seamless viewing experience by delivering content from places closer to the end users [16].

Additionally, CDNs provide dynamic content acceleration, which enhances the delivery of dynamic content like live streams and interactive media in addition to

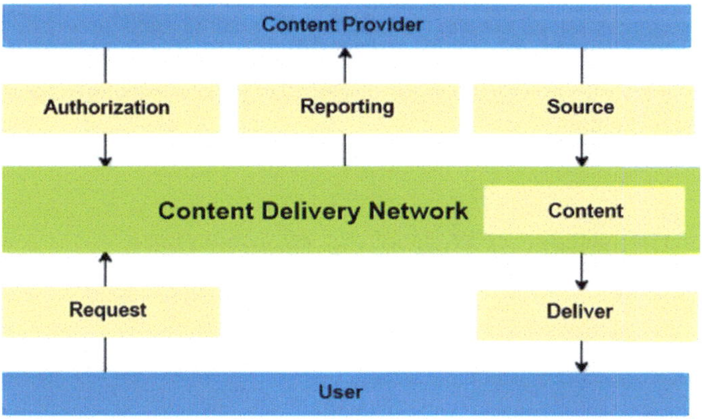

Fig. 3 CDN architecture

static information like movies and photographs. For applications where low latency and high dependability are critical, such as interactive gaming or live sports broadcasting, this functionality is essential. In order to preserve the integrity and privacy of multimedia material during transmission, CDNs also incorporate a number of security measures, such as SSL encryption, secure token authentication, and protection against Distributed Denial of Service (DDoS). These security precautions aid in defending against online attacks that can impede the delivery of material or jeopardize private data (Fig. 3).

4.4 Structure and Functionality of CDNs

A network of strategically placed computers, sometimes referred to as edge servers or points of presence (PoPs), makes up CDNs. The primary source of the content is the origin server, from which these servers store copies of the content. Among the main features of CDNs are.

Content Caching: Storing static content locally for end users is one of CDNs' main uses. Together with other static assets like CSS and JavaScript files, this also includes multimedia files like audio, video, and image files. CDNs reduce the amount of distance that data must travel, which lowers load times and increases access speeds. They do this by storing cached copies at several PoPs throughout the world.

Load balancing: To balance the load, many servers receive incoming traffic from CDNs. This improves the speed and effectiveness of content delivery by preventing any one server from becoming overloaded and guaranteeing that users are served by the server nearest to them. During high-traffic events like live streaming sporting events or product launches, load balancing is especially crucial.

Content Optimization: Content Delivery Networks (CDNs) frequently have functionalities that optimize the transmission of material. These functionalities may include file compression, device-specific image scaling, and media file conversion into more streaming-friendly formats. These improvements aid in lowering the needed bandwidth and enhance end users' playback, particularly for those with sluggish Internet connections.

4.5 The Advantages of Multimedia Content Delivery via CDNs

There are several benefits of using CDNs, particularly when delivering multimedia content:

Reduced delay and Faster Load Times: CDNs drastically cut down on the delay involved in retrieving data from faraway servers by caching material closer to the end users. This is especially crucial for streaming audio and video, as buffering can negatively impact the user experience. Smoother playback is a result of faster load times, and this is essential for user retention and happiness.

High availability and scalability: Content Delivery Networks (CDNs) are built to manage enormous volumes of traffic by dividing the load across several servers. When there is a sudden surge in spectators, as happens with major sporting events or viral material, scalability becomes critical because it prevents traditional servers from becoming overwhelmed. Even in situations of heavy demand, CDNs guarantee that the content is accessible and available.

Enhanced Security: A range of security capabilities offered by CDNs aid in the protection of user information and content. Among these is DDoS prevention, which lessens the effect of attacks known as distributed denial of service, which try to overload servers with traffic. Additionally, secure token authentication is a feature of CDNs that guarantees that only authorized users can access certain content. Furthermore, data is secured while transmission using SSL/TLS encryption to prevent tampering and eavesdropping [17].

Enhanced Redundancy and Reliability: Content Delivery Networks (CDNs) offer redundancy by distributing multiple copies of content among many servers. Because of this redundancy, the material will continue to be supplied from other servers in the network even in the event of a server failure. For real-time apps and live streaming services, where downtime can result in severe consumer displeasure, this stability is especially important.

4.6 CDN Applications for Multimedia Distribution

Applications where effective multimedia content distribution is essential employ CDNs. Among the important use cases are.

Video Streaming: Sites like YouTube, Netflix, and Hulu rely heavily on Content Delivery Networks (CDNs). They facilitate the buffer-free, high-quality delivery of video material to a worldwide audience. A seamless watching experience is ensured by adaptive bitrate streaming, a technology that is enabled by CDNs. It modifies the video stream's quality in real time dependent on the user's network conditions.

Live Streaming: CDNs allow millions of viewers worldwide to watch live events in real time, including concerts, sports, and seminars. By using Content Delivery Networks (CDNs), latency is reduced and content delivery is guaranteed, even during periods of high demand.

Online Gaming: Response times and latency are important factors in online gaming. By caching game data closer to the player, facilitating quicker game update downloads, and more effectively distributing content, CDNs assist in minimizing latency. As a result, playing games becomes more responsive.

E-commerce: CDNs are used by e-commerce companies to quickly distribute multimedia material, such as product photos and videos. In e-commerce, quick loading times are essential since they have a direct effect on conversion rates and customer experience. Additionally, during sales events or new product launches, CDNs assist in managing traffic surges.

Corporate and Educational Content: Organizations that offer training videos, webinars, and other multimedia resources to employees and students in different locations do so by means of Content Delivery Networks (CDNs). CDNs ensure that material is available and provides a consistent experience for users, regardless of their location.

4.7 Encoding and Transcoding of Media

Multimedia content must be offered in several formats and resolutions due to the variety of devices and network conditions. Here's when transcoding and media encoding are useful. The process of converting unprocessed multimedia data into a format appropriate for dissemination and storage is known as encoding. AAC and MP3 are common audio codecs, and H.264 and H.265 (HEVC) for video. The media's compatibility, file size, and quality are all impacted by the codec selection. For instance, H.265 provides more compression efficiency than H.264, enabling high-quality video at lower bitrates—a benefit for lower storage and bandwidth costs. Conversely, transcoding is the process of changing the bitrate or format of media that has already been encoded. Adaptive bitrate streaming (ABR), which modifies a video stream's quality in real time according to the viewer's network circumstances, depends on this procedure. ABR is used by technologies such as HTTP Live Streaming (HLS) and Dynamic Adaptive Streaming over HTTP (DASH) to provide users with the greatest quality experience possible, irrespective of their device capabilities or Internet speed. To ensure a seamless viewing experience for both, a viewer with a faster connection would receive a 480p stream, while someone on a high-speed connection might receive a 1080p broadcast.

Scalable encoding and transcoding solutions are offered by cloud-based services as AWS Elemental MediaConvert, Google Cloud Transcoder, and Azure Media Services. These services offer features like automatic captioning and content protection, as well as supporting a large number of formats and automating the conversion process. For content providers that have to swiftly prepare media for several platforms and devices, this automation is very helpful. Moreover, these services allow for smooth workflows from content intake to delivery by integrating with other cloud infrastructure.

4.8 Processing Data in Real Time

Real-time data processing is essential to multimedia applications in order to provide responsive and timely services. This field has seen a revolution because to serverless computing platforms like AWS Lambda, Azure Functions, and Google Cloud Functions, which let developers run code in response to events without having to worry about managing servers. This approach is perfect for real-time tasks like sentiment analysis, object detection, and scene recognition, where functions can be triggered by incoming data streams. One example of a real-time task would be live video analysis.

Data streaming technologies that can manage massive volumes of streaming data from numerous sources, like Amazon Kinesis and Apache Kafka, also offer real-time processing. Real-time data collection, processing, and analysis are made possible by these systems. This allows for the deployment of applications such as live content moderation, in which media is screened and marked for objectionable content as it is posted. In interactive media experiences like virtual reality settings or live sports analytics, where data is processed instantly to improve user engagement, real-time analytics can also be used.

4.9 Apache Hadoop in Multimedia Data Processing

Multimedia data is known for its enormous volume and diversity, and Apache Hadoop provides a robust framework for the distributed processing and storing of huge information. With its scalable architecture, each machine can provide local processing and storage, ranging from a single server to thousands of units. This section examines the Hadoop Distributed File System (HDFS) and MapReduce, two of its fundamental building blocks, and how to use them to process multimedia data, including audio, video, and image files.

4.9.1 Distributed File System for Hadoop (HDFS)

The foundation of Hadoop's storage system, HDFS is designed to store big files in a distributed setting. Because of its fault-tolerant architecture, data may be reliably stored even in the case of hardware malfunctions. This is especially crucial for multimedia data, which frequently needs high durability and dependability and might be rather large (Fig. 4)

.

Scalability: By dividing up files over a cluster of computers, HDFS can grow to contain petabytes of data. Every file is divided into blocks, each of which is often 128 MB in size. Each block is then copied among a number of nodes, generally three. High data reliability is ensured by this replication, which makes sure that even in the event of a node failure, the data may still be accessed from another node.

High Throughput: HDFS is best suited for batch processing activities because it is made for high-throughput data access rather than low-latency data access. A limited number of huge files work better than a large number of little files, and it is optimized for massive datasets. This is especially helpful for multimedia applications that often involve big data volumes, such as streaming videos or storing images.

Fault Tolerance and Data Integrity: The system automatically duplicates data from failing nodes and uses checksums to periodically check for data corruption. The integrity of multimedia files, which must be preserved for correct processing and playback, depends on this capability.

Because HDFS can effectively store and handle massive amounts of data, it is a great option for applications like audio libraries, video archives, and image repositories. It offers a strong basis for expansive multimedia applications since it can manage a wide range of data formats and support a variety of data processing and analytics functions.

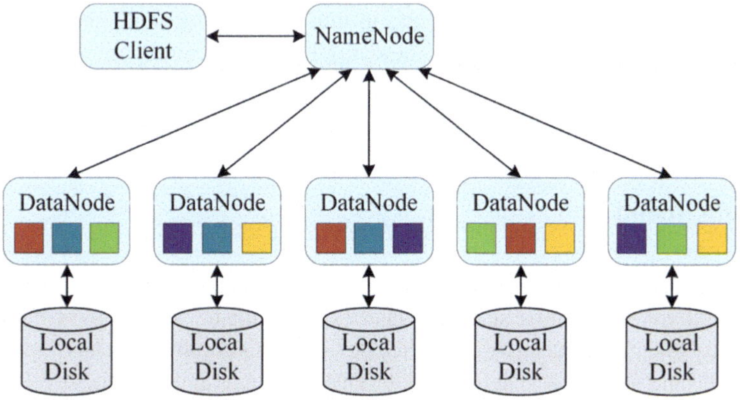

Fig. 4 Hadoop distributed file system

4.9.2 Map-Reduce

A distributed cluster of computers may process huge datasets more easily by using the MapReduce programming model and processing approach. It simplifies the intricacy of parallel processing, hence aiding developers in the creation of distributed applications.

Parallel Processing: The Map phase and the Reduce phase are how MapReduce splits the data processing task into two parts. The input data is divided into separate portions during the map phase, which is then processed concurrently by several mappers. After processing a certain amount of data, each mapper produces key-value pairs. In a video analysis assignment, for example, each mapper may process a video segment in order to extract features like faces or objects.

Data Aggregation and Analysis: The framework conducts a summary action, like counting occurrences or average values, in the Reduce phase after aggregating the outputs from the Map phase based on the key values. For jobs like creating metadata or indexing multimedia content, this stage is essential. For instance, the Reduce step might aggregate these results to index the entire video by the objects present, after mapping video segments to recognized objects.

Fault Tolerance: MapReduce is built to elegantly manage hardware malfunctions. Hadoop can transfer the task to another node in case of a node failure during the Map or Reduce phase, guaranteeing the successful completion of the operation. Processing massive multimedia files, which can be resource- and time-intensive, requires resilience.

The ease of use and efficiency of MapReduce in batch processing massive amounts of data is one of its main benefits. It can be applied to multimodal situations for the following purposes.

Video and Image Analysis: Taking metadata out of photos and movies, like tags or keywords, to build searchable databases.

Audio processing is the process of examining big audio datasets to find features like sentiment analysis or voice recognition.

Analyzing usage logs from multimedia apps in order to spot trends, determine what content is popular, and comprehend user behavior is known as log analysis.

5 Conclusion

In summary, the combination of cloud computing and the Internet of Things (IoT) has the potential to revolutionize several industries due to the unprecedented volume and complexity of data produced by connected devices. This combination has allowed for on-demand and scalable access to computational resources as well as the development of a stable framework for the management and analysis of large volumes of data. The addition of multimedia IoT, which includes image, video, and audio data, expands on the capabilities of these technologies and creates new opportunities for rich, immersive applications in a variety of industries, including smart cities, healthcare, agriculture, and industrial automation.

The integration of cloud computing within the IoT ecosystem addresses critical challenges, including the need for scalable storage and processing capabilities, the complexity of real-time data analytics, and the requirement for cost-effective solutions. Cloud platforms offer the flexibility to scale resources in response to fluctuating data volumes, enabling organizations to optimize operations without the burden of maintaining expensive on-premises infrastructure. Additionally, the combination of advanced data analytics and machine learning capabilities provided by cloud services facilitates the extraction of actionable insights, thereby enhancing operational efficiency and decision-making. The combination of cloud computing with IoT is not without difficulties, despite these important benefits. To effectively utilize the potential of these technologies, issues including latency, bandwidth restrictions, data privacy concerns, and compatibility must be resolved. As data volume and sensitivity continue to rise, it is imperative to ensure secure and compliant data management processes.

As we move toward the future, continuous improvements in edge computing, artificial intelligence, and data security will play a vital role in overcoming these difficulties and unlocking the full potential of cloud-enabled IoT systems. The continued evolution of this landscape will drive innovation, improve quality of life, and create new opportunities for businesses and consumers alike. For researchers, practitioners, and policymakers, managing this complex and rapidly changing environment will require a collaborative and forward-thinking strategy, ensuring that the benefits of emerging technologies are realized while avoiding associated dangers. As a result, the combination of cloud computing with IoT not only constitutes a noteworthy technological development but also acts as a trigger for more extensive social and economic change.

References

1. Botta, A., de Donato, W., Persico, V., Pescapé, A.: Integration of cloud computing and Internet of Things: a survey. Future Gener. Comput. Syst. **56**, 684–700 (2016)
2. Yuriyama, M., Kushida, T.: Sensor-cloud infrastructure—physical sensor management with virtualized sensors on cloud computing. In: 2010 13th International Conference on Network-Based Information Systems, Takayama, Japan, pp. 1–8 (2010)
3. Bandyopadhyay, S., Sengupta, M., Maiti, S., Dutta, S.: Role of middleware for Internet of Things: a study. Int. J. Comput. Sci. Eng. Surv. **2**(3), 94–105 (2011)
4. Aazam, M., Khan, I., Alsaffar, A.A., Huh, E.N.: Cloud of Things: integrating Internet of Things and cloud computing and the issues involved. In: Proceedings of the 11th IEEE International Bhurban Conference on Applied Sciences and Technology (IBCAST), Islamabad, Pakistan, pp. 414–419 (2014)
5. Atzori, L., Iera, A., Morabito, G.: The Internet of Things: a survey. Comput. Netw. **54**(15), 2787–2805 (2010)
6. Fortino, G., Trunfio, P. (eds.) Internet of Things Based on Smart Objects: Technology, Middleware and Applications. Springer (2014)

7. Stergiou, C., Psannis, K.E., Kim, B.G., Gupta, B.: Secure integration of IoT and cloud computing. Future Gener. Comput. Syst. **78**, 964–975 (2018)
8. Lopez, P.G., Krishnamurthy, D.R.S., Carvallo, J.P., Brocanelli, M., Jimenez-Peris, R.: Elasticity in cloud computing: state of the art and research challenges. IEEE Trans. Serv. Comput. **11**(2), 430–445 (2018)
9. Khan, R., Khan, S.U., Zaheer, R., Khan, S.: Future internet: the Internet of Things architecture, possible applications and key challenges. In: 2012 10th International Conference on Frontiers of Information Technology, Islamabad, Pakistan, pp. 257–260 (2012)
10. Hassan, M.M., Song, B., Huh., E.N.: A framework of sensor–cloud integration opportunities and challenges. In: 2011 International Conference on Ubiquitous Information Management and Communication, pp. 618–626, Seoul, South Korea, (2011)
11. Mahmud, M.S., Bell, R.G., Hossain, A.Z.M.A.: Cloud computing and IoT-based health monitoring systems: a survey. J. Netw. Comput. Appl. **162**, 1–23 (2020)
12. Xu, L., He, W., Li, S.: Internet of Things in industries: a survey. IEEE Trans. Ind. Inf. **10**(4), 2233–2243 (2014)
13. de Assis Alves Neto, M.D., da Silva, F.G., Villas, L.A., Loureiro, A.A.: Internet of Things platform for smart cities: a survey. Sensors **18**(9):1–30 (2018)
14. Baker, T., Aljohani, B., Nasir, Q., Asim, M., Hameed, R.H., Buyya, M.: A survey and taxonomy on energy and resource efficient techniques in cloud computing systems. Computing **102**, 1–27 (2020)
15. Alam, K.M., Saini, M., Saddik, A.E.: tNote: a social network of vehicles under IoT paradigm. In: 2015 IEEE International Conference on Multimedia & Expo Workshops (ICMEW), pp. 1–6, Torino, Italy (2015)
16. Luan, T.H., Gao, L., Li, Z., Xiang, Y., Wei, G., Sun, L.: Fog computing: focusing on mobile users at the edge (2015). arXiv preprint arXiv:1502.01815
17. Prabhu, S.R.: Fog Computing: Concepts. Wiley-IEEE Press, Frameworks and Technologies (2019)

Extended Reality and Immersive Multimedia for Gaming Applications

Sanchit, Bharat Bhushan, Mustafa Al-Asadi, and Khursheed Aurangzeb

Abstract In daily work life, XR refers to VR, AR, and MR, which is becoming more common. These technologies have been developing since 1957. Cross-platform, cross-media story, and brand worlds are the growing fields of XR application, these are few examples which explain that game industry features beyond gaming. But the growing amount of user participation is the main challenge in creating such an application which satisfied the need of user without compromising the experience. XR is also changing multimedia, by combining real-time interaction and push the limits of immersive experience. New technologies in this field deal with great amount of user interaction in real world. For this, new hardware and software have entered the consumer market, and entirely new technologies are being created at affordable pricing. This paper presents the information about the current technologies used in reality spectrum, how they contribute in gaming and different modes for users to interact and immerse. It also explores devices like VR headsets to AR-enabled glasses and MR resultant combination of virtual and physical space. This paper also highlights the role of XR in multimedia, including holographic video, remote rendering, and media streaming. Finally, it provides detailed information about the hardware, software, and devices which are used for developing in this field and recent advancement.

Keywords Extended reality · Virtual reality · Augmented reality · Multimedia data · Remote rendering

---

Sanchit (✉) · B. Bhushan
Department of Computer Science and Engineering, School of Engineering and Technology, Sharda University, Greater Noida, India
e-mail: sanchitsinha14@gmail.com

M. Al-Asadi
The University of Mashreq, Baghdad, Iraq
e-mail: mustafa.alasadi@uom.edu.iq

K. Aurangzeb
Department of Computer Engineering, College of Computer and Information Sciences, King Saud University, Riyadh, Saudi Arabia

R. Sharma et al. (eds.), *Multimedia Technologies in the Internet of Things Environment, Volume 4*, Studies in Big Data 173, https://doi.org/10.1007/978-981-96-4356-1_11

1 Introduction

Virtual Reality (VR) and Augmented Reality (AR) technologies are incredibly popular, as the software and hardware of these technologies evolve and become more widely available for variety of applications. Their ability to enhance user experience on playing games, simulation, training, learning, or other tasks is the reason why they have been applied in various fields. VR delivers a simulated environment where one can go in a virtual world with a VR headset and interact with it. Although VR cannot relate with real-world environment which is considered as a problem, which is solved by AR which presented new methods of visualization to enable computer-generated content in real world [1]. But AR lacks in interaction between virtual and real-world content. For tackling this challenge, Mixed Reality (MR) is introduced which merges the virtual and real worlds in such a manner that real-world object can make interaction with a virtual object to provide a scenario to the user. These three domains VR, AR, and MR complete the Extended Reality (XR) spectrum, covering the full spectrum of real-and-virtual combined environment [2]. Ultimate display is presented by Sutherland which would replicate reality in all available senses, almost 50 years ago. It considered as the start of all reality mimicking technologies we are using nowadays. Many research organizations have been developed in these past 50 years but only limited systems have matured outside laboratory-based prototypes. Morton Heiling invented the Sensorama Simulator in 1957, it is a little bigger than an arcade cabinet in a game room and is enhanced by sensations as wind, scents, vibrations, stereo audio, and feedback on the handlebars. Ivan Sutherland created the first AR helmet with AR lenses, known as the "Sword of Damocles," in 1966 [5].

Achieving smooth integration between virtual and real-world elements is one of the biggest issues in the field of XR. Although VR transports viewers to an entirely digital world, its usefulness is limited since it frequently lacks relevance to real-life situations. AR, on the other hand, attempts to establish interactive depth between the two areas by superimposing digital content on top of the real world. By interactively combining actual and virtual things, MR aims to close this gap, but it is vulnerable by expensive development costs, complicated technology, and the requirement for advanced hardware and software systems [3, 4]. In addition, user engagement is still a challenge because it is difficult to guarantee intuitive interaction without sacrificing the immersive experience. Some of these issues have been resolved by improvements in software and hardware technologies, which have made XR solutions easier to use and more accessible. The user experience has been revolutionized by gadgets like portable VR headsets, AR glasses with enhanced field of view, and MR platforms with advanced spatial mapping features. Developers are now able to create applications that seamlessly blend virtual and real environments thanks to development tools like ARKit, ARCore, and open-source VR platforms [5]. Furthermore, creative ways to designing user interfaces and manufacturing cost reductions have helped make XR technology more accessible to a larger audience. These advancements are opening the door for XR applications in fields including gaming, education, and healthcare.

This paper provides an overview of current advancement in XR, focusing on the technology, application, software, and hardware in gaming. The paper also highlights how XR has changed the game industry. Further, the study looks at cross-media and cross-platform tools and applications. This paper also covers how besides revolutionizing gaming, XR technologies are changing multimedia experiences. Applications of XR technology offer immersive storytelling capabilities by combining 3D geometric rendering with realistic multimedia streams of 360-degree video and volumetric material. Advanced XR media formats, such as holographic video and object-based 3D audio, improve the engagement of the users by producing realistic interactions. The potential of XR has been further enhanced by the idea of cloud rendering, which processes high-quality virtual environments in real time while preserving the portability of the device. These developments illustrate how XR overlaps increasingly with multimedia, integrating it into next-generation digital platforms. Finally, the paper discusses recent technological developments, including advance VR headset and AR glasses, which make XR easier and more affordable [6].

The paper is organized as follows. Section 2 presents an overview of the technologies and methodology used in each category. Section 3 is about how integration of VR and AR in gaming, and how it helps to develop the gaming industry and games produced by using these technologies. Section 4 presents brief description of hardware, software, tools, and device development in each domain, and also provides information about some devices which are available for public. Section 5 explains how XR is useful in multimedia streaming and development. Section 6 highlights the major advancements.

2 Reality Spectrum

The term XR or cross reality is a collection of cutting-edge technologies that involve use of 3D models or simulations across immersive, virtual, and physical platforms. These technologies include VR, AR, and MR. These technologies complete the reality spectrum. These are the technologies that link computer technology (such as information overlay) to the real world for the purpose of extending experiences beyond the reality. The preponderance of distinct terms for slightly different technologies indicates the growth area in this field [7]. Four major reality spectrum techniques are explained in the subsection below.

2.1 Virtual Reality

The term VR is described as 3D computer-generated environment that user can interact with. Computers powerful enough of projecting 3D information into display are used to access VR. Giving the user the feeling that they are a part of the video by immersing them in the surrounding visual display; this is the goal of the original technologies. The concept of the ultimate display lets user feel immersed using numerous input and output devices in a computer-generated environment. Although VR was targeted toward gaming in its market introduction, there are many other potential applications [8]. Some existing VR application have many sectors and fields like healthcare, exercise, simulation, education and training. VR can be divided into two categories named immersive and non-immersive. Details about these two categories is explained in the subsection below.

2.1.1 Non-Immersive

Through a combination of screens surrounding the user, VR presents the user with virtual information. Driving and flight simulations are some examples of this, where user is surrounded by multiple screens creating simulation of driver seat or cockpit without being fully immersed. On a standard display a simulated or recorded VR scene can be played non-immersively. On a computer, the viewing direction can be manually changed using a mouse and/or keyboard and on tablet or smartphones it can be changed by tilting and rotating. Many VR viewing and recording systems are currently available on market [8, 9].

2.1.2 Immersive

Using a head-mounted display (HMD) to track a user's positions and show the VR data based on the position of user; such wearables are example of immersive VR. User can experience 360 degrees of virtual environment. When someone think about the word VR they usually think about the immersive experience. A panoramic image lets user experience as if he/she is inside the environment. A stereoscopic 3D effect is created by displaying slightly different image to right and left eye over HMD which increases the immersion. Newer model of VR headsets operates independently with all data processing integrated into the viewer, whereas the original models required powerful hardware [9].

Currently, there are several open platforms for VR software development. While hardware manufactures like Oculus and High-Tech Computer (HTC) have released their software development kit (SDKs), on other hand, PlayStation does not plan to open their interface for public. Devices having their own SDKs are closely bound with the hardware, and for cross-device development standardization is necessary. Open-source software and open-source hardware combines with a strong desire for standardization is given be open-source virtual reality (OSVR53) [10].

2.2 *Augmented Reality*

Interfaces in AR are a blend of both virtual content and real content. Its basic functionality is to create links among the real world and electronic information generated by a device, which user could interact directly or with some inputs. By adding layers of digital information AR aims to expand and digitally integrate the physical environment in real time. User can view various information (graphics, video, audio, symbols, letters, numbers) overlaying in real world with help of various technologies [11]. AR can be divided into four categories, namely, marker-based, marker-less, projections, and overlaps. Details about these four categories is explained in the subsection below.

2.2.1 Marker-Based and Marker-Less AR

In marker-based AR technology, a camera is usually connected to a device marker as a real-space visual object. Therefore, the marker can project a definite and unique outcome. Simple models that look easily recognizable, such as QR codes, serve as markers in apps due to being easily recognizable and requiring no special processing powers. It is also referred as image recognition.

However, the most widely used technology in AR applications is marker-less technology. In these situations, data is provided based on the position of the electronic device using technologies like accelerometer, speed meter, Global Positioning System (GPS), digital compass, etc. It is utilized to find precise location and address mapping. It is also known as geolocation [12].

2.2.2 Projections and Overlaps

Apps that make use of the projection approach instead make use of actual objects that are projected with artificial light from the outside world. By projecting light onto a real-world surface and detecting human contact with it, applications built on this kind of technology enable human connection. The absence of a marker in this kind of AR indicates the requirement for more processing capacity, which is useful for handling virtual items. Therefore, for fully enjoying this, the device needs greater processing power. In overlaps, the technology is superimposing an object with an enlarged view in real world, replacing it partially or completely. Without determining or recognizing the object it won't replace original object with augmented one. That's why object recognition is vital in AR overlay [13].

2.3 Mixed Reality

MR is among the mechanism that has attracted interest recently. Its environment is unusual because it blends actual and virtual items, is registered in three dimensions, and is interactive in real time. It may enable smooth interaction among the actual and simulated worlds. For these reasons, MR encounters are realistic, accurate, and captivating. A system that is primarily artificial with some real-world imagery added, like texture mapping video onto virtual objects, is called augmented virtuality (AV). AV and AR are combined to form MR. In an AR environment, real and computer-generated elements are combined, registered in three dimensions, and communicating in real time [14].

By defining an MR system as an interactive system that blends digital and physical elements, this definition of AR is expanded. While an MR system also handles inputs from the user and the real world, an AR system focuses mostly on the system's output (3D, 2D, and visual). In addition to enabling smooth interaction between virtual and real environments, MR offers a physical interface metaphor for manipulating objects [15]. The interface also provides for a fluid transition between virtuality and reality. Users of MR applications are diverse these days. Due to the growing prevalence of MR applications across a wide range of industries, including entertainment, the military, healthcare, and education, operators are moving from creators and researchers to a diverse range of consumers. MR is still a relatively new technology, given the limited commercial and public applications of this technology. But it existed for almost 50 years in research field with application in various domains, such as technical support, entertainment, and military. Furthermore, MR systems are expensive to develop and design [16].

2.4 Extended Reality

XR denotes to mixed, virtual, and augmented (MR, VR, AR) reality. XR is an umbrella term for diverse kinds of technologies. Like VR, AR, and MR there is another term called augmented virtuality (AV) which can be defined as the ability to interactively explore a virtual representation obtained from the real environment. One could think AV is subset of MR that merges virtual object into the real world. In most cases, this can be reached by streaming videos from physical environment (by webcam) or by using 3D scanning of items, videos, or real-world objects and draped in virtual objects, giving the impression that the virtual world ID is somewhat realistic while keeping its flexibility [17, 18]. Two central features of XR are immersion and interactivity. VR and XR treated as first and foremost as a medium and virtual environment or virtual world as the content of this medium. VR/XR medium characterized by immersion and interaction. Other characteristics are virtual world, creators, participants, i.e., the content [19]. Details about these two categories are explained in the subsection below.

2.4.1 Immersion

Immersion denotes to the sense or feeling of being in a surrounding which can be achieved either physically or mentally. The physical means help in the sense of entering a virtual environment as content of VR, whereas physically approached immersion is the elimination of the real world in order to sense with a digital world. System can range from totally immersive to non-immersive, with different levels of immersion. VR that is immersive through the hardware: VR accessed through headsets is immersive, but VR viewed in desktop is not. Because of their lack of immersion, non-immersive systems like regular monitor or 3D glasses make one wonder if they are even VR system at all [20].

2.4.2 Interactivity

Interactivity or interaction can refer either as users' interaction with each other through the medium or user interaction with the medium (virtual object in real world or virtual environment, XR). Interactivity may relate to simple navigational possibilities, even with the last option of user-medium interaction: that the user is not limited to preset viewpoints but can position themselves in the environment and alter perspectives. Interactivity can be theorized in variety of ways [20]. Rather than combining these diverse conceptualizations, the contrasts between these technologies shall be highlighted. While some defined interactivity as a technological feature that is immediate interaction with virtual objects, other defines it as human-to-human behavior, whether or not an XR medium was used. In several instances, interactivity was completely disregarded since XR was viewed as a way to access physical resource, such information that was displayed digitally, without the need to change or interact with them [8]. Table 1 summarizes spectrum types and working methods below.

3 Reality Blending Gaming

Since gaming is regarded as a form of expression and communication, it is frequently associated with the entertainment sector, along with films and videos. New gaming experiences that are beyond traditional screen-based gaming are made possible by immersive technologies. These technologies take gaming to a more immersive, interactive, and frequently more realistic level by providing unique means of interacting with and engaging with virtual environments [21]. By giving players additional methods to engage with the game world and one another, these technologies expand the possibilities of multiplayer and social online gaming. Further, details about immersive gaming are explained in the subsection below.

Table 1 Spectrum types and working

Spectrum	Type	Explanation
VR	Non-immersive [7, 8]	• Combination of screens surrounding • Driving and flight simulations
	Immersive [8]	• Track a user's position and present VR information • 360 degrees of virtual environment
AR	Marker-based AR [9]	• Uses a camera combined with device marker as a real-world visual object • It is also referring as image recognition
	Marker-less AR [9]	• This technology provides data based on position of electronic device • It's also known as geolocation
	Projections [9]	• Apps that make use of the projection approach instead make use of actual objects that are projected with artificial light from the outside world
	Overlaps [9]	• This technology overlaying an object with an enlarged view in real world, replacing it partially or completely
MR [10, 11]		• Blends actual and virtual items • Registered in three-dimensional objects • Smooth interaction between virtual and real environment
XR	Immersion [11, 12]	• It refers to the sense or feeling of being in an environment which can be achieved either physically or mentally
	Interactivity [12]	• Users' interaction with each other through the medium or user interaction with the medium

3.1 Aspects of Gaming

VR initially marketed toward gaming. VR provides a 3D virtual space to user to interact with multiple other users through VR HMDs. Term social VR is an evolving sociotechnical organization which is changing the way of people to socialize, interact, and meet online. In most of the online and virtual world games, avatars play an important part. While exploring virtual world in games avatars act as the link of virtual assets. Avatar is a representation of a user's physical self on Internet and let him/her to experience the activities (crafting, personalizing, manipulation) and adventure of virtual world [22].

AR games combine outdoor actives with gaming experience because of which it is often described as new leisure activities. AR gaming refers to link of media and outdoor time into leisure time which made AR gaming feel bulky [13]. Skill-based computer games, real-world sports, and paintball are examples of physical games. Real-world games are mostly physical, and with AR the player is free to use their whole body in the game. Puzzle games can also be projected into real world using AR. Two most famous types of games are discussed in the subsections below [23].

3.1.1 Strategy Gaming

Players engage in turn-based strategy games with militaries of miniature characters in battle games. Games in which conflict are resolved using a set of directions that typically calls for dice to randomize outcomes, and decisions by consensus to resolve ambiguous situation. Particularly with younger players, the resolution of these ambiguities can occasionally be contentious, which reduced enjoyment of game. Thus, it is preferred to remove ambiguities wherever probable. The game situations are similar to war but some uncertainties like measurements may not be known, the game rules might not be interpreted clearly, or the player does not understand where those characters are on the map. These circumstances are less likely to happen when the rules of the game have been well written. Face-to-face communication support in AR eliminates the ambiguities in game simulation and additional visual and audio effect might be added.

The intricacy of game rules is another drawback of strategy games. The desire to recreate historical states with realistic-looking soldiers and terrain with properly recreated rules of combat is one of the many reasons why people play battle games. Since players have to solve the rules manually, the rules must be straightforward, making them highly abstract from how units and weapons really behave on the battleground. Both sets of issues can be resolve in a battle game with AR enhancement, however some compromise might be required. A proposed strategy game may either replace the player's characters entirely or use them as physical layer on top of a simulated game layer [24].

3.1.2 Role Playing Games (RPGs)

Role playing games are those games where the player chooses character and its role in a fictional world or scenario. Players describe their character's activities inside the game scenario rather than acting out the role of their characters, which sets it apart from improvisational drama and live role playing. In RPGs, one player is the game's master (GM), who manages game's mechanics, game world, and non-player characters. Depending on their style of play, participants can influence the game's outcome to varying degrees, however the referee typically fabricates the general storyline.

While the material of computer role-playing games (CRPGs) is comparable to the typical RPGs, the playing style is very different. Character and plot puzzles, tactical fighting, and occasionally easy games skill are the main features of CRPGs. Typical RPGs are very social and personal, whereas CRPGs are typically played alone or over a network with few communication channels due to the constraints of computer-based social interaction. Real-world RPGs can have some inadequacies. Players can communicate directly face to face. AR is able to integrate virtual entities with real world at nominal cost of ordinary face-to-face communication, and with increase in support software it can further alleviate these problems. Each player in a hybrid AR RPG may use an interface device to handle their data [25].

3.2 *VR and Gaming Integration*

Other than gaming and entertainment, VR can be applied to a wide range of fields and industries which helps to improve gaming experience as well. Simulation aspect is one of the largely applicable and widely used applications of VR, and according to the user's needs it can be created and customized. There are two main types of simulation: immersive and non-immersive. Applications for VR simulations are numerous and include prototyping, developing, testing tools and items, and training simulations. Driving simulators, pilot flight simulators, and combat simulators for military personnel are a few of the often-utilized VR simulations in the industrial sector [26]. Further, explain in subsection below.

3.2.1 Driving Simulation

Driving simulation is one of the famous major uses of simulations by VR for gaming as well as driving training industry. These simulations allow user to place in hard driving scenarios eliminating real threat with the help of VR. This simulator can be helpful in several purposes, such as training trainee drivers in low-stress driving environment and collecting data on driving behavior. The driver can be put into a simulated vehicle within an environment similar to the real one. Following the simulation results, drivers can get feedback on how to improve their actions while driving and be made aware of their mistakes.

Driving simulations have capacity to gather real-time information on how user reacts to different conditions. This data can be utilized in variety of ways, like improving safety measures for automobile, creating training modules, developing a better user experience for drivers, or even for research and development of autonomous vehicles (AVs). With VR simulations consistently offering chances for secure and effective data collecting and user testing, AV has been a developing technology that will continue to grow and progress further [27].

3.2.2 Product Design and Prototyping

VR has facility to look at 3D models in virtual environment in a way that is challenging to present on screen, this is one of the most useful applications of VR. For evaluation and testing purposes, prototypes designed for goods can be modeled and shown in a virtual space. Presenting a virtual prototype or part without having to invest a lot of money, material, effort, or time in creating the prototype in real life is a major benefit of displaying these models in VR. With the help of simulation, VR can show how the product reacts under dissimilar scenarios, and it also helps to create a suitable scenario for the product. VR simulation can be used to show the effect of different communications among the real-world subject and the prototype. Based on the outcomes of the simulated interaction, this can assist the prototype designers in identifying any aspects of the prototype that require improvement. Prototyping could proceed more quickly, prototype production waste could be decreased, and the functions of the prototype could be better understood thanks to the view of the product in a virtual environment and modify the VR design for a quicker turnaround and faster results [28].

3.2.3 Education

Although VR has not been widely implemented within education, there are many encouraging researches and examples of how VR may be useful in learning. By keeping student interested in what is happening within the VR atmosphere, using VR can help them pay more attention. The majority of teenagers find it difficult to focus in class, particularly when they believe the subject being covered are unrelated to them. Students are more engaged and interested in what they are studying when they use innovative technology like VR, which immerses them in a virtual world.

By enabling students to participate in meaningful experience, VR also gives them the chance to create and practice their own knowledge. Through immersive engagement in instructional activities, students are able to better understand the subject. Students can securely and effectively experience diverse locations and environments through VR, which enable them to learn and explore variety of concepts. This may include a view of moon, space, or underwater that is not possible to visit in real life [29].

3.3 AR and MR Gaming Integration

With AR interfaces, users can interact with digital content, including 3D objects that can be superimposed in real time on their environment. Nintendo and Niantic released the augmented reality game "Pokémon Go" in 2016, and within a week, the number of downloads skyrocketed. Another AR app that was taken advantage of by the popular Nintendo home system, the Switch, was "Mario Kart Home Circuit." Mario Kart makes the most of AR technologies. The real-and-virtual GoKarts interact in real time because of this game's usage of sensors and cameras, which enable mapping an AR circuit based on the surrounding location. Some other popular AR games are discussed below:

- Harry Potter Wizards Unite: The Wizarding World is brought to life in this game from the makers of Pokémon GO, which allows players to cast spells, locate magical creatures, and fight evil forces in the real world.
- Minecraft Earth: Despite being formally retired in 2021, Minecraft Earth was an entertaining AR game that lets users construct buildings, collect resources, and explore the real-world Minecraft realm.

Applications based on AR technology offer weak interactions with images, movies, and other types of content. However, MR-based game applications can overcome the drawbacks of the standard system. Interaction with game components incorporated in real-world scenes is made possible via MR. For instance, the user can move a gaming agent on board in a strategic game by using vocal commands. The agent's movement in actual space was then visible to us because of MR. In the past, VR- and AR-based game applications just showed the game on a computer or smartphone screen, without allowing players to interact with the game's objects in real time [30].

4 Requirements

XR demands specific hardware, software, and tools to deliver immersive and interactive experience. The requirement of XR can depend on the type of technology. Some requirements are discussed in subsection below.

4.1 Hardware

The traditional way of classifying hardware into input and output device categories. Most of the time, we're looking at hybrid devices, like an HMD, which naturally gives input data with its extra tracking functions. Only gadgets that offer active sensory stimulation beyond the stimulation of simply holding the item are taken into consideration as displays. Every technology needs respective hardware to work properly [31]. Further explanation is given in subsection below.

4.1.1 VR Hardware

At unprecedented speed, both input and output devices are being developed and designed. At present, various existing VR hardware are still at the prototype level. Some of the hardware tools that are being used in VR gaming development are mentioned in the subsections below:

- Visual Display: In the late 1960s, research on HMDs which displays computer-generated content began, which uses set of cathode-ray tube (CRT)-based optical see-through mechanical tracking and relay optics. OLED and LCD are current display technologies [33].
- Haptic Devices: Haptic devices are present across different fields. One of the approaches are vests which includes vibrotactile elements, and on the other hand there is hybrid approach which implements as a controller. In vests, early models like Auralizer from 1990s transfer the sonic waves into vibration by using audio

output produced by the system. In the scientific VR community, haptic displays are common area often shared with visual display [33, 34].

- Multisensory Display: Displays which stimulate other senses, like olfactory or tactile feedback. In market, olfactory displays that are worn on the body have been proposed as an accessory to future HMD solutions or in combination with the HMD's display component. Exotic displays from the haptic and olfactory domains were brought to market by the recent development. Warm and chilled air can be blown in the user's face thanks to a water mist created by an ultrasonic ionizing device. The gadget uses vibration engines to provide haptic feedback and a microphone for communication with the outside world [34].

4.1.2 AR Hardware

Real-time service is the key feature of AR. The development of wireless networks, tracking technology, high-end camera quality, sensors, and smartphone allowed operation of AR even in mobile surroundings. Most of the AR hardware components are IoT devices which can be alienated into.

- Display in AR: Users can interact with AR systems using display devices. HMDs, monitors, and wearable technology (clothing, gloves, and glasses) are a few examples. HMD uses one or more cameras that are based on holography and optics (diffraction and reflection). Another gadget that gives the user more information is the Head-Up Display (HUD), which is compact and light [35].
- Sensors in AR: The position of a generic person or generic object can be determined thanks to sensors, which are highly helpful in tracking mechanisms. This operation is required for the visual recording of the physical environment and its digital information in order to create a fusion of images of the digital and physical worlds. The tracking precision of the AR system is increased by a variety of tracking tools and methods (such as GPS and ultrasound) with varying ranges and settings [36, 37].
- Input Devices in AR: Users can interact with AR systems through input devices. The AR interface serves as a bridge between these gadgets and the AR systems. The input can take many different forms, including speaking, touching, blinking, and gesturing. Using movements that are translated into orders, the user of the Vomar app can rearrange the furniture in his home [38].

4.1.3 MR Hardware

Three types of HMDs are often used in the MR domain: see-through displays with optical combiners, video see-through displays, and traditional VR HMDs. The display, tracking devices, and input devices are the primary hardware elements for MR.

- Input Devices in MR: Digital information displayed over the real environment can be manipulated using input devices. Input devices for Mixed Reality Regeneration Concept (MRRC) were keyboard, fiducial marker, and web camera [38].
- Tracking Devices: In basic terms, tracking makes MR technology possible by generating the combined virtual-real environment. MRRC uses real-time, vision-based camera tracking technology, which includes a two-dimensional fiducial marker system and a webcam. The foundation of all vision-based tracking techniques is the tracking and identification of specific characteristics in pictures [32].
- Displays in MR: MRRC was screen-based display as it is cost-effective. Even though it is less immersive, this is the easiest MR setup as it removes HMD issues [32] (Table 2).

Table 2 Hardware advancements

Spectrum	Hardware	Description
VR	Visual display [33]	• HMD which displays computer-generated content • Uses set of CRT-based optical see-through mechanical tracking and relay optics
	Haptic devices [33, 34]	• Vests which include vibrotactile elements • Transfer the sonic waves into vibration by using audio output
	Multisensory display [34]	• Displays which stimulate other senses, like olfactory or tactile feedback • An ultrasonic ionizing system is used to generate water mist
AR	Display [36]	• Display devices allow user to interact with AR system
	Sensors [36, 37]	• Generic person or generic object can be determined • Highly helpful in tracking mechanisms
	Input devices [38]	• Users can interact with AR systems • The input types and nature: touching to speaking, gesticulating to blinking
MR	Tracking [38]	• Tracking makes MR technology possible • MRRC uses real-time, vision-based camera
	Input devices [32]	• It is possible to change digital data that is shown over the actual surroundings • The webcam, fiducial marker, and keyboard served as the MRRC's input devices
	Displays [32]	• MR setup as it removes HMD issues • MRRC was screen-based display as it is cost-effective

4.2 *Tools and Applications*

Throughout VR, AR, and MR the software environment is shaped differently for all realities. It has an important role and remarkable position with its role to ensure contact between user and software environment.

4.2.1 VR Tools

Game Engines: Unity and unreal engines are the currently established games engines which support the Oculus Rift. Mobile development is also supported by the current version of Unity for example plugin communication and GearVR with HTC Vive is available. Developers need to apply some tricks to decrease complexity because now VR devices rendering requirements increased [35, 36].

Multi-user Experiences: Multi-user environment is one of the most important topics in VR. HMDs supporting community platforms are increasing consequently in numbers. High fidelity is one of the most important topics and it is seen as spiritual successor to Second Life [36].

Aerospace: Since 1986, NASA has used VR in aerospace for telerobotics. VR has been used to train astronauts for the past 25 years. Another early use case was the use of virtual technologies to explore planetary settings [37].

4.2.2 AR Tools

AR Markup Language: AR required specific software according to the application, however by using ARML (augmented reality markup language) the software uses cameras and tracking devices to take advantage of the actual world's coordinates in order to send the location data to an XML file. By defining their relationship, the functional block of ARML enables fusion between the virtual and real worlds, enabling the integration of virtual objects into the real world [38].

ARKit: For AR development in iOS, Apple created a development kit called ARKit. Apple launched ARKit in 2017 with its iPhone lineup. Apps built with ARKit are in different categories: games, business, education, productivity, etc. [39].

ARCore: ARCore is a development kit for Android apps developed by Google. After launch of ARKit Google wrapped up Tango (previous AR platform from Google) and released with a completely new AR SDK [39].

4.2.3 MR Tools

Mixed Reality Regenerative Concept (MRRC): MRRC was created to introduce biomedical science students to the regenerative concept and in vitro tissue engineering procedures through the use of MR technology. Offline and real-time software technologies are used to develop MRRC. Tools which come under this offline software technologies are used to prepare the content in the augmentation as well as the MR environment [40].

A suitable software architecture required integration of specific components to develop an MR system [41].

This architecture consists of three parts: the network, client, and server. Communication between the server and the client to execute the MR application is enabled by network. Data management and processes execution is done by server [42, 43]. Table 3 summarizes the tools and application with these features.

4.3 Devices

Devices play a critical role in immersing users in digital worlds or enhancing real-world experiences with digital content. The primary categories of XR devices are given below:

Table 3 Tools and application in different spectrums

Spectrum	Tools and application	Description
VR	Game engines [25, 26]	• Supports the oculus rift • Mobile development is also supported
	Multi-user experiences [26]	• HMDs supporting community platforms
	Aerospace [27]	• VR is used to train astronauts • Planetary environment's exploration using virtual technology
AR	AR markup language [28]	• Fusion between virtual and real world is possible by functional block of ARML
	ARKit [29]	• For AR development in iOS, Apple created a development kit called ARKit
	ARCore [29]	• ARCore is a development kit for Android developed by Google
MR	Mixed reality regenerative concept (MRRC) [30]	• MRRC was developed for Biomedical Science students by using MR technology
	Architecture [31–33]	• Suitable architecture while integration components facilitate are needed • This architecture consists of three parts: the network, client, and server

Mobile HMDs: The mobile HMDs use smartphone as a whole for the display and processing of data. They provide a simple casing, which keeps the smartphone at a specified distance from lens.

Stationary HMDs: The stationary HMDs are equipped with additional sensors besides their optical tracking. They have accelerometers, magnetometers, and gyroscopes, and they use sensor fusion to integrate the data with the optical tracking. These devices are constructed using mobile phones' existing display technology.

Smart glasses: These glasses represent an HMD device in shape of glasses that allows to view web pages and application in AR [43, 44].

Some popular devices:

- Google Cardboard: The Google Cardboard is the most basic viewer, which does not even let user to strap the gadget on.
- GearVR: GearVR is a high-quality smartphone holder developed by Samsung. It provides an additional touchpad on side of the case.
- PlayStation VR: PlayStation VR is a VR headset developed by Sony. It supports the display of 360-degree photo and video.
- HTC Vive: HTC Vive is a virtual and MR headset produced by HTC Corporation. It makes use of Fresnel lenses to reduce weight and size of the display.
- Google Glass: Google Glass is an AR headset which looks exactly like simple glasses. It provides Google search results in front of the eyes and has gesture control.
- Microsoft HoloLens: Microsoft HoloLens is an AR/MR headset developed by Microsoft. It supports 3D computer-aided design and gaming.
- Meta Quest: Meta Quest is a MR headset developed by Reality Labs. It can be used on its own without a computer.
- Apple Vision Pro: Apple Vision Pro is an MR headset developed by Apple. Currently, it is the most advanced headset in the market. It can be used on its own and has a large application library (Table 4).

5 Multimedia

Recently, the development of technology has seen the rise of the sector in terms of XR products and solutions and thus raise demands for fresh or improved architectural concepts. Often, XR applications connect to both 3D geometric rendering and multimedia paradigms, a requirement that may be very complicated. An alternative approach to XR is integrating photo-realistic multimedia streams such as 360-degree and volumetric media in the immersive environment. However, technological gaps and paradigm conflicts could limit the development of a broad portfolio of products and solutions at the intersection of the two paradigms in XR [45]. The concept of cloud gaming has long been proven, and it allows users to access high-performance gaming computations from anywhere with a low-powered machine. Still, this is an excellent stepping point for XR remote rendering where several adjustments concerning

Table 4 Spectrum devices and features

Device	Features	Picture
Google Cardboard	• Basic VR viewer • Low cost	
GearVR	• High-quality smartphone holder • Additional touchpad	
Playstation VR	• 360-degree photo and video	
HTC Vive	• VR and MR headset • Fresnel lenses	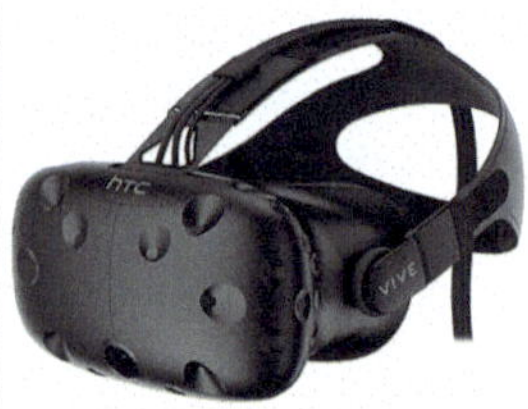
Google Glass	• AR headset looks like simple glasses • Gesture control	
Microsoft HoloLens	• AR/MR headset • 3D computer-aided design and gaming	

(continued)

Table 4 (continued)

Device	Features	Picture
Meta Quest	• MR headset • Used on its own without a computer	
Apple Vision Pro	• Most advanced MR headset • Large application library	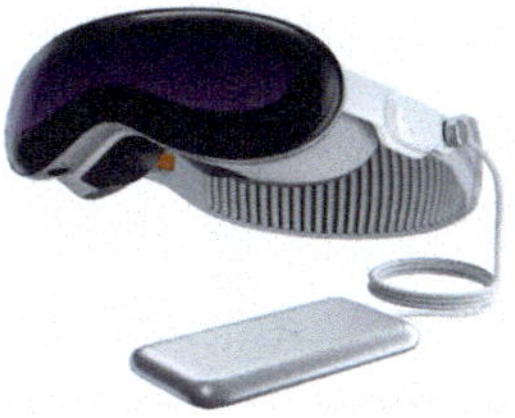

multimedia and gaming infrastructures need to be considered as the demands are much higher [2].

5.1 XR Immersive Media Formats and Transmission

Text, audio, and video modalities have always formed the core of multimedia transmission. The quality of 2D video delivery has already been greatly enhanced by recent developments in immersive media, which have led to advancements in both types of video research (supporting HDR, 4 K, etc.). There have been many advances within the audio domain for both the home theater experience and immersive media with forms of pre-existing modality modes: MPEG-H object-based 3D audio and Dolby Atmos [46].

5.1.1 Holographic Video

There are a number of methods for enabling spatial expressiveness in holographic video, from video-first methods, which extend video concepts with functionality to provide a sense of volume on the rendered media, to 3D model-based methods, which take inspiration from the gaming industry and express 3D models and their spatiotemporal relationships [47]. Traditional static content, like meshes, is given a time dimension in the 3D-based technique so that it can be viewed as holographic video content [48]. Since Point Clouds may directly communicate the collected content, they are currently chosen over Meshes for such purposes. Unlike 3D-based approaches, which aim to depict the entire scene in 3D, video-based approaches take into account the variations in frames across viewing angles when expressing the holography. Multiview video, which combines frames from many viewing angles into a single container and uses prediction to achieve compression efficiency both in the time domain and between views, is an early example of such an attempt [49]. A holographic experience or a more conventional method of using multiview video to replicate or improve a 3D capture representation can be produced, depending on the rendering device. Finally, the latest advancement in video holography is immersive video compression and transmission. The phrase immersive video refers to holographic video that does not fit into any of the previously described categories [50].

5.1.2 Tactile Data

Complex patterns can be created on tactile displays to reach multiple levels of information complexity. The user must translate the combination of stimuli to the intended meaning when these patterns are not self-explanatory or intuitive. Through its haptic feedback devices, a user will be able to discern vast volumes of data with sufficient training. Humans are highly suited to this, although it does require training. Depending on the modality, adjustments in a number of parameters can produce patterns. In short, the way and resolution in which the changes in stimuli are seen determines the ability to form patterns [51].

5.2 Remote Rendering

Reusing existing technologies like user interfaces, interaction techniques, and 3D rendering—all of which are frequently employed by the gaming industry—is necessary for XR advances. With an increasing quantity of geometry, texturing, lighting, animation, and physics, the virtual worlds that are produced are typically complicated. These worlds are also dynamic, changing based on the user's interactions and progress [52]. Game engines have included spatial data structures to organize the models and components in a virtual environment and enhance rendering efficiency. Bounding volume hierarchies (BVHs), scene graphs, and spatial partitioning are the most popular. However, the way these spatial data structures are implemented varies depending on the game engine or 3D application [53]. Therefore, it is essential to have a uniform file format for sharing virtual worlds between users and platforms. Individual 3D objects can already be described in a variety of forms. The glTF format, created by Kronos Group, was adopted and is now mostly utilized in web-based engines for comprehensive scene description and animation. A common scene description format will facilitate content exchange and deployment across engines and architectures, given the wide variety of XR devices now on the market [54].

5.2.1 Media Production

Game engines have expanded to various media channels, such as television, because of their growing functionality. TV series like West World and The Mandalorian utilize Unreal Engine to create in-camera visual effects (VFX), a novel technique for producing visual effects in real time while live action films are being filmed. The method requires realistic VFX rendering based on the camera's position and low-latency camera tracking using optical, feature, and inertial tracking techniques. Since the process of building a real-time virtual environment is already a type of XR, these tracking and rendering techniques can also be used in XR applications [55].

5.2.2 Ray Tracing

For developing a real-time graphics system, a significant amount of processing power, particularly when attempting to use ray-tracing algorithms to simulate lighting accurately, is required. This is made feasible by a new generation of graphics cards with dedicated ray-tracing cores from Nvidia and AMD, and games like Metro Exodus and Cyberpunk 2077 demonstrate that real-time raytracing is now widely available. These games, however, rely on completely controllable virtual worlds and prerendered 3D models. Numerous unknowns, such as location, lighting, occlusion, orientation, and interactivity, present a barrier in XR applications [56].

5.3 Media Streaming

Despite their increasing popularity, powerful graphics systems are still rather large and not appropriate for head-mounted or portable devices. Cloud and gaming companies have been searching for ways to fill this gap and develop remote rendering and gaming solutions. Microsoft hopes to provide comparable solutions with their Azure7 cloud offerings, while Nvidia's CloudXR6 offers a framework for delivering XR content to untethered devices. CloudXR makes it possible for mobile XR systems to produce advanced visual techniques from distant servers. A server runs a virtual environment in a gaming engine, and mobile devices get the video stream from virtual cameras in the scene [57]. The position and orientation of the XR devices are stored in a different connection to the server. Low network latency and compression algorithms are essential for split rendering to operate well, but when done right, they can open up new possibilities for realistic XR experiences. The ability to create complex photo-realistic 3D content remotely without requiring a device is an example of such an architecture. Remote rendering is obviously necessary and beneficial for supporting lightweight AR devices with low battery life. One major issue, though, is finding a balance between the energy required to transfer real-time 3D assets between the end device and the cloud and the shift of rendering resources [58].

6 Recent Advancement

XR is revolutionary in many sectors, such as entertainment, healthcare, manufacturing, and education. With advance graphics, real-time interaction, and responsive controls, immersive experiences of VR and AR in gaming increase significantly. More comfortable and lighter headsets are now available in the market, making it possible for user to engage for longer sessions. Some advances in multimedia and technology are explained below.

6.1 *VR Gaming*

Recent innovations in VR have been focused on higher immersion and availability. Improvement in eye-tracking and foveated rendering has improved performance and produced crisper images. Haptic feedback and hand tracking continue to advance, creating even more natural interaction. Haptic feedback continues to become more realistic. Additionally, wireless VR is getting better due to faster bandwidth technologies, making it possible for PC VR to broadcast smoothly to standalone (independent) headsets. These advances make VR more immersive, usable, and accessible in multimedia. Some of the applications in VR are given below:

Powell et al. [59] presented a serious game for immersive VR HMD-based long-term upper extremity exercise, which provides guideline into designing VR systems that are not just entertaining but also clinically effective for rehabilitation. Carroll et al. [60] presented effect of VR gaming player experience in significance of strategy and racing games, which helps to clarify how VR technology and game design interact, providing helpful advice for developing engaging VR gaming experiences that suit to a wide range of player preferences. Mayor et al. [61] presented ScrumVR, a VR serious video game, designed to teach the key aspects of the most widespread agile methodology for software development, which shows how VR may change the educational process by making difficult subjects like Scrum more approachable and interesting. Alam et al. [62] proposed comparison between engagement, performance, and usability of traditional seated VR with a room-scale variation within serious games. Liang et al. [63] presented efficient transfer of safety knowledge and enhanced interactive safety training by designing and developing a serious game based on VR technology, which highlights how VR serious games might improve preparedness and safety awareness in risky situations.

6.2 *AR Gaming*

The aim of recent advances in AR is to improve user experience and real-world integration. Current devices contain better optics, grater resolution, and clearer spatial mapping that move AR overlay closer to the feel of real world. AR glasses are becoming lighter and more comfortable, with improved field of view (FOV). Advanced AI and machine learning allow better recognition of object and interaction with the environment in real time. As platforms like ARKit and ARCore develop further, they make it possible for AR experiences in education, gaming, and multimedia apps to be more seamless. Some of the applications in AR are given below:

Feijs et al. [64] presented the design guidelines of different mechanics of four AR serious games inspired by the psychological needs within self-determination, and they designed and developed those games themselves. Ylardia et al. [65] proposed course of history using AR with gamification in timeline activities with primary and secondary students, which highlights how AR has the ability to revolutionize history education by making it an exciting and engaging subject that encourages stronger ties to the past. Du et al. [66] proposed mobile gesture interaction method for AR gaming, which demonstrates how gesture detection technology has advanced, opening the door to more fluid and engaging AR gaming experiences. Gaydos et al. [67] presented a new hybrid AR environment for conceptual understanding, which shows how hybrid AR may be used to develop interactive, participatory learning experiences that encourage teamwork and active participation, going beyond conventional pedagogical approaches. von der Putten et al. [68] proposed collaborative AR game which used both subjective and objective behaviors, known as TimeWarp which highlights how AR has the ability to produce captivating cooperative experiences, highlighting the role that presence and interaction play in the popularity of AR games such as TimeWarp.

6.3 MR Gaming

The smooth merging of virtual and real worlds is the main goal of recent advancement in MR. MR devices continue to move to new heights with advances such as improved dimensional awareness, better pass-through capability, and more immersive experience of holographic constructs. Cloud computing supports scalability for MR, making complex simulations and real-time collaboration possible. These advances expanding MR application in design, education, healthcare, multimedia, etc. Some of the applications in MR are given below.

Hang et al. [69] presented a serious game for learning traditional handicraft using MR, which demonstrates the importance of MR as a cutting-edge teaching tool in higher education by highlighting its potential in education, especially for technical and craftsmanship. Kiran et al. [70] proposed adaptive MR robot games (AMRRG) representing an approach to integrating consumer robots into public spaces for entertainment, which highlights the entertainment value and user-centric design of adaptive MR-based robotic games and demonstrates how they might improve customer experiences. Doong et al. [71] explored the learning effect of game-based learning system developed by MR and location, which highlights the potential of location-based MR games as a cutting-edge teaching tool, providing guidance on how to develop and use them to improve learning results. Kim et al. [72] presented the benefits of a new type of treatment, eye-contact game which uses. Mr. Prabhakaran et al. [73] addressed the challenge of finding optimized in-home first-person perspective camera setup for MR gaming system, which addresses technical issues with camera positioning to improve accessibility and user satisfaction while helping to produce high-quality, immersive MR gaming systems (Table 5).

Table 5 Summary of recent advancement

Spectrum	References	Year	Contribution
VR	Powell et al. [59]	2022	• Serious game for immersive VR • VR HMD-based long-term upper extremity exercise
	Carroll et al. [60]	2019	• VR gaming player experience • Strategy and racing games
	Mayor et al. [61]	2023	• Presents ScrumVR • Agile methodology for software development
	Alam et al. [62]	2020	• Traditional seated VR with a room-scale
	Liang et al. [63]	2019	• Interactive safety training • Designing and developing a serious game
AR	Feijs et al. [64]	2022	• Different mechanics of four AR serious games • Psychological needs within self-determination
	Ylardia et al. [65]	2024	• Course of history using AR • Timeline activities with primary and secondary students
	Du et al. [66]	2022	• Mobile gesture interaction method for AR gaming
	Gaydos et al. [67]	2018	• Tests a new hybrid AR environment
	von der Putten et al. [68]	2012	• Studied collaborative AR game
MR	Yongxin Hang et al. [69]	2023	• Serious game for learning traditional handicraft
	Ajmeera Kiran et al. [70]	2024	• Adaptive mixed reality robot games (AMRRG) • Integrating consumer robots into public spaces for entertainment
	Ji-Liang Doong et al. [71]	2015	• Effect of game-based learning system developed by MR
	Kim et al. [72]	2020	• Benefits of a new type of treatment • Eye-contact game which uses MR
	Prabhakaran et al. [73]	2024	• First-person perspective camera setup for MR gaming system

7 Conclusion

XR which includes VR, AR, and MR is transforming a number of industries, including gaming and multimedia. From the very first version which was in 1957, the XR technology has evolved and become more user-friendly and realistic. AR superimposes digital features on the real world, VR immerses users in fully virtual environments, and MR bridges the gap by interactively combining virtual and real-world things. Despite their transformative potential, some problems in their wider acceptance include limited access, high cost, and technical complexity. With the help of software tools like ARKit and ARCore and hardware innovations like portable VR headsets and improved AR glasses, usability and affordability have significantly improved. Because of these advancements, XR is now more widely available and may be used in a variety of industries, including healthcare, education, and gaming. By combining real-time interactivity and creative storytelling, XR has completely changed the gaming business and made games more realistic and engaging. XR is transforming multimedia beyond gaming, providing applications such as immersive 360-degree media and holographic video. Potential of XR is further enhanced by cloud rendering technologies, which enable high-quality virtual worlds while preserving device portability. This paper highlighted the current state of XR, its applications, and potential integration of advanced multimedia technologies. With continuous advancements in hardware, software, and interactive methodologies, XR will change digital experiences and create next-generation applications in gaming, multimedia, and other fields.

References

1. Rebbani, Z.: Auto-fit two movable projections into a single 3D display screen for virtual and augmented reality scenes. Int. J. Adv. Trends Comput. Sci. Eng. **9**(4), 5308–5314 (2020). https://doi.org/10.30534/ijatcse/2020/163942020
2. Rokhsaritalemi, S., Sadeghi-Niaraki, A., Choi, S.-M.: A review on mixed reality: current trends, challenges and prospects. Appl. Sci. **10**(2), 636 (2020). https://doi.org/10.3390/app10020636
3. Perks, M.: 11. regulating in-game monetization : implications of regulation on games production. Game Prod. Stud. 217–234 (2021). https://doi.org/10.1515/9789048551736-013
4. Young, C.: 7. unity production: capturing the everyday game maker market. Game Prod. Stud. 141–158 (2021). https://doi.org/10.1515/9789048551736-009.
5. Jeon, S., Kim, J.: Effects of augmented-reality-based exercise on muscle parameters, physical performance, and exercise self-efficacy for older adults. Int. J. Environ. Res. Public Health **17**(9), 3260 (2020). https://doi.org/10.3390/ijerph17093260
6. Campo-Prieto, P., Cancela, J.M., Rodríguez-Fuentes, G.: Immersive virtual reality as physical therapy in older adults: present or future (systematic review). Virtual Real. **25**(3), 801–817 (2021). https://doi.org/10.1007/s10055-020-00495-x
7. Lissa, M., Bhuvaneswari, V.: Augmented reality. Innovating with Augmented Reality, pp. 1–27 (2021).https://doi.org/10.1201/9781003175896-1
8. Aschauer, A., Reisner-Kollmann, I., Wolfartsberger, J.: Creating an open-source augmented reality remote support tool for industry: challenges and learnings. Proced. Comput. Sci. **180**, 269–279 (2021). https://doi.org/10.1016/j.procs.2021.01.164

9. Santi, G.M., Ceruti, A., Liverani, A., Osti, F.: Augmented reality in industry 4.0 and future innovation programs. Technologies **9**(2), 33 (2021). https://doi.org/10.3390/technologies9020033
10. Stanney, K., Lawson, B.D., Rokers, B., Dennison, M., Fidopiastis, C., Stoffregen, T., Weech, S., Fulvio, J.M.: Identifying causes of and solutions for cybersickness in immersive technology: reformulation of a research and development agenda. Int. J. Hum. Comput. Interact. **36**(19), 1783–1803 (2020). https://doi.org/10.1080/10447318.2020.1828535
11. LaRocco, M.: Developing the 'best practices' of virtual reality design: industry standards at the frontier of emerging media. J. Vis. Cult. **19**(1), 96–111 (2020). https://doi.org/10.1177/1470412920906255
12. Marino, E., Barbieri, L., Colacino, B., Fleri, A.K., Bruno, F.: An augmented reality inspection tool to support workers in Industry 4.0 environments. Comput. Indus. **127**, 103412 (2021). https://doi.org/10.1016/j.compind.2021.103412
13. Sendari, S., Firmansah, A., Aripriharta.: Performance analysis of augmented reality based on VUFORIA using 3D marker detection. In: 2020 4th International Conference on Vocational Education and Training (ICOVET) (2020). https://doi.org/10.1109/icovet50258.2020.9230276
14. Williams, R., Erkoyuncu, J.A., Masood, T., Vrabic, R.: Augmented reality assisted calibration of digital twins of mobile robots. IFAC-PapersOnLine **53**(3), 203–208 (2020). https://doi.org/10.1016/j.ifacol.2020.11.033
15. Ficarra, B.: Virtual reality, augmented reality, and mixed reality. Emerging Technologies for Nurses (2020). https://doi.org/10.1891/9780826146519 0004
16. Ong, S., Siddaraju, V. K.: Shared experiences. Beginning Windows Mixed Reality Programming, 189–218 (2021). https://doi.org/10.1007/978-1-4842-7104-9_9
17. Lai, Z.-H., Tao, W., Leu, M.C., Yin, Z.: Smart augmented reality instructional system for mechanical assembly towards worker-centered intelligent manufacturing. J. Manuf. Syst. **55**, 69–81 (2020). https://doi.org/10.1016/j.jmsy.2020.02.010
18. Baroroh, D.K., Chu, C.-H., Wang, L.: Systematic literature review on augmented reality in smart manufacturing: collaboration between human and computational intelligence. J. Manuf. Syst. **61**, 696–711 (2021). https://doi.org/10.1016/j.jmsy.2020.10.017
19. Schumann, M., Fuchs, C., Kollatsch, C., Klimant, P.: Evaluation of augmented reality supported approaches for product design and production processes. Proced. CIRP **97**, 160–165 (2021). https://doi.org/10.1016/j.procir.2020.05.219
20. Hakim, R., Dijaya, R.: Augmented Reality Application Design for 3-Dimensional Model Visualization Product Packaging Design (2023). https://doi.org/10.21070/ups.697
21. Arjoranta, J., Koskimaa, R., Siitonen, M.: Immersive gaming as journalism. Immersive Journalism as Storytelling, pp. 137–146 (2021). https://doi.org/10.4324/9780429437748-15
22. Roehl, B.: Building a social VR App. VR Developer Gems, pp. 289–320 (2019). https://doi.org/10.1201/b21598-16
23. Serious Game for Flood Safety Training, N.: Applying Serious Game Elements to Enhance the Flood Safety Training Apps (Preprint) (2020). https://doi.org/10.2196/preprints.24386.
24. Gilad, B.: War gaming: virtual reality, real lessons. Strategy Leadersh. **38**(6), 38–42 (2010). https://doi.org/10.1108/10878571011088050
25. Bycer, J.: The foundation of CRPG Design. Game Des. Deep Dive, 11–27 (2023). https://doi.org/10.1201/9781003331599-3
26. Ng, E., Bakr, A.A., Cai, Y.: Integration of wearable devices in VR-enhanced aircraft maintenance training. Gaming Media and Social Effects, 281–290 (2023). https://doi.org/10.1007/978-981-99-4958-8_12
27. Gironacci, I.: XR management training simulator supported by content-based scenario recommendation. In: 2022 IEEE International Conference on Artificial Intelligence and Virtual Reality (AIVR), pp. 104–108 (2022). https://doi.org/10.1109/aivr56993.2022.00021
28. Liou, F.F.: Product prototyping. Rapid Prototyping and Engineering Applications, pp. 19–84 (2019). https://doi.org/10.1201/9780429029721-2
29. Kim, H.-Y., Kim, S.-S., Sim, S.-B.: Exploring the applicability of effective physical education classes using virtual (VR) sports : targeting physical education teachers. Korean J. Sports Sci. **33**(2), 387–396 (2024). https://doi.org/10.35159/kjss.2024.4.33.2.387

30. Wetzel, R., Blum, L., Broll, W., Oppermann, L.: Designing mobile augmented reality games. Handbook of Augmented Reality, pp. 513–539 (2011). https://doi.org/10.1007/978-1-4614-0064-6_25
31. Quandt, M., Beinke, T., Freitag, M.: User-centered evaluation of an augmented reality-based assistance system for maintenance. Proced. CIRP **93**, 921–926 (2020). https://doi.org/10.1016/j.procir.2020.03.053
32. Nishchyk, A., Geentjens, W., Medina, A., Klein, M., Chen, W.: An augmented reality game for helping elderly to perform physical exercises at home. Lecture Notes in Computer Science, pp. 233–241 (2020). https://doi.org/10.1007/978-3-030-58796-3_28
33. Rahouti, A., Lovreglio, R., Datoussaïd, S., Descamps, T.: Prototyping and validating a non-immersive virtual reality serious game for healthcare fire safety training. Fire Technol. **57**(6), 3041–3078 (2021). https://doi.org/10.1007/s10694-021-01098-x
34. Harley, D.: Palmer luckey and the rise of contemporary virtual reality. Converg. Int. J. Res. New Media Technol. **26**(5–6), 1144–1158 (2019). https://doi.org/10.1177/1354856519860237
35. Peddie, J.: Overview of augmented reality system organization. Augmented Reality, pp. 89–100 (2023). https://doi.org/10.1007/978-3-031-32581-6_4
36. Dalle Mura, M., Dini, G.: An augmented reality approach for supporting panel alignment in car body assembly. J. Manuf. Syst. **59**, 251–260 (2021). https://doi.org/10.1016/j.jmsy.2021.03.004
37. Huang, R., Sun, M.: Network algorithm real-time depth image 3D human recognition for augmented reality. J. Real-Time Image Proc. **18**(2), 307–319 (2020). https://doi.org/10.1007/s11554-020-01045-z
38. Mourtzis, D., Angelopoulos, J., Panopoulos, N.: A framework for automatic generation of augmented reality maintenance & repair instructions based on convolutional neural networks. Proced. CIRP **93**, 977–982 (2020). https://doi.org/10.1016/j.procir.2020.04.130
39. Flavián, C., Ibáñez-Sánchez, S., Orús, C.: The impact of virtual, augmented and mixed reality technologies on the customer experience. J. Bus. Res. **100**, 547–560 (2019). https://doi.org/10.1016/j.jbusres.2018.10.050
40. Cheng, J.C., Chen, K., Chen, W.: State-of-the-art review on mixed reality applications in the AECO industry. J. Constr. Eng. Manag. **146**(2) (2020). https://doi.org/10.1061/(asce)co.1943-7862.0001749.
41. Aruanno, B., Garzotto, F.: Memholo: mixed reality experiences for subjects with alzheimer's disease. Multimed. Tools Appl. **78**(10), 13517–13537 (2019). https://doi.org/10.1007/s11042-018-7089-8
42. Brito, C., Alves, N., Magalhães, L., Guevara, M.: Bim mixed reality tool for the inspection of heritage buildings. ISPRS Annal. Photogr. Remote Sens. Spat. Inf. Sci. IV-2/W6, 25–29 (2019). https://doi.org/10.5194/isprs-annals-iv-2-w6-25-2019
43. Pepe, A., Trotta, G.F., Mohr-Ziak, P., Gsaxner, C., Wallner, J., Bevilacqua, V., Egger, J.: A marker-less registration approach for mixed reality–aided maxillofacial surgery: a pilot evaluation. J. Digit. Imagin. **32**(6), 1008–1018 (2019). https://doi.org/10.1007/s10278-019-00272-6
44. Chen, S.S.-C., Duh, H.: Interface of mixed reality: from the past to the future. CCF Trans. Pervasive Comput. Interact. **1**(1), 69–87 (2019). https://doi.org/10.1007/s42486-018-0002-8
45. Flotyński, J.: E-XR: Explorable extended reality environments. Knowledge-Based Explorable Extended Reality Environments, pp. 85–96 (2020). https://doi.org/10.1007/978-3-030-59965-2_5
46. Kim, S., Billinghurst, M., Kim, K.: Multimodal interfaces and communication cues for remote collaboration. J. Multimodal User Interfaces **14**(4), 313–319 (2020). https://doi.org/10.1007/s12193-020-00346-8
47. Poller, P., Chikobava, M., Hodges, J., Kritzler, M., Michahelles, F., Becker, T.: Back-end semantics for multimodal dialog on XR devices. In: 26th International Conference on Intelligent User Interfaces, pp. 75–77 (2021). https://doi.org/10.1145/3397482.3450719
48. Numan, N., Haar, F. ter, Cesar, P.: Generative RGB-D face completion for head-mounted display removal. In: 2021 IEEE Conference on Virtual Reality and 3D User Interfaces Abstracts and Workshops (VRW), pp. 109–116 (2021). https://doi.org/10.1109/vrw52623.2021.00028

49. Broxton, M., Flynn, J., Overbeck, R., Erickson, D., Hedman, P., Duvall, M., Dourgarian, J., Busch, J., Whalen, M., Debevec, P.: Immersive light field video with a layered mesh representation. ACM Trans. Gr. **39**(4) (2020). https://doi.org/10.1145/3386569.3392485
50. Salahieh, B., Bhatia, S., Boyce, J.: Multi-pass renderer in MPEG test model for immersive video. 2019 Picture Coding Symposium (PCS) (2019). https://doi.org/10.1109/pcs48520.2019.8954515
51. Ramirez, E.J.: Virtual experience, real harm. The Ethics of Virtual and Augmented Reality, pp. 70–98 (2021). https://doi.org/10.4324/9781003042228-4
52. Ratcliffe, J., Soave, F., Hoover, M., Ortega, F.R., Bryan-Kinns, N., Tokarchuk, L., Farkhatdinov, I.: Remote XR studies: exploring three key challenges of remote XR experimentation. In: Extended Abstracts of the 2021 CHI Conference on Human Factors in Computing Systems, pp. 1–4 (2021). https://doi.org/10.1145/3411763.3442472
53. Paradis, D.J., Segee, B.: Remote rendering and rendering in virtual machines. In: 2016 International Conference on Computational Science and Computational Intelligence (CSCI), pp. 218–221 (2016). https://doi.org/10.1109/csci.2016.0048
54. Zellmann, S.: Augmenting Image Warping-Based Remote Volume Rendering with Ray Tracing (2020). https://doi.org/10.36227/techrxiv.12568583
55. Medoff, N.J., Kaye, B.K.: XR: Inside of media, inside the mind. Now Media: The Evolution of Electronic Communication, pp. 211–235 (2021). https://doi.org/10.4324/9781003020721-8
56. Swinford, C., Lee, N.: Ray tracing in video games. Encyclopedia of Computer Graphics and Games, pp. 1–9 (2023). https://doi.org/10.1007/978-3-319-08234-9_469-1
57. Chung, J.-M.: XR and Multimedia Video Technologies. Emerging Metaverse XR and Video Multimedia Technologies, pp. 183–228 (2022). https://doi.org/10.1007/978-1-4842-8928-0_5
58. Meardi, G., Mockford, K.: Optimized pixel streaming for ultra-low latency cloud gaming and VR/XR applications. SMPTE 2023 Media Technology Summit, pp. 14–15 (2023). https://doi.org/10.5594/m002023
59. Elor, A., Powell, M., Mahmoodi, E., Teodorescu, M., Kurniawan, S.: Gaming beyond the novelty effect of immersive virtual reality for physical rehabilitation. IEEE Trans. Games **14**(1), 107–115 (2022). https://doi.org/10.1109/TG.2021.3069445
60. Carroll, M., Osborne, E., Yildirim, C.: Effects of VR gaming and game genre on player experience. In: 2019 IEEE Games, Entertainment, Media Conference (GEM), pp. 1–6. New Haven, CT, USA (2019). https://doi.org/10.1109/GEM.2019.8811554
61. López-Fernández, D., Mayor, J., Pérez, J., Gordillo, A.: Learning and motivational impact of using a virtual reality serious video game to learn scrum. IEEE Trans. Games **15**(3), 430–439 (2023). https://doi.org/10.1109/TG.2022.3213127
62. Shewaga, R., Uribe-Quevedo, A., Kapralos, B., Alam, F.: A comparison of seated and room-scale virtual reality in a serious game for epidural preparation. IEEE Trans. Emerg. Topics Comput. **8**(1), 218–232 (2020). https://doi.org/10.1109/TETC.2017.2746085
63. Liang, Z., Zhou, K., Gao, K.: Development of virtual reality serious game for underground rock-related hazards safety training. IEEE Access **7**, 118639–118649 (2019). https://doi.org/10.1109/ACCESS.2019.2934990
64. Li, J., Spek, E.V.D., Hu, J., Feijs, L.: Extracting design guidelines for augmented reality serious games for children. IEEE Access **10**, 66660–66671 (2022). https://doi.org/10.1109/ACCESS.2022.3184775
65. Lázaro Carrascosa, C., Ylardia, I.P., Paredes-Velasco, M., García-Suelto, M.d.C.N. : Game-based learning with augmented reality for history education. IEEE Revista Iberoamericana de Tecnologias del Aprendizaje **19**, 14–23 (2024). https://doi.org/10.1109/RITA.2024.3368348
66. Du, G. et al.: A mobile gesture interaction method for augmented reality games using hybrid filters. IEEE Trans. Instrum. Meas. **71**(9507612), 1–12 (2022). https://doi.org/10.1109/TIM.2022.3181305
67. Oh, S., So, H.-J., Gaydos, M.: Hybrid augmented reality for participatory learning: the hidden efficacy of multi-user game-based simulation. IEEE Trans. Learn. Technol. **11**(1), 115–127 (2018). https://doi.org/10.1109/TLT.2017.2750673

68. von der Pütten, A.M., et al.: Subjective and behavioral presence measurement and interactivity in the collaborative augmented reality game TimeWarp☆. Interact. Comput. **24**(4), 317–325 (2012). https://doi.org/10.1016/j.intcom.2012.03.004
69. Hang, Y., Wang, H., Sang, Z., Huang, R., Ye, L.: The impact of mixed reality serious games on mortise and tenon learning in college students. Comput. Educ. X Real. **3**(100042), ISSN 2949–6780 (2023). https://doi.org/10.1016/j.cexr.2023.100042. (https://www.sciencedirect.com/science/article/pii/S2949678023000363)
70. Kiran, A., Refonaa, J., Nabeel, M., Navaprakash, N., Reddy, V.L., Lalitha, R.V.S.: Adaptive mixed reality robotic games for personalized consumer robot entertainment. Entertain. Comput. **52**(100825), ISSN 1875–9521 (2025). https://doi.org/10.1016/j.entcom.2024.100825. (https://www.sciencedirect.com/science/article/pii/S1875952124001939)
71. Doong, J.-L., Lai, C.-H., Chuang, K.-H., Hsu, C.-C.: Learning effects of location based mixed reality game: a pilot study. Proced. Manuf. **3**, 1603–1607 (2015). ISSN 2351–9789. https://doi.org/10.1016/j.promfg.2015.07.448. (https://www.sciencedirect.com/science/article/pii/S2351978915004497)
72. Kim, S., Ryu, J., Choi, Y., Kang, Y., Li, H., Kim, K.: Eye-contact game using mixed reality for the treatment of children with attention deficit hyperactivity disorder. IEEE Access **8**, 45996–46006 (2020). https://doi.org/10.1109/ACCESS.2020.2977688
73. Chung, Y.-Y., Prabhakaran, B.: Optimizing camera setup for in-home first-person rendering mixed-reality gaming system. IEEE Trans. Games **16**(2), 396–408 (2024). https://doi.org/10.1109/TG.2023.3277399

Multi-model Fusion for Prediction and Segmentation of Brain Tumor Using CNNs

R. Thilagavathy, M. Gayathri, R. Deepalakshmi, Raghvendra Kumar, and Ahmed Alkhayyat

Abstract Healthcare has undergone a transformation due to medical imaging technology, which offers detailed insights into the human body. The proposed model aims to advance the analysis of brain tumors by utilizing a complex method that combines CT and MRI -Scan data into an intuitive Flask-based web application. Pioneer-based registration guarantees accurate alignment of various patient pictures, creating a common coordinate system for thorough anatomical comparisons. We use transfer learning to improve the analytical capabilities of the VGG-19 CNN architecture, allowing for more nuanced analysis. The next step, Image Fusion, uses the complimentary capabilities of data from both CT and MRI scans to maximize the segmentation accuracy. By isolating areas of interest, the Watershed transformation makes it easier to categorize data more precisely. Furthermore, a CNN predicts the existence of brain tumors, expediting the process of diagnosis and treatment, and eventually supporting a patient-centered and effective healthcare paradigm. These developments improve accuracy and accessibility in medical procedures while also streamlining the challenging assessment of brain tumors yielding an accuracy of 92.16%.

R. Thilagavathy · M. Gayathri
Department of Computing Technologies, School of Computing, SRM Institute of Science and Technology, Kattankulathur, Tamil Nadu, India
e-mail: thilagar2@srmist.edu.in

M. Gayathri
e-mail: gayathrm2@srmist.edu.in

R. Deepalakshmi
Department of Computer Science and Engineering, SRM Institute of Science and Technology, Tiruchirapalli, Tamil Nadu, India
e-mail: deepadinesh1018@gmail.com

R. Kumar (✉)
Department of Computer Science and Engineering, GIET University, Gunupur, India
e-mail: raghvendra@giet.edu

A. Alkhayyat
College of Technical Engineering, The Islamic University, Najaf, Iraq
e-mail: ahmedalkhayyat85@iunajaf.edu.iq

R. Sharma et al. (eds.), *Multimedia Technologies in the Internet of Things Environment, Volume 4*, Studies in Big Data 173, https://doi.org/10.1007/978-981-96-4356-1_12

Keywords CT-scan · MRI-scan · CNN · Flask framework · VGG-19 · Image fusion · Image segmentation · Watershed transformation

1 Introduction

The critical need for precise and efficient brain tumor analysis is addressed by our endeavor. Through our Flask-based web application, customers can easily submit CT and MRI scans to begin this advanced analysis. A keystone of our approach is landmark-based registration, which guarantees accurate picture alignment for comprehensive comparisons across various patient datasets. We use a CNN in transfer learning to further improve the tumor segmentation accuracy. Our system can better analyse brain images by utilising prior knowledge thanks to this advanced technique. The solution that we have developed is the new Image Fusion approach, which combines data from CT and MRI images. This combination improves the accuracy of tumor segmentation and offers more in- depth information for better diagnostic results in medical settings.

2 Related Works

The topic of illness prediction has seen a great deal of research employing a variety of machine learning approaches and algorithms, which may find use in the medical and healthcare sectors. Utilizing their suggested methodologies and findings, this study examines a few of those research publications investigations. The authors [1] showcase that brain tumors, ranking 10th in causes of death for both adults and children, vary in type and have low survival rates. Accurate classification is vital, achieved through multiclass classification methods [2]. Magnetic Resonance Imaging (MRI) is the preferred tool for tumor identification. Convolutional Neural Networks (CNNs) are the most effective approach for image classification, particularly in this context. A CNN model successfully classified brain images into four classes (no tumor, glioma, meningioma, and pituitary tumor) with 84% accuracy rate.

It has been highlighted [3] that the development of quantitative image analysis, especially radiomics, has become essential for forecasting the course of treatment for brain malignancies such as glioblastoma multiforme (GBM). Texture, shape and signal intensity are just a few of multiple quantitative characteristics that are evaluated in order to provide a thorough insight of the pathophysiology. The paper highlights how inconsistent manual segmentation in brain tumor analysis may be addressed by Convolutional Neural Networks (CNNs). It delves further into the cutting- edge science of radiomics with the goal of identifying quantitative characteristics that may be used to forecast crucial clinical outcomes like therapeutic response and survival [4]. This two-pronged strategy demonstrates how cutting-edge technologies may improve and automate brain tumor segmentation accuracy with 81.19% accuracy rate.

The authors [5] examine cutting-edge deep learning techniques for brain tumor segmentation and highlight how well they work to overcome obstacles. Deep learning shows up as a solution for quantitative analysis because of its outstanding performance. A critical study of current issues in the field of medical image analysis rounds out the debate. Notable designs such as UNet models [6] and ensemble approaches have potential, but they also come with a number of drawbacks that need to be carefully considered, such class imbalance difficulties and pre-processing and weight initialization [7].

It is found that due to time limits and possible confusion with strokes, glioma, the most prevalent CNS tumor, presents difficulties when manually segmenting from MRI. Brain tumor segmentation in MRI is one of the most challenging jobs, yet deep learning allows automation despite class imbalances [8]. To address these issues, a deep multi-task learning framework integrating a V-Net-based distance transform decoder and a multi-depth fusion module is presented. With an average Dice score of 78%, the model delivers high- quality segmentation on BraTS datasets, demonstrating its potential for precise and automated brain tumor segmentation [9, 10].

The authors [11] employ hybrid models that combine aspects from well-known architectures such as SegNet, U-Net, and ResNet18 are proposed and include U-SegNet, Seg-UNet, and Res-SegNet. These models' skip connections and depth modifications are intended to improve accuracy. The hybrid design continuously produces greater accuracy in brain tumor segmentation, as demonstrated by evaluation using the BraTS dataset, indicating their potential to improve segmentation algorithms [12].

3 Proposed Approach

The complementary strengths of CT and MRI are leveraged in a Flask web application as part of our proposed approach to brain tumor analysis. The registration procedure is initiated by users submitting their MRI and CT scans, which is simple and relies on landmarks for precise alignment across different patient datasets. By employing past knowledge, we may boost accuracy by applying transfer learning with a Convolutional Neural Network (CNN). The ultimate result of our approach is the innovative Image Fusion process, which enhances tumor segmentation accuracy by merging CT and MRI data. Through optimization and improvement of brain tumor analysis, this comprehensive approach aims to improve patient-centered medical processes and diagnostic outcomes.

The same procedure is shown Fig. 1 sequentially in the pipeline and architecture diagrams below. When the user accesses the web platform, everything starts. A simple user interface invites people to upload their MRI and CT scans. The Flask backend activates upon picture upload. It initiates the processing pipeline and acknowledges user input. The supplied photographs are taken by the landmark- based registration module, which makes sure they align precisely to create the ideal environment for successful fusion. The images are sent to the fusion module after they have been aligned. After the images are broken down using wavelet decomposition [13] and the components are combined using the VGG-19 model, a composite image with more diagnostic information is produced.

The Watershed algorithm is then used to segment the fused image. The deep learning model is used to examine the identified locations and determine how likely they are to be tumor regions [14]. The results are finally shown to the user. Potential tumor locations are highlighted on the image by the algorithm along with scores that represent the likelihood of the tumor being malignant (Fig. 2).

A. Motivation

The pressing need for accurate neuroimaging instruments, particularly for brain tumor analysis, is what motivated this project. A precise diagnosis and treatment plan are sometimes impossible to achieve with current approaches [15]. A foundation is provided by the combination of MRI and CT, although difficulties with

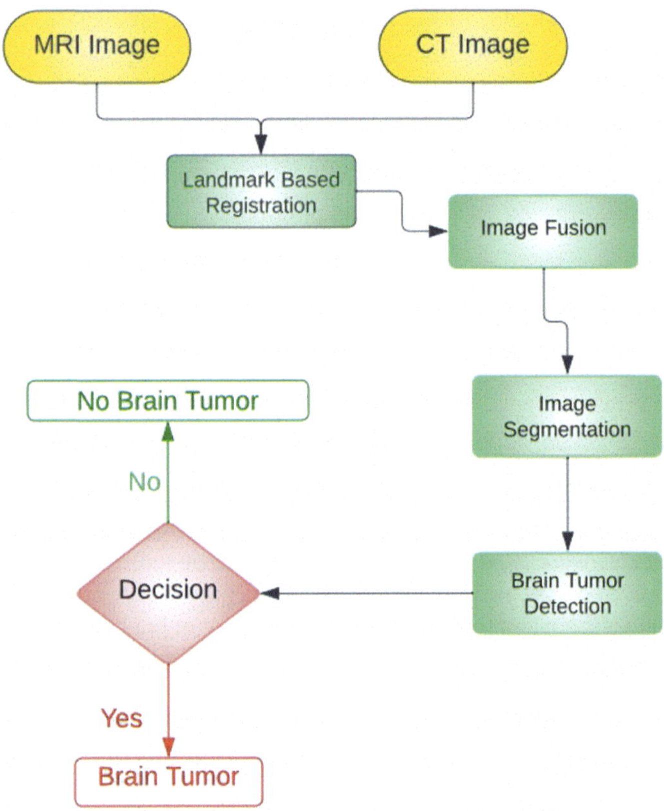

Fig. 1 Flow diagram of Brain Tumor Prediction

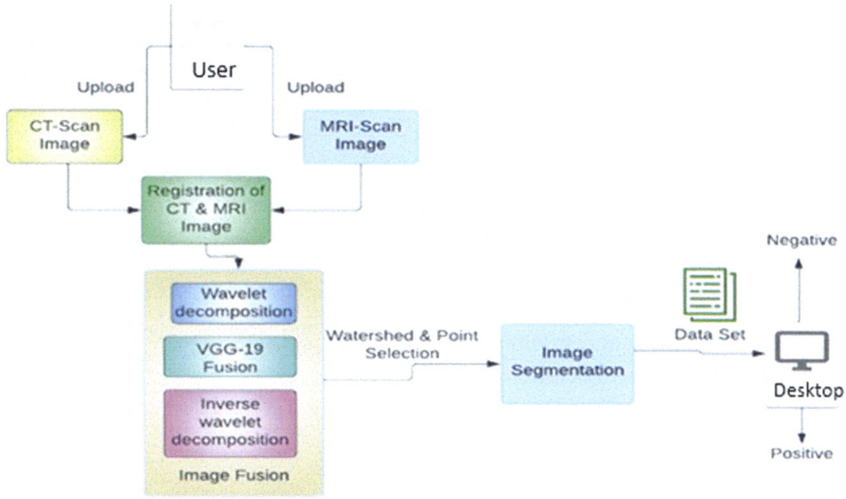

Fig. 2 Architecture diagram of Brain Tumor Prediction

multimodal data alignment and interpretation still exist. Our research uses cutting-edge methods like landmark-based registration and convolutional neural networks (CNNs) because of how crucial correct tumor analysis is. By addressing a major gap in existing healthcare procedures, the objective is to improve the efficacy and precision of brain tumor identification. Our research uses cutting-edge technology with the goal of using them to improve healthcare in a larger way, beyond only current clinical applications. The possibility for better diagnosis and the start of a new age of efficient and individualized patient care lies in the combination of deep learning with sophisticated imaging [16]. The aim is to improve the overall quality of life for those who are impacted by these difficult medical problems, and the ultimate motivation is to have a significant influence on brain tumor diagnosis and prognosis.

B. Brain Tumor Detection Through CNN

The way CNN works in case of brain tumor detection can be broken down into following layers:

C. Convolution Layer

Convolution operations are used by a CNN's convolution layer to extract features from the input data. Filters are deployed to capture local patterns so that the network may learn hierarchical representations on its own. This layer is critical to recognize components such as edges and textures for efficient feature extraction from visual input.

D. ReLU Layer

By using the ReLU activation function, the Rectified Linear Unit (ReLU) layer produces non-linearity. The network's capacity to comprehend complex patterns and gradients is enhanced, which quickens the rate of convergence during training. With ReLU, negative values are replaced with zero to more effectively depict complicated linkages in data.

E. **Pooling Layer**

Pooling improves computational efficiency, highlights the most important characteristics, and strengthens the translational invariance of the network. It can decrease spatial dimensions without compromising the integrity of the data.

F. **Fully Connected Layer**

One popular method is max pooling, which down-samples the data by selecting the largest value from a set of nearby pixels.

The completely connected layer integrates high- level characteristics from previous layers for the final classification or regression. The network may learn intricate correlations throughout the whole input space since each neuron in this layer is connected to every other neuron in the layer that came before it. This layer transforms collected data into judgments or predictions, completing the network's processing pipeline.

4 Model Training

A. **Image Registration**

- **Landmark-Based Registration**

Aligning many medical imaging modalities, such CT and MRI pictures, accurately can be challenging to do. Our solution to this problem is Landmark-Based Registration, which identifies and compares certain locations, or "landmarks," on both images. Correct overlay is ensured by comparing the photos to a single anatomical reference and reducing disparities to create a unified portrayal.

- **Transfer Learning with VGG-19 CNN**

In image processing tasks, Modern neural networks, and Convolutional Neural Networks in particular, have shown remarkable efficacy in image processing applications [17]. With the pre- trained parameters of the VGG-19 CNN, we include an additional level of precision in the registration process. This connection allows the system to accomplish accurate image alignment by using data from vast databases.

- **YCbCr Color Format Conversion**

Traditional RGB formats are still in use, however they might not be the greatest at preserving and recording finely nuanced brightness nuances. To get around this, our system converts these images to the YCbCr format. This conversion emphasizes the brightness component (Y) while taking into consideration the nuances of chrominance (Cb and Cr). This stage is necessary for a lossless, high-quality representation of the diagnostic scans.

- **Discrete Wavelet Transform Application**

Our method allows us to analyze the finer details of the images by extracting information about the frequency and spatial distribution of the images using the Discrete Wavelet Transform (DWT). DWT effectively splits image characteristics into discrete frequency bands, laying the foundation for the best possible fusion procedure in the next module.

B. **Image Fusion**

- **Wavelet Decomposition**

The challenging issue of merging CT and MRI data is handled using a particular mathematical technique called Wavelet Decomposition in image processing. Every image is decomposed into approximation coefficients that capture broad characteristics and detail coefficients that display finer details using this approach. Removing layers to show a more complex image is analogous to this. By combining macrostructure and fine characteristics, these coefficients allow for intelligent image fusion.

- **Fusion Leveraging VGG-19**

We have incorporated the well-known VGG-19 CNN model into our fusion method, improving upon previous approaches. After processing the deconstructed coefficients from CT and MRI images, VGG-19 examines each set of deconstructed coefficients (LL, LH, LV, LD) independently. By leveraging its depth and complexity to find and include the best components from both modalities, VGG-19 produces a richer representation. Deep learning ensures high diagnostic value and increases fusion accuracy (Fig. 3).

- **Inverse Wavelet Transformation**

The final step in the fusion process is the Inverse Wavelet Transformation. By meticulously re-layering coefficients, it reassembles features from CT and MRI images into a single, seamlessly formed image. This intricate process that preserves data integrity results in a harmonic fusion of both methodologies.

C. **Image Segmentation**

- **Watershed Algorithm**

Inspired by geographic concepts, the Watershed Algorithm sees an image as a topographic landscape where the intensities of the pixels correspond to heights. Like natural watersheds, it draws boundaries, splitting images into distinct catchment basins. This method is crucial for accurate point of interest identification and segmentation since these basins represent potential tumor sites in our investigation.

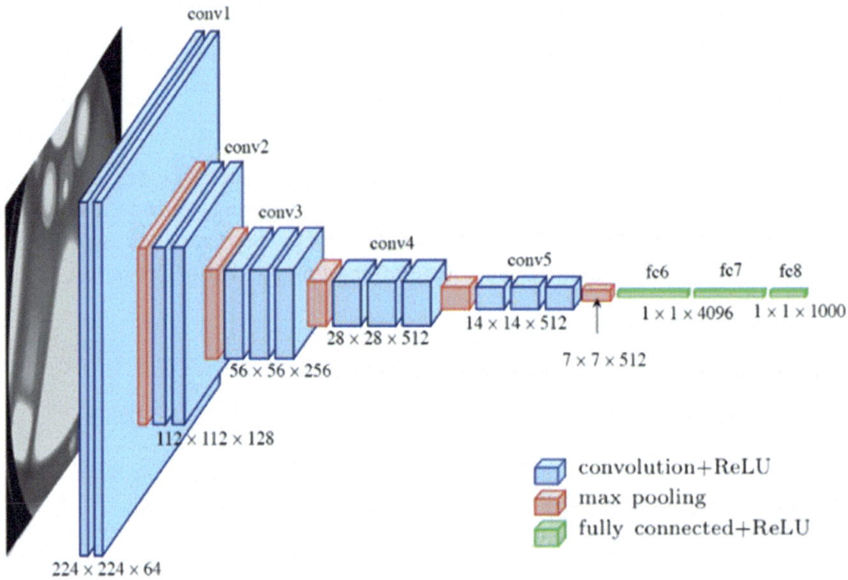

Fig. 3 Network architecture of VGG-19 model: convolution is denoted by Conv, and completely linked by FC

- **Marker Selection**

Our fusion method integrates deep learning capabilities with the widely recognized VGG-19 Convolutional Neural Network (CNN) model. In contrast to traditional fusion methods, which might just make use of basic averaging techniques or simpler integration methods, using the VGG-19 introduces a new level of complexity and detail. The model acts upon the decomposed coefficients by independently evaluating each set (LL, LH, LV, and LD) from the CT and MRI images. The VGG-19 network finds and incorporates the best components from each modality, making it simpler to generate a richer, more comprehensive representation. Its depth and intricacy allow it to capture and understand details that simpler fusion approaches could overlook.

- **Transformation and Growth Based on Marked Points**

The Watershed Algorithm begins segmenting the image after the markers are positioned. The areas around each marker enlarge, mimicking a situation of water filling, and creating distinct catchment basins. As these basins become closer to one another, expansion causes their borders to become harder where they meet. This laborious processing method ensures that every pixel in the image is recognized and assigned to a specific location. The segmented image that results is tapestry-like, with each separate zone denoting a different anatomical or pathological component.

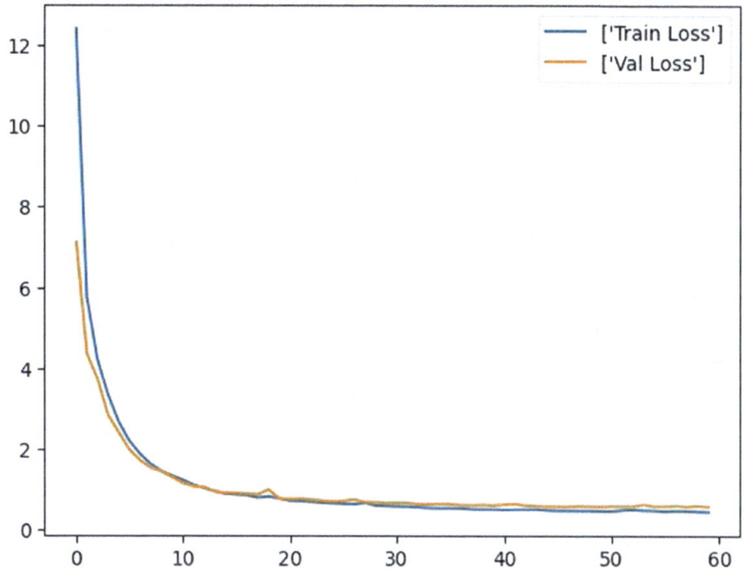

Fig. 4 Training loss of the proposed model

5 Validation of Accuracy and Losses of VGG-19 CNN

During the training phase, the difference between the actual target values and the model's predictions is measured by the training loss metric. With higher performance shown by the values, it offers insight into how well the model is learning from the training set (Fig. 4).

The percentage of correctly categorized occurrences that the model classified during training is known as training accuracy. It is used as a performance statistic to evaluate how well the model learns patterns from the training set of data and makes precise predictions (Fig. 5).

6 Results

He dataset used in this study was obtained from Kaggle and includes more than 150 individuals with both positive and negative brain tumor conditions. It also includes a combination of fusion images, segmentation images, and CT/MRI scan data. The dataset offers helpful information for developing the prediction model (Figs. 6 and 7).

Our study's procedure begins with landmark-based registration, which makes sure that the images are oriented in the same coordinate system before moving on to image fusion, which is required for MRI and CT scans of the same patient. As so, the

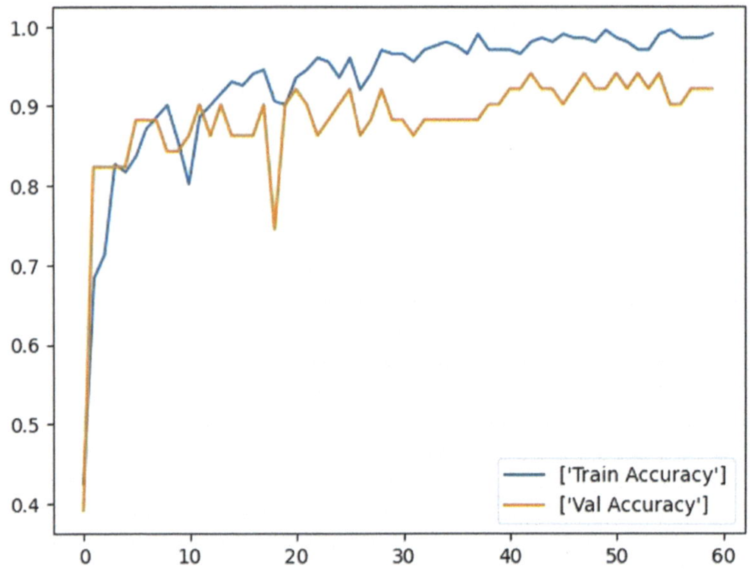

Fig. 5 Training accuracy of the proposed model

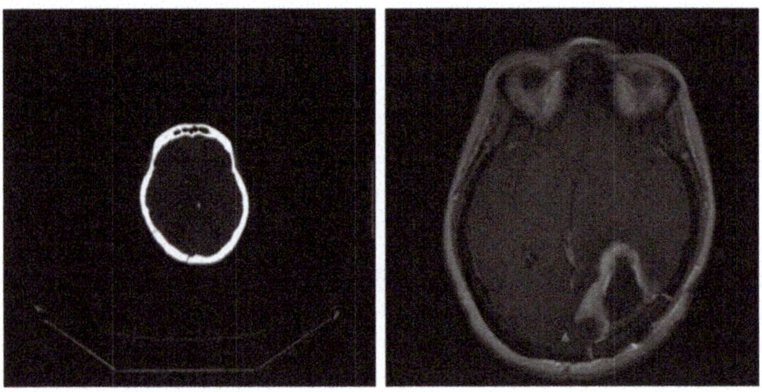

Fig. 6 Dataset sample: positive CT and MRI scan

comparable pixels in both photos represent the same anatomical regions of the brain. Because it enables complementary information from many modalities, including contrast, texture, and structure, to be merged together, picture alignment is crucial for image fusion. Additionally, image alignment lessens the faults and artifacts that result from image misalignment, such as distortion, ghosting, and blurring. Landmark-based registration can achieve high precision and long-term reliability for the merging of brain images (Fig. 8).

In the next stage, we use the Discrete Wavelet Transform (DWT) to break the registered images into four sub-bands: LL (low-low frequencies), LH (low-high frequencies), HL (high-low frequencies), and HH (high-high frequencies). Next,

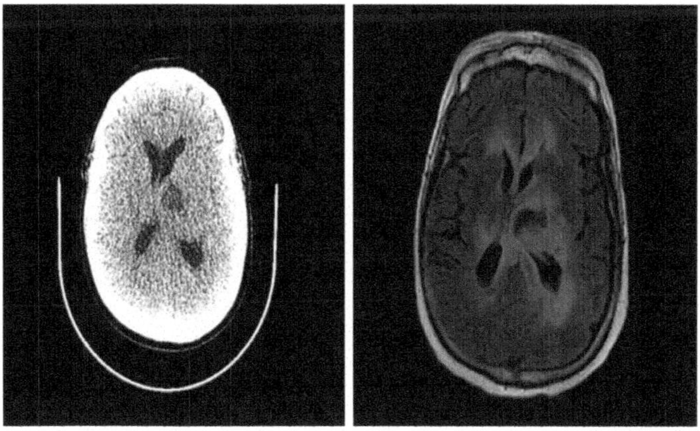

Fig. 7 Dataset sample: negative CT and MRI scan

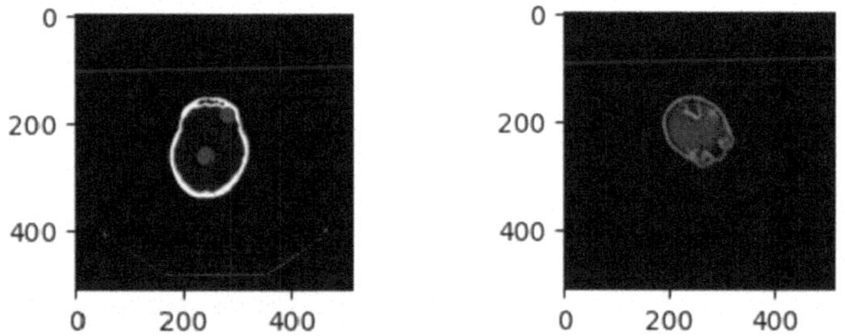

Fig. 8 Landmark-based registration

by carefully pairing the two pictures, the matched sub-bands are supplied into the VGG-19 neural network. The output of the last convolutional layer is a fused sub-band. The inverse Discrete Wavelet Transform is then used to rebuild the combined picture. During this reconstruction process, the data from the four sub-bands is joined to make a logical whole. Figure 9 illustrates the utility of the image fusion approach by showing the successful generation of the fused picture from the registered photos.

Next, we do the segmentation stage using the Watershed Algorithm, a region-based technique based on picture morphology. When it comes to brain tumor segmentation, the Watershed Algorithm's ability to distinguish between touching or overlapping items in an image is widely recognized. The fused picture is subjected to this method, yielding the segmented image, as seen visually in Fig. 10.

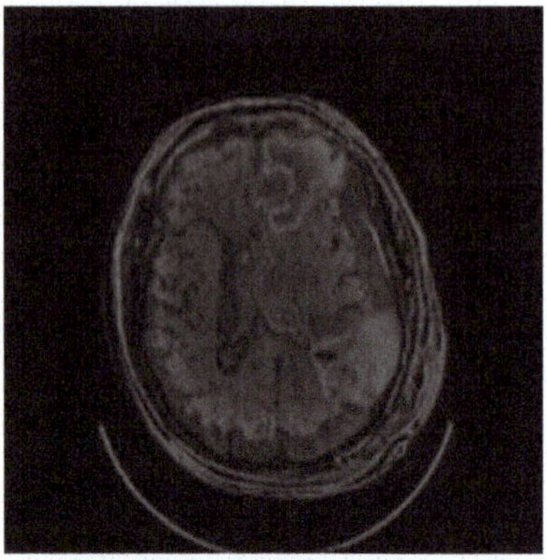

Fig. 9 Fused image

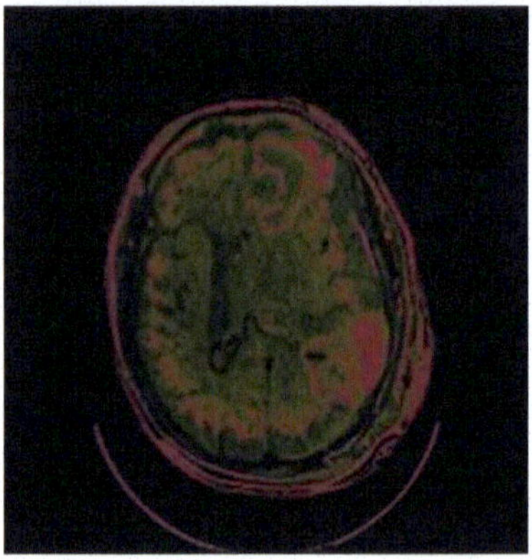

Fig. 10 Segmented image

The last figure, represents our study endeavor's culmination and visually represents the critical stage of brain tumor identification (Fig. 11). The significant task of determining if brain tumors are present in the separated parts of this visually appealing image is taken on by our classification system, which is based on the Regional Convolutional Neural Network (RCNN). The graphical representation demonstrates the interplay between deep learning, picture segmentation, and diagnostic accuracy and serves as evidence of the effectiveness of our proposed technique (Fig. 12). The outcome demonstrated here demonstrates how quickly and accurately our system can provide assessments, offering invaluable insights into the healthcare sector (Table 1).

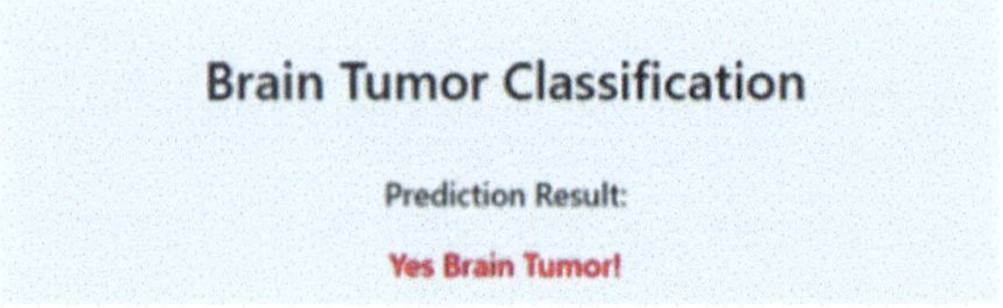

Fig. 11 Prediction

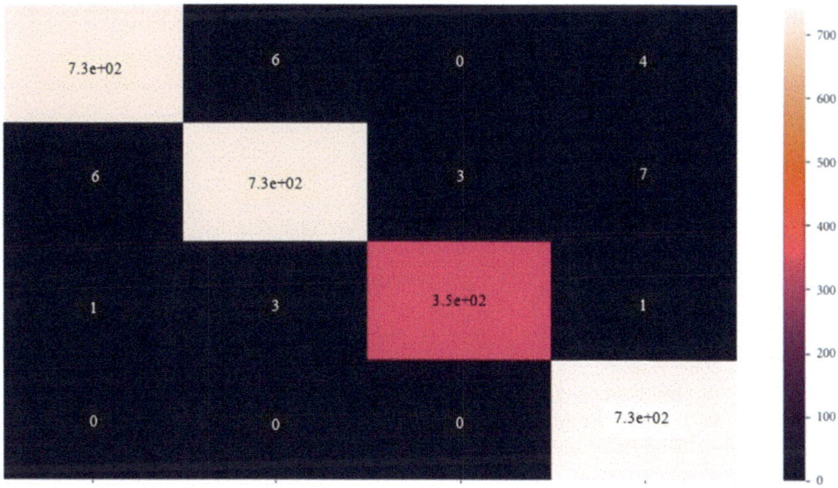

Fig. 12 Confusion matrix of proposed model

Table 1 Comparison with previous work

Related work	Method	Accuracy (%)
[1]	CNN	84
[3]	CNN, GMB	81.19
[9]	CNN, V-Net, BraTS	78
Proposed model	VGG-19 CNN, image fusion, watershed transformation	92.16

7 Conclusion

Our innovative concept is a significant step forward in the healthcare industry as it combines the capabilities of CT and MRI using cutting-edge algorithms. This initiative provides a robust technique aimed at enhancing the prognosis and diagnosis of brain tumors. A Flask-based web application that provides a seamless approach for users to merge and harmonize CT and MRI scans is the core component of our strategy. By facilitating a more comprehensive and in-depth analysis of potential brain tumor sites, this integration enhances the overall efficacy of diagnostic procedures. Our solution is based on the powerful VGG-19 Convolutional Neural Network, which leverages deep learning for sophisticated analysis. Our method ensures high precision tumor detection in potentially tumorous locations and emphasizes the accuracy of image analysis when paired with the Watershed transformation, with the accuracy coming out to be 92.12%. Our approach goes beyond just a straightforward diagnostic tool and shows the potential for combining state-of-the-art technology with medical procedures, emphasizing a commitment to better patient care outcomes. In navigating the intersection of technology and healthcare, our research demonstrates the transformative potential that arises when innovation combines with the need to improve patient well-being.

References

1. Tiwari, P., Pant, B., Elarabawy, M.M., Abd-Elnaby, M., Mohd, N., Dhiman, G., Sharma, S.: CNN based multiclass brain tumor detection using medical imaging. Comput. Intell. Neurosci. **21**(2022), 1830010 (2022). https://doi.org/10.1155/2022/1830010.PMID:35774437;PMCID: PMC9239800
2. Bakas, S., Reyes, M., Jakab, A., Bauer, S., Rempfler, M., Crimi, A., Shinohara, R.T.: Identifying the best machine learning algorithms for brain tumor segmentation, progression assessment, and overall survival prediction in the BRATS challenge. arXiv preprint arXiv:1811.02629.
3. Bhandari, A., Koppen, J., Agzarian, M.: Convolutional neural networks for brain tumor segmentation. IEEE Trans. Med. Imaging (2016)
4. Isensee, F., Kickingereder, P., Wick, W., Bendszus, M., Maier-Hein, K.H.: Brain tumor segmentation and radiomics survival prediction: contribution to the BRATS 2017 challenge. In: Crimi, A., Bakas, S., Kuijf, H., Keyvan, F., Reyes, M., van Walsum, T.: Brainlesion: Glioma, Multiple Sclerosis, Stroke and Traumatic Brain Injuries: 4th International Workshop BrainLes 2018 Held in Conjunction with MICCAI 2018 Granada Spain September 16 2018 Revised Selected Papers Part I, pp. 287–297. Springer International Publishing, Cham (2019)
5. Magadza, T., Viriri, S.: Deep learning for brain tumor segmentation: a survey of state-of-the-art. J. Imaging **7**(2), 19 (2021). https://doi.org/10.3390/jimaging7020019
6. Islam, M., VS, V., Jose, V.J.M., Wijethilake, N., Utkarsh, U., Ren, H.: Brain tumor segmentation and survival prediction using 3D attention UNet. arXiv preprint arXiv:2104.00985 (2021)
7. Menze, B.H., Jakab, A., Bauer, S., Kalpathy-Cramer, J., Farahani, K., Kirby, J., Burren, Y., Porz, N., Slotboom, J., Wiest, R., Lanczi, L., Gerstner, E., Weber, M.A., Arbel, T., Avants, B.B., Ayache, N., Buendia, P., Collins, D.L., Cordier, N., Corso, J.J., Criminisi, A., Das, T., Delingette, H., Demiralp, C., Durst, C.R., Dojat, M., Doyle, S., Festa, J., Forbes, F.: The multimodal brain tumor image segmentation benchmark (BRATS). IEEE Trans. Med. Imaging **34**(10), 1993–2024 (2015)
8. Sun, L., Zhang, S., Chen, H., Luo, L.: Brain tumor segmentation and survival prediction using multimodal MRI scans with deep learning. Front. Neurosci. **13**, 810 (2019)
9. Huang, H., Yang, G., Zhang, W., Xu, X., Yang, W., Jiang, W., Lai, X.: A deep multi-task learning framework for brain tumor segmentation. Front. Oncol. **4**(11), 690244 (2021). https://doi.org/10.3389/fonc.2021.690244.PMID:34150660;PMCID:PMC8212784
10. Singh, N., Prasath, V.B.S., Mukherjee, S.: Multimodal medical image fusion using discrete cosine transform and principal component analysis for brain tumor detection. In: 2018 IEEE International Conference on Computational Intelligence and Computing Research (ICCIC), pp. 1–6 (2018)
11. Daimary, D., Bora, M.B., Amitab, K., Kandar, D.: Brain tumor segmentation from MRI images using hybrid convolutional neural networks. Procedia Comput. Sci. **167**:2419–2428
12. Byra Reddy, G.R., Prasanna Kumar, H.: A review of multimodal medical image fusion. Int. J. Biomed. Eng. Technol. **34**(2), 119–132 (2020)
13. Prasath, V.B.S., Singh, N., Mukherjee, S.: Multimodal medical image fusion using wavelet transform and fuzzy logic for brain tumor detection. In: 2017 IEEE International Conference on Computational Intelligence and Computing Research (ICCIC), pp. 1–6 (2017)
14. Rathore, S., et al.: Brain tumor segmentation and classification using deep learning. Int. J. Adv. Comput. Sci. Appl. **10**(5), 34–40 (2019)
15. Liu, Y., Wang, Z., Li, J., Li, X., Zhang, J.: Flexible fusion network for multi-modal brain tumor segmentation. IEEE J. Biomed. Health Inform. **27**(7), 3349–3359 (2023)

16. Zhou, S.K., Greenspan, H., Shen, D.: Deep learning for medical image analysis. Acad. Radiol. **24**(11), 1284–1299 (2017)
17. Zheng, Y., Yang, C., Merkulov, A.: Breast cancer screening using convolutional neural network and follow-up digital mammography, vol. 4 (2018). https://doi.org/10.1117/12.2304564

The manufacturer's authorised representative in the EU is Springer Nature Customer Service Centre GmbH, Europaplatz 3, 69115 Heidelberg, Germany. If you have any concerns regarding our products, please contact ProductSafety@springernature.com

Printed and bound by CPI Group (UK) Ltd, Croydon, CR0 4YY
30/09/2026
02225298-0006